The Essentials of
STATISTICS

Changbuk Jung (T.A)

Spring 2007 (PLS 201)

Political Science Dept.
M. S. U.

The Essentials of
STATISTICS

A Tool for Social Research

Joseph F. Healey

Christopher Newport University

THOMSON

WADSWORTH

Australia • Brazil • Canada • Mexico • Singapore • Spain
United Kingdom • United States

THOMSON
WADSWORTH

The Essentials of Statistics: A Tool for Social Research
Joseph F. Healey

Acquisitions Editor: *Bob Jucha*
Assistant Editor: *Katia Krukowski*
Editorial Assistant: *Kristin Marrs*
Technology Project Manager: *Dee Dee Zobian*
Marketing Manager: *Michelle Williams*
Marketing Assistant: *Gregory Hughes*
Marketing Communications Manager: *Linda Yip*
Project Manager, Editorial Production:
Matt Ballantyne
Creative Director: *Rob Hugel*

Print Buyer: *Karen Hunt*
Permissions Editor: *Bob Kauser*
Production Service: *Scratchgravel Publishing
Services*
Copy Editor: *Pat Tompkins*
Illustrator: *Lotus Art*
Cover Designer: *Yvo*
Cover Image: © *Digital Vision/Getty Images*
Compositor: *Interactive Composition Corporation*
Text and Cover Printer: *West Group*

Thomson Higher Education
10 Davis Drive
Belmont, CA 94002-3098
USA

For more information about our
products, contact us at:
Thomson Learning Academic Resource Center
1-800-423-0563

For permission to use material from this text
or product, submit a request online at
http://www.thomsonrights.com.
Any additional questions about permissions
can be submitted by e-mail to
thomsonrights@thomson.com.

Library of Congress Control Number: 2006904375

ISBN 0-495-00975-X

Brief Contents

Detailed Contents

PART III   **BIVARIATE MEASURES OF ASSOCIATION / 251**

Preface

Sociology and the other social sciences (including political science, social work, public administration, criminal justice, urban studies, and gerontology) are research-based disciplines, and statistics are part of their everyday language. To join the conversation, you must be literate in the vocabulary of research, data analysis, and scientific thinking. Fluency in statistics will help you understand the research reports you encounter in everyday life and the professional research literature of your discipline. You will also be able to conduct quantitative research, contribute to the growing body of social science knowledge, and reach your full potential as a social scientist.

Although essential, learning (and teaching) statistics can be a challenge. Students in statistics courses typically bring with them a wide range of mathematical backgrounds and an equally diverse set of career goals. They are often puzzled about the relevance of statistics for them and, not infrequently, there is some math anxiety to deal with.

This text introduces statistical analysis for the social sciences while addressing these challenges. The text is an abbreviated version of *Statistics: A Tool for Social Research* and presents only the most essential material from the larger text. It makes minimal assumptions about mathematical background (the ability to read a simple formula is sufficient preparation for virtually all of the material in the text), and it includes a variety of special features to help students analyze data successfully. The theoretical and mathematical explanations are kept at an elementary level, as is appropriate in a first exposure to social statistics. The text has been written especially for sociology and social work programs but is sufficiently flexible to be used in any program with a social science base.

GOAL OF THE TEXT AND CHANGES IN THE ESSENTIALS VERSION

The goal of this text is to develop basic statistical literacy. The statistically literate person understands and appreciates the role of statistics in the research process, is competent to perform basic calculations, and can read and appreciate the professional research literature in his or her field as well as any research reports encountered in everyday life. These three aspects of statistical literacy provide a framework for discussing the additional features of this text.

1. An Appreciation of Statistics. A statistically literate person understands the relevance of statistics for social research, can analyze and interpret the meaning of a statistical test, and can select an appropriate statistic for a given purpose and a given set of data. This textbook develops these qualities, within the constraints imposed by the introductory nature of the course, in the following ways:

- *The relevance of statistics.* Chapter 1 includes a discussion of the role of statistics in social research and stresses the usefulness of statistics in analyzing and manipulating data and answering research questions. Each example problem is framed in the context of a research situation. A question is posed and then, with the aid of a statistic, answered. The relevance of statistics for answering questions is thus stressed throughout the text. This central theme of usefulness is further reinforced by a series of boxes labeled "Application," each of which illustrates some specific way statistics can be used to answer questions.

    Almost all end-of-chapter problems are labeled by the social science discipline or subdiscipline from which they are drawn: SOC for sociology, SW for social work, PS for political science, CJ for criminal justice, PA for public administration, and GER for gerontology. By identifying problems with specific disciplines, students can more easily see the relevance of statistics to their own academic interests. (Not incidentally, they will also see that the disciplines have a large subject matter in common.)

- *Interpreting statistics.* For most students, interpretation—saying what statistics mean—is a big challenge. The ability to interpret statistics can be developed only by exposure and experience. To provide exposure, I have been careful, in the example problems, to express the meaning of the statistic in terms of the original research question. To provide experience, the end-of-chapter problems almost always call for an interpretation of the statistic calculated. To provide examples, many of the Answers to Odd-Numbered Computational Problems in the back of the text are expressed in words as well as numbers.

- *Using statistics: Ideas for research projects.* Appendix E offers ideas for independent data-analysis projects for students. The projects require students to use SPSS to analyze the GSS, the data set provided with this text. The projects can be assigned at intervals throughout the semester or at the end of the course. Each project provides an opportunity for students to practice and apply their statistical skills and, above all, to exercise their ability to understand and interpret the meaning of the statistics they produce.

2. Computational Competence.   Students should emerge from their first course in statistics with the ability to perform elementary forms of data analysis—to execute a series of calculations and arrive at the correct answer. To be sure, computers and calculators have made computation less of an issue today. Yet computation is inseparable from statistics, and, because social science majors frequently do not have strong quantitative backgrounds, I have included a number of features to help students cope with these challenges:

- *Step-by-step boxes* presenting computational procedures are provided for each statistic. These appear immediately following the introduction of the statistic.

- *Extensive problem sets* are provided at the end of each chapter. For the most part, these problems use fictitious data and are designed for ease of computation.
- *Solutions* to odd-numbered computational problems are provided so that students may check their answers.
- *SPSS for Windows* is incorporated throughout the text to give students access to the computational power of the computer.

3. The Ability to Read the Professional Social Science Literature. The statistically literate person can comprehend and critically appreciate research reports written by others. The development of this quality is a particular problem at the introductory level because (1) the vocabulary of professional researchers is so much more concise than the language of the textbook, and (2) the statistics featured in the literature are more advanced than those covered at the introductory level. To help bridge this gap, I have been careful always to express the meaning of each statistic in terms of answering a social science research question.

Additional Features. A number of other features make the text more meaningful for students and more useful for instructors:

- *Readability and clarity.* The writing style is informal and accessible to students without ignoring the traditional vocabulary of statistics. Problems and examples have been written to maximize student interest and to focus on issues of concern and significance. For the more difficult material (such as hypothesis testing), students are first walked through an example problem before being confronted by formal terminology and concepts. Each chapter ends with a summary of major points and formulas and a glossary of important concepts. A list of frequently used formulas inside the front and back covers along with a glossary of symbols in Appendix H can be used for quick reference.
- *Organization and coverage.* The text is divided into four parts, with most of the coverage devoted to univariate descriptive statistics, inferential statistics, and bivariate measures of association. The distinction between description and inference is introduced in the first chapter and maintained throughout the text. In selecting statistics for inclusion, I have tried to strike a balance between the essential concepts with which students must be familiar and the amount of material students can reasonably be expected to learn in their first (and perhaps only) statistics course, while bearing in mind that different instructors will naturally wish to stress different aspects of the subject. Thus, the text covers a full gamut of the usual statistics, with each chapter broken into subsections so that instructors may choose the particular statistics they wish to include. In this edition, the text has been shortened and streamlined by moving some infrequently used techniques and statistical procedures to the companion website.
- *Learning objectives.* Learning objectives are stated at the beginning of each chapter. These are intended to serve as "study guides" and to help students identify and focus on the most important material.

- *Review of mathematical skills.* A comprehensive review of all of the mathematical skills that will be used in this text is presented in the Prologue. Students who are inexperienced or out of practice with mathematics may want to study this review early in the course and refer to it as needed. A self-test is included so students may check their level of preparation for the course.

- *Explicitly linked statistical techniques and end-of-chapter problems.* After a technique is introduced, students are directed to specific problems for practice and review. The "how-to-do-it" aspects of calculation are reinforced immediately and clearly.

- *Progressively organized end-of-chapter problems.* Simpler problems with small data sets are presented first. Often, explicit instructions or hints accompany the first several problems in a set. The problems gradually become more challenging and require more decision making by the student (e.g., choosing the most appropriate statistic for a certain situation). Thus, each problem set develops problem-solving abilities gradually and progressively. Some of the more challenging problems (e.g., using the largest data sets) have been moved to the website.

- *Computer applications.* To help students take advantage of the power of the computer, this text integrates SPSS, the leading social science statistical software. Appendix F provides an introduction to SPSS. The demonstrations at the ends of chapters explain how to use the statistical package to produce the statistics presented in the chapter. Student exercises analyzing data with SPSS are also included. The student version of SPSS is available as a supplement to this text, and the student version of MicroCase is available online at the website.

- *Realistic, up-to-date data.* The database for computer applications in the text is a shortened version of the 2004 General Social Survey. This database will give students the opportunity to practice their statistical skills on "real-life" data. The database is described in Appendix G and is available in SPSS format at the website for this text. Also available at the website are other data sets and additional exercises and projects.

- *Instructor's Manual/Test Bank.* The *Instructor's Manual* includes chapter summaries, a test item file of multiple-choice questions, answers to even-numbered computational problems, and step-by-step solutions to selected problems. In addition, the *Instructor's Manual* includes cumulative exercises (with answers) that can be used for testing purposes.

- *ExamView® computerized testing.* Quickly create customized tests that can be delivered in print or online. ExamView's simple "what you see is what you get" interface allows you easily to generate tests of up to 250 items. (Contains all the Test Bank questions electronically.)

- *Study Guide.* Easy to read and easy to understand, the *Study Guide* uses examples from contemporary social problems to help build students' understanding of the concepts presented in the text. Each chapter of the *Study Guide* includes learning objectives, chapter summaries, multiple-choice questions, work problems with detailed answers and explanations, and data analysis exercises for SPSS and MicroCase®. The *Study Guide* also includes a feature on reading statistics.

- *Companion website.* Access the companion website through www. thomsonedu.com/sociology. Free access! Click on the companion website to find useful student learning resources for each chapter of the text, including tutorial practice quizzes with feedback, web links, Internet exercises, flash cards of the text's glossary, crossword puzzles, and more. For instructors, the site includes a password-protected *Instructor's Manual.*

The text has been thoroughly reviewed for clarity and readability. As with previous editions, my goal is to provide a comprehensive, flexible, and student-oriented text that will provide a challenging first exposure to social statistics.

ACKNOWLEDGMENTS

This text has been in development, in one form or another, for over twenty years. An enormous number of people have made contributions, both great and small, to this project and, at the risk of inadvertently omitting someone, I am bound at least to attempt to acknowledge my many debts.

This edition reflects the thoughtful guidance of Bob Jucha of Wadsworth, and I thank him for his many contributions. Much of the integrity and quality of this book is a direct result of the very thorough (and often highly critical) reviews that have been conducted over the years. I am consistently impressed by the sensitivity of my colleagues to the needs of the students, and for their assistance in preparing this essentials edition, I would like to thank the following reviewers: Marina Adler, University of Maryland, Baltimore; Michael G. Bisciglia, Louisiana State University; Robert M. Carini, University of Louisville; Keith Carroll, Benedictine University; James Cassell, Henderson State University; Nicole Dash, University of North Texas; Tamela Eitle, University of Miami; Kenneth Fernandez, University of Nevada, Las Vegas; Percy Galimbertti, Texas A&M University; Aleta Top Gustavson, Southern Illinois University, Carbondale; Jennifer Keene, University of Nevada, Las Vegas; Kirk McClure, University of Kansas; John David Rausch, Jr., West Texas A&M University; Alden Roberts, Texas Tech University; and James R. Rogers, Texas A&M University, College Station. Whatever failings are contained in the text are, of course, my responsibility and are probably the results of my occasional decisions not to follow the advice of my colleagues.

I would like to thank the instructors who made statistics understandable to me (Professors Satoshi Ito, Noelie Herzog, and Ed Erikson) and all of my colleagues at Christopher Newport University for their support and encouragement (especially Professors F. Samuel Bauer, Robert Durel, James Forte, Marcus Griffin, Ruth Kernodle, Marion Manton, Timothy Marshall, Cheryl Mathews, Lea Pellet, Virginia Purtle, and William Winter). I would be very remiss if I did not acknowledge the constant support and excellent assistance of Mrs. Iris Price, and I thank all of my students for their patience and thoughtful feedback. Also, I am grateful to the Literary Executor of the late Sir Ronald A. Fisher, F.R.S., to Dr. Frank Yates, F.R.S., and to Longman Group Ltd., London, for permission to reprint Appendixes B, C, and D, from their book *Statistical Tables for Biological, Agricultural and Medical Research* (6th edition, 1974).

Finally, I want to acknowledge the support of my family and rededicate this work to them. I have the extreme good fortune to be a member of an extended family that is remarkable in many ways and that continues to increase in size. Although I cannot list everyone, I would like to especially thank the older generation (my mother, Alice T. Healey), the next generation (my sons Kevin and Christopher, my daughters-in-law Jennifer and Jessica), the new members (my wife Patricia Healey and Christopher, Katherine, and Jennifer Schroen), and the youngest generation (Benjamin and Caroline Healey).

The Essentials of
STATISTICS

Prologue
Basic Mathematics Review

You will probably be relieved to hear that this text, your first exposure to statistics for social science research, is not particularly mathematical and does not stress computation per se. Although you will encounter many numbers to work with and numerous formulas to use, the major emphasis is on understanding the role of statistics in research and the logic by which we attempt to answer research questions empirically. In addition, the example problems and many of the homework problems have been intentionally simplified so that the computations will not unduly distract you from the task of understanding the statistics themselves.

On the other hand, you may regret to learn that there is, inevitably, some arithmetic that you simply cannot avoid if you want to master this material. It is likely that some of you haven't had any math in a long time, others have convinced themselves that they cannot do math under any circumstances, and still others are just rusty and out of practice. All of you will find that mathematical operations that might seem complex and intimidating can be broken down into simple steps. If you have forgotten how to cope with some of these steps or are unfamiliar with these operations, this prologue will ease you into the skills you will need to do all of the computations in this textbook.

CALCULATORS AND COMPUTERS

A calculator is a virtual necessity for this text. Even the simplest, least expensive model will save you time and effort and is definitely worthwhile. However, I recommend that you consider investing in a more sophisticated calculator with memories and preprogrammed functions, especially the statistical models that can compute means and standard deviations automatically. Calculators with these capabilities are available for less than $20.00 and will be worth the small effort it will take to learn to use them.

In the same vein, several computerized statistical packages (or statpaks) commonly available on college campuses can further enhance your statistical and research capabilities. The most widely used of these is the Statistical Package for the Social Sciences (SPSS). This program comes in a student version, which is available bundled with this text (for a small fee).[1] Statistical packages such as SPSS are many times more powerful than even the most sophisticated handheld calculators, and it will be well worth your time to learn how to use them because they will eventually save you time and effort. Appendix F introduces SPSS, and *SPSS for Windows* demonstrations and exercises appear at the end of most chapters to show you how to use the program to generate and

---

[1]Another statpak, called MicroCase, is available free and can be downloaded from the website for this text. All of the computer exercises in this text are available in MicroCase format at the website.

interpret the statistics covered in the text. Many other programs may be available to you to help you accomplish the goal of generating accurate statistical results with a minimum of effort and time. Even spreadsheet programs such as Microsoft Excel, which is included in many versions of Microsoft Office, have some statistical capabilities. You should be aware that all of these programs (other than the simplest calculators) will require some effort to learn but the rewards will be worth the effort.

In summary, you should find a way at the beginning of this course—with a calculator, a statpak, or both—to minimize the tedium and hassle of mere computing. This will permit you to devote maximum effort to the truly important goal of increasing your understanding of the meaning of statistics in particular and social science research in general.

VARIABLES AND SYMBOLS

Statistics are a set of techniques by which we can describe, analyze, and manipulate **variables.** A variable is a trait that can change value from case to case or from time to time. Examples of variables would include height, weight, level of prejudice, and political party preference. The possible values or scores associated with a given variable might be numerous (for example, income) or relatively few (for example, gender). I will often use symbols, usually the letter X, to refer to variables in general or to a specific variable.

Sometimes we will need to refer to a specific value or set of values of a variable. This is usually done with the aid of subscripts. So, the symbol X_1 (read "X-sub-one") would refer to the first score in a set of scores, X_2 ("X-sub-two") to the second score, and so forth. Also, we will use the subscript i to refer to all the scores in a set. Thus, the symbol X_i ("X-sub-eye") refers to all of the scores associated with a given variable (for example, the test grades of a particular class).

OPERATIONS

You are familiar with the four basic mathematical operations of addition, subtraction, multiplication, and division and the standard symbols $(+, -, \times, \div)$ used to denote them. The latter two operations can be symbolized in a variety of ways. For example, the operation of multiplying some number a by some number b may be symbolized in (at least) six different ways:

$a \times b$
$a \cdot b$
$a * b$
ab
$a(b)$
$(a)(b)$

In this text, we will commonly use the "adjacent symbols" format (that is, ab), the conventional times sign $(\times)$, or adjacent parentheses to indicate multiplication. On most calculators and computers, the asterisk (*) is the symbol for multiplication.

The operation of division can also be expressed in several different ways. In this text, we will use either of these two methods:

$$a/b \quad \text{or} \quad \frac{a}{b}$$

Several of the formulas with which we will be working require us to find the square of a number. To do this, simply multiply the number by itself. This operation is symbolized as X^2 (read "X squared"), which is the same thing as $(X)(X)$. If X has a value of 4, then

$$X^2 = (X)(X) = (4)(4) = 16$$

or we could say that "4 squared is 16."

The square root of a number is the value that, when multiplied by itself, results in the original number. So the square root of 16 is 4 because (4)(4) is 16. The operation of finding the square root of a number is symbolized as:

$$\sqrt{X}$$

A final operation with which you should be familiar is summation, or the addition of the scores associated with a particular variable. When a formula requires the addition of a series of scores, this operation is usually symbolized as ΣX_i. Σ is the uppercase Greek letter sigma and stands for "the summation of." So the combination of symbols ΣX_i means "the summation of all the scores" and directs us to add the value of all the scores for that variable. If four people had family sizes of 2, 4, 5, and 7, then the summation of these four scores for this variable could be symbolized as:

$$\Sigma X_i = 2 + 4 + 5 + 7 = 18$$

The symbol Σ is an operator, just like the $+$ or $\times$ signs. It directs us to add all of the scores on the variable indicated by the X symbol.

There are two other common uses of the summation sign and, unfortunately, the symbols denoting these uses are not, at first glance, sharply different from each other or from the symbol used above. A little practice and some careful attention to these various meanings should minimize the confusion. The first set of symbols is ΣX_i^2, which means "the sum of the squared scores." This quantity is found by *first* squaring each of the scores and *then* adding the squared scores together. A second common set of symbols will be $(\Sigma X_i)^2$, which means "the sum of the scores, squared." This quantity is found by *first* summing the scores and *then* squaring the total.

These distinctions might be confusing at first, so let's see if an example helps to clarify them. Suppose we had a set of three scores: 10, 12, and 13. So:

$$X_i = 10, 12, 13$$

The sum of these scores would be indicated as:

$$\Sigma X_i = 10 + 12 + 13 = 35$$

The sum of the squared scores would be:

$$\Sigma X_i^2 = (10)^2 + (12)^2 + (13)^2 = 100 + 144 + 169 = 413$$

Take careful note of the order of operations here. First, the scores are squared one at a time and then the squared scores are added. This is a completely different operation from squaring the sum of the scores:

$$(\Sigma X_i)^2 = (10 + 12 + 13)^2 = (35)^2 = 1,225$$

To find this quantity, first the scores are summed and then the total of all the scores is squared. The value of the sum of the scores, squared (1,225) is not the same as the value of the sum of the squared scores (413). The following chart indicates the operations associated with each set of symbols:

Symbols	Operations
ΣX_i	Add the scores.
ΣX_i^2	First square the scores and then add the squared scores.
$(\Sigma X_i)^2$	First add the scores and then square the total.

OPERATIONS WITH NEGATIVE NUMBERS

A number can be either positive (if it is preceded by a + sign or by no sign at all) or negative (if it is preceded by a − sign). Positive numbers are greater than zero, and negative numbers are less than zero. It is very important to keep track of signs because they will affect the outcome of virtually every mathematical operation. This section will briefly summarize the relevant rules for dealing with negative numbers. First, adding a negative number is the same as subtraction. For example:

$$3 + (-1) = 3 - 1 = 2$$

Second, subtraction changes the sign of a negative number:

$$3 - (-1) = 3 + 1 = 4$$

Note the importance of keeping track of signs here. If you neglected to change the sign of the negative number in the second expression, you would arrive at the wrong answer.

For multiplication and division, you should be aware of various combinations of negative and positive numbers. For purposes of this text, you will rarely have to multiply or divide more than two numbers at a time, and we will confine our attention to this situation. Ignoring the case of all positive numbers, this leaves several possible combinations. A negative number times a positive number results in a negative value:

$$(-3)(4) = -12$$

or

$$(3)(-4) = -12$$

A negative number multiplied by a negative number is always positive:

$$(-3)(-4) = 12$$

Division follows the same patterns. If there is a single negative number in the calculations, the answer will be negative. If both numbers are negative, the answer will be positive. So:

$$(-4)/(2) = -2$$

and

$$(4)/(-2) = -2$$

but

$$(-4)/(-2) = 2$$

Negative numbers do not have square roots because multiplying a number by itself cannot result in a negative value. Squaring a negative number always results in a positive value (see the multiplication rules above).

ACCURACY AND ROUNDING OFF

A possible source of confusion in computation involves the issues of accuracy and rounding off. People work at different levels of accuracy and precision and, for this reason alone, may arrive at different answers to problems. This is important because, if you work at one level of precision and I (or your instructor or your study partner) work at another, we can arrive at solutions that are at least slightly different. You may sometimes think you've gotten the wrong answer when all you've really done is round off at a different place in the calculations or in a different way.

There are two issues here: when to round off and how to round off. In this text, I have followed the convention of working in as much accuracy as my calculator or statistics package will allow and then rounding off to two places of accuracy (two places beyond or to the right of the decimal point) only at the very end. If a set of calculations is lengthy and requires the reporting of intermediate sums or subtotals, I will round the subtotals off to two places also.

In terms of how to round off, begin by looking at the digit immediately to the right of the last digit you want to retain. If you want to round off to 100ths (two places beyond the decimal point), look at the digit in the 1,000ths place (three places beyond the decimal point). If that digit is 5 or more, round up. For example, 23.346 would round off to 23.35. If the digit to the right is less than 5, round down. So, 23.343 would become 23.34.

Let's look at some more examples of how to follow the rounding rules stated above. If you are calculating the mean value of a set of test scores and your calculator shows a final value of 83.459067, and you want to round off to two places beyond the decimal point, look at the digit three places beyond the decimal point. In this case the value is 9 (greater than 5), so we would round the second digit beyond the decimal point up and report the mean as 83.46. If the value had been 83.453067, we would have reported our final answer as 83.45.

FORMULAS, COMPLEX OPERATIONS, AND THE ORDER OF OPERATIONS

A mathematical formula is a set of directions, stated in general symbols, for calculating a particular statistic. To "solve a formula," you replace the symbols with the proper values and then manipulate the values through a series of calculations. Even the most complex formula can be rendered manageable if it is broken down into smaller steps. Working through these steps requires some knowledge of general procedure and the rules of precedence of mathematical operations. This is because the order in which you perform calculations may affect your final answer. Consider the following expression:

$$2 + 3(4)$$

Note that if you do the addition first, you will evaluate the expression as:

$$5(4) = 20$$

but if you do the multiplication first, the expression becomes:

$$2 + 12 = 14$$

Obviously, it is crucial to complete the steps of a calculation in the correct order.

The basic rules of precedence are to find all squares and square roots first, then do all multiplication and division, and finally complete all addition and subtraction. So the following expression:

$$8 + 2 \times 2^2/2$$

would be evaluated as:

$$8 + 2 \times 4/2 = 8 + 8/2 = 8 + 4 = 12$$

The rules of precedence may be overridden when an expression contains parentheses. Solve all expressions within parentheses before applying the rules stated above. For most of the complex formulas in this text, the parentheses control the order of calculations. Consider the following expression:

$$(8 + 2) - 4(3)^2/(8 - 6)$$

Resolving the parenthetical expressions first, we would have:

$$(10) - 4 \times 9/(2) = 10 - 36/2 = 10 - 18 = -8$$

Without the parentheses, the same expression would be evaluated as

$$8 + 2 - 4 \times 3^2/8 - 6 = 8 + 2 - 4 \times 9/8 - 6 = 8 + 2 - 36/8 - 6$$
$$= 8 + 2 - 4.5 - 6 = 10 - 10.5 = -.5$$

A final operation you will encounter in some formulas in this text involves denominators of fractions that themselves contain fractions. In this situation, solve the fraction in the denominator first and then complete the division. For example,

$$\frac{15 - 9}{6/2}$$

would become:

$$\frac{15 - 9}{6/2} = \frac{6}{3} = 2$$

When you confront complex expressions such as these, don't be intimidated. If you're patient with yourself and work through them step by step, beginning with the parenthetical expression, you can manage even the most imposing formulas.

EXERCISES

Try the problems below as a "self-test" on the material presented in this review. If you can handle these problems, you're ready to do all of the arithmetic in this text. If you have difficulty with any of these problems, please review the appropriate section of this prologue. You might also want to use this section as an opportunity to become more familiar with your calculator. Answers are given on the next page, along with some commentary and some reminders.

1. Complete each of the following:
  a. $17 \times 3 =$
  b. $17(3) =$
  c. $(17)(3) =$
  d. $17/3 =$

e. $(42)^2 =$

f. $\sqrt{113} =$

2. For the set of scores (X_i) of 50, 55, 60, 65, and 70, evaluate each of the expressions below:

$\Sigma X_i =$

$\Sigma X_i^2 =$

$(\Sigma X_i)^2 =$

3. Complete each of the following:
   a. $17 + (-3) + (4) + (-2) =$
   b. $15 - 3 - (-5) + 2 =$
   c. $(-27)(54) =$
   d. $(113)(-2) =$
   e. $(-14)(-100) =$
   f. $-34/-2 =$
   g. $322/-11 =$

4. Round off each of the following to two places beyond the decimal point:
   a. 17.17532
   b. 43.119
   c. 1,076.77337
   d. 32.4651152301
   e. 32.4751152301

5. Evaluate each of the following:
   a. $(3 + 7)/10 =$
   b. $3 + 7/10 =$
   c. $\dfrac{(4 - 3) + (7 + 2)(3)}{(4 + 5)(10)} =$
   d. $\dfrac{22 + 44}{15/3} =$

ANSWERS TO EXERCISES

1. a. 51   **b.** 51   **c.** 51
(The obvious purpose of these first three problems is to remind you that there are several different ways of expressing multiplication.)

   d. 5.67 (Note the rounding off.)   **e.** 1,764

   f. 10.63

2. The first expression translates to "the sum of the scores," so this operation would be:

$\Sigma X_i = 50 + 55 + 60 + 65 + 70 = 300$

The second expression is the "sum of the squared scores." So:

$\Sigma X_i^2 = (50)^2 + (55)^2 + (60)^2 + (65)^2 + (70)^2$

$\Sigma X_i^2 = 2,500 + 3,025 + 3,600 + 4,225 + 4,900$

$\Sigma X_i^2 = 18,250$

The third expression is "the sum of the scores, squared":

$(\Sigma X_i)^2 = (50 + 55 + 60 + 65 + 70)^2$

$(\Sigma X_i)^2 = (300)^2$

$(\Sigma X_i)^2 = 90,000$

Remember that ΣX_i^2 and $(\Sigma X_i)^2$ are two completely different expressions with very different values.

3. a. 16   **b.** 19 (Remember to change the sign of -5.)

   c. $-1,458$ **d.** -226 **e.** 1,400

   f. 17   **g.** -29.27

4. a. 17.18   **b.** 43.12   **c.** 1,076.77

   d. 32.47   **e.** 32.48

5. a. 1   **b.** 3.7 (Note again the importance of parentheses.)

   c. 0.31   **d.** 13.2

1 Introduction

LEARNING OBJECTIVES

By the end of this chapter, you will be able to:

1. Describe the limited but crucial role of statistics in social research.
2. Distinguish among three applications of statistics (univariate descriptive, bivariate descriptive, and inferential) and identify situations in which each is appropriate.
3. Identify and describe three levels of measurement and cite examples of variables from each.

1.1 WHY STUDY STATISTICS?

Students sometimes approach their first course in statistics with questions about the value of the subject. What, after all, do numbers and statistics have to do with understanding people and society? In a sense, this entire book will attempt to answer this question, and the value of statistics will become clear as we move from chapter to chapter. For now, the importance of statistics can be demonstrated, in a preliminary way, by briefly reviewing the research process as it operates in the social sciences. These disciplines are scientific in the sense that social scientists attempt to verify their ideas and theories through research. Broadly conceived, **research** is any process by which information is systematically and carefully gathered for the purpose of answering questions, examining ideas, or testing theories. Research is a disciplined inquiry that can take numerous forms. Statistical analysis is relevant only for research projects in which information is represented as numbers. Numerical information is called **data,** and the sole purpose of statistics is to manipulate and analyze data. **Statistics,** then, are a set of mathematical techniques that social scientists use to organize and manipulate data so that they can answer questions and test theories.

What is so important about learning how to manipulate data? On one hand, some of the most important and enlightening works in the social sciences do not use any statistical techniques. There is nothing magical about data and statistics. The mere presence of numbers guarantees nothing about the quality of a scientific inquiry. On the other hand, data can be the most trustworthy information available to the researcher and, consequently, deserve special attention. Data that have been carefully collected and thoughtfully analyzed are the strongest, most objective foundations for building theory and enhancing understanding. Without a firm base in data, the social sciences would lose the right to the name *science* and would be of far less value.

Thus, the social sciences rely heavily on data analysis to advance knowledge. Let me be very clear about one point: It is never enough merely to gather data (or, for that matter, any kind of information). Even the most objective and carefully collected numerical information does not and cannot speak for itself. The researcher must be able to use statistics effectively to organize, evaluate, and analyze the data. Without a good understanding of the principles of statistical analysis, the researcher will be unable to make sense of the data. Without the appropriate application of statistical techniques, the data will remain mute and useless.

Statistics are an indispensable tool for the social sciences. They provide the scientist with some of the most useful techniques for evaluating hypotheses and testing theory. The next section describes the relationships among theory, research, and statistics in more detail.

1.2 THE ROLE OF STATISTICS IN SCIENTIFIC INQUIRY

Figure 1.1 graphically represents the role of statistics in research. The diagram is based on the thinking of Walter Wallace and illustrates how the knowledge base of any scientific enterprise grows and develops. One point the diagram makes is that scientific theory and research continually shape each other. Statistics are one of the most important means by which research and theory interact. Let's take a closer look at the wheel.

Because the figure is circular, it has no beginning or end, and we could begin our discussion at any point. For the sake of convenience, let's begin at the top and follow the arrows clockwise around the circle. A **theory** is an explanation of the relationships between phenomena. People naturally (and endlessly) wonder about problems in society (such as prejudice, poverty, child abuse, or serial murders) and, in their attempt to understand these phenomena, they develop explanations ("lack of education causes prejudice"). This kind of informal "theorizing" about society is no doubt very familiar to you. A major difference between our informal, everyday explanations of social phenomena and scientific theory is that the latter is subject to rigorous testing. Let's take the problem of racial prejudice as an example to illustrate how the research process works.

What causes racial prejudice? A theory called the *contact hypothesis* provides one possible answer to this question. The social psychologist Gordon Allport stated this theory more than 40 years ago, and it has been tested on a number of occasions since that time.[1] The theory links prejudice to the

FIGURE 1.1   "THE WHEEL OF SCIENCE"

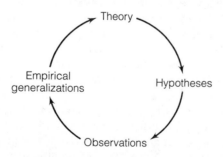

Reprinted by permission of Transaction Publishers.

[1]Allport, Gordon, 1954. *The Nature of Prejudice.* Reading, MA: Addison-Wesley. For recent attempts to test this theory, see: McLaren, Lauren. 2003. "Anti-Immigrant Prejudice in Europe: Contact, Threat Perception, and Preferences for the Exclusion of Migrants." *Social Forces,* 81: 909–937; Pettigrew, Thomas, 1997. "Generalized Intergroup Contact Effects on Prejudice." *Personality and Social Psychology Bulletin.* 23:173–185; and Sigelman, Lee and Welch, Susan, 1993. "The Contact Hypothesis Revisited: Black-White Interaction and Positive Racial Attitudes." *Social Forces.* 71:781–795.

volume and nature of interaction between members of different racial groups. Specifically, the hypothesis asserts that contact situations in which different groups have equal status and are engaged in cooperative behavior will reduce prejudice on all sides. The greater the extent to which contact is equal and cooperative, the more likely people will see each other as individuals and not as representatives of a particular group. For example, the contact hypothesis predicts that members of a racially mixed athletic team who cooperate with each other to achieve victory would tend to experience a decline in prejudice. On the other hand, when different groups compete for jobs, housing, or other valuable resources, prejudice will increase.

The contact hypothesis is not a complete explanation of prejudice, of course, but it illustrates a sociological theory. This theory offers an explanation for the relationship between two social phenomena: (1) prejudice and (2) equal-status, cooperative contact between members of different groups. People who have little contact will be more prejudiced, and those who experience more contact will be less prejudiced.

Before moving on, let's examine theory in more detail. The contact hypothesis, as with most theories, is stated in terms of causal relationships between variables. A **variable** is any trait that can change values from case to case. Examples of variables are gender, age, income, or political party affiliation. In any specific theory, some variables will be identified as causes and others will be identified as effects or results. In the language of science, the causes are called **independent variables** and the effects or result variables are called **dependent variables.** In our theory, contact is the independent variable (or the cause) and prejudice is the dependent variable (the result or effect). In other words, we are arguing that lack of equal-status contact is a cause of prejudice or that an individual's level of prejudice depends on the extent to which he or she participates in equal-status, cooperative contacts with other groups.

So far, we have a theory of prejudice and an independent and a dependent variable. What we don't know yet is whether the theory is true or false. To find out, we need to compare our theory with the facts: We need to do some research. The next steps in the process would be to define our terms and ideas more specifically and exactly. One problem we often face in doing research is that scientific theories are too complex and abstract to be fully tested in a single research project. To conduct research, one or more hypotheses must be derived from the theory. A **hypothesis** is a statement about the relationship between variables that, while logically derived from the theory, is much more specific and exact than the theory.

For example, to test the contact hypothesis, we would have to say exactly what we mean by prejudice and we would need to describe "equal-status, cooperative contact" in great detail. There has been substantial research on the effect of contact on prejudice, and we would consult the research literature to develop and clarify our definitions of these concepts.

As our definitions develop and the hypotheses take shape, our next step in the research process involves deciding exactly how we will gather our data. We must decide how cases will be selected and tested, how exactly the variables will be measured, and a host of related matters. Ultimately, these plans will lead to the observation phase (the bottom of the wheel of science), where we actually measure social reality. Before we can do this, we must have a very clear idea of what we are looking for and a well-defined strategy for conducting the search.

To test the contact hypothesis, we would begin with people from different racial or ethnic groups. We might place some subjects in situations that required them to cooperate with members of other groups and other subjects in situations that feature intergroup competition. We would need to measure levels of prejudice before and after each type of contact. We might do this by administering a survey that asked subjects to agree or disagree with statements such as "Greater efforts must be made to racially integrate the public school system" or "Skin color is irrelevant and people are just people." Our goal would be to see if the people exposed to the cooperative contact situation actually become less prejudiced.

Now, finally, we come to statistics. As the observation phase of our research project comes to an end, we will be confronted with a large collection of numerical information or data. If our sample consisted of 100 people, we would have 200 completed surveys measuring prejudice: 100 completed before the contact situation and 100 filled out afterward. Try to imagine dealing with 200 completed surveys. If we had asked each respondent just five questions to measure his or her prejudice, we would have a total of 1,000 separate pieces of information to deal with. What do we do? We must have some systematic way to organize and analyze this information, and at this point, statistics will become very valuable. Statistics will supply us with many ideas about "what to do" with the data, and we will begin to look at some of the options in the next chapter. For now, let me stress two points about statistics.

First, statistics are crucial. Statistics enable social scientists to conduct **quantitative research:**[2] research based on the analysis of numerical information or data. Researchers use statistical techniques to organize and manipulate data so that they can test hypotheses, shape and refine theories, and improve our understanding of the social world. Second, and somewhat paradoxically, the role of statistics is limited. As Figure 1.1 makes clear, scientific research proceeds through several mutually interdependent stages, and statistics become directly relevant only at the end of the observation stage. Before any statistical analysis can be legitimately applied, the preceding phases of the process must be successfully completed. If the researcher has asked poorly conceived questions or has made serious errors of design or method, then even the most sophisticated statistical analysis is valueless. As useful as they can be, statistics cannot substitute for rigorous conceptualization, detailed and careful planning, or creative use of theory. Statistics cannot salvage a poorly conceived or designed research project. They cannot make sense out of garbage.

On the other hand, inappropriate statistical applications can limit the usefulness of an otherwise carefully done project. Only by successfully completing *all* phases of the process can a quantitative research project hope to contribute to understanding. A reasonable knowledge of the uses and limitations of statistics is as essential to the education of the social scientist as is training in theory and methodology.

As the statistical analysis comes to an end, we would begin to develop empirical generalizations. While we would be primarily focused on assessing our theory, we would also look for other trends in the data. Assuming that we found

---

[2]Social science researchers also do *qualitative research* or research in which information is expressed in a form other than numbers. Interviews, participant observation, and content analysis are examples of research methodologies that are often qualitative.

that equal-status, cooperative contact reduces prejudice in general, we might go on to ask if the pattern applies to males as well as females, to the well educated as well as the poorly educated, to older respondents as well as to the younger. As we probed the data, we might begin to develop some generalizations based on the empirical patterns we observe. For example, what if we found that contact reduced prejudice for younger respondents but not for older respondents? Could it be that younger people are less "set in their ways" and have attitudes and feelings that are more open to change? As we developed tentative explanations, we would begin to revise or elaborate our theory.

If we change the theory to take account of these findings, however, a new research project designed to test the revised theory is necessary and the wheel of science would begin to turn again. We (or perhaps some other researchers) would go through the entire process once again with this new—and, hopefully, improved—theory. This second project might result in further revisions and elaboration that would require still more research projects, and the wheel of science would continue turning as long as scientists were able to suggest additional revisions or develop new insights. Ideally, every time the wheel turned, our understandings of the phenomena under consideration would improve.

This description of the research process does not include white-coated, clipboard-carrying scientists who, in a blinding flash of inspiration, discover some fundamental truth about reality and shout, "Eureka!" The truth is that, in the normal course of science, we rarely can say with absolute certainty that a given theory or idea is definitely true or false. Rather, evidence for (or against) a theory will gradually accumulate over time, and ultimate judgments of truth will likely be the result of many years of hard work, research, and debate.

Let's briefly review our imaginary research project. We began with an idea or theory about intergroup contact and racial prejudice. We imagined some of the steps necessary to test the theory and took a quick look at the various stages of the research project. We wound up back at the level of theory, ready to begin a new project guided by a revised theory. We saw how theory can motivate a research project and how our observations might cause us to revise the theory and, thus, motivate a new research project. Wallace's wheel of science illustrates how theory stimulates research and how research shapes theory. This constant interaction between theory and research is the lifeblood of science and the key to enhancing our understandings of the social world.

The dialog between theory and research occurs at many levels and in multiple forms. Statistics are one of the most important links between these two realms. Statistics permit us to analyze data, to identify and probe trends and relationships, to develop generalizations, and to revise and improve our theories. As you will see throughout this text, statistics are limited in many ways. They are also an indispensable part of the research enterprise. Without statistics, the interaction between theory and research would become extremely difficult, and the progress of our disciplines would be severely retarded. *(For practice in describing the relationship between theory and research and the role of statistics in research, see problems 1.1 and 1.2.)*

1.3 THE GOALS OF THIS TEXT

In the preceding section, I argued that statistics are a crucial part of how scientific investigations are carried out and that, therefore, some training in statistical analysis is a crucial component in the education of every social scientist. In this section, I will address the questions of how much training is necessary and what the purposes of that training are.

First, this textbook takes the point of view that statistics are tools. They can be very useful as part of the process by which we increase our knowledge of the social world, but they are not ends in themselves. Thus, we will not take a "mathematical" approach to the subject. Statistical techniques will be presented as a set of tools that can help answer important questions. This emphasis does not mean that we will dispense with arithmetic entirely, of course. This text includes enough mathematical material so that you can develop a basic understanding of why statistics "do what they do." Our focus, however, will be on how these techniques are applied in the social sciences.

Second, all of you will soon become involved in advanced coursework in your major fields of study, and you will find that much of the literature used in these courses assumes at least basic statistical literacy. Furthermore, many of you, after graduation, will find yourselves in positions—either in a career or in graduate school—where some understanding of statistics will be very helpful or perhaps even required. Very few of you will become statisticians per se (and this text is not intended for the preprofessional statistician), but you must have a grasp of statistics to read and critically appreciate your own professional literature. As a student in the social sciences and in many careers related to the social sciences, you simply cannot realize your full potential without a background in statistics.

Within these constraints, this textbook is an introduction to statistics as they are used in the social sciences. The general goal here is to develop an appreciation—a "healthy respect"—for statistics and their place in research. You should emerge from this experience with the ability to use statistics intelligently and to know when other people have done so. You should be familiar with the advantages and limitations of the more commonly used statistical techniques, and you should know which techniques are appropriate for a given set of data and a given purpose. Lastly, you should develop sufficient statistical and computational skills and enough experience in the interpretation of statistics to be able to carry out some elementary forms of data analysis by yourself.

1.4 DESCRIPTIVE AND INFERENTIAL STATISTICS

As noted earlier, the general function of statistics is to manipulate data to answer research questions. This section introduces the two general classes of statistical techniques that, depending on the research situation, are available to accomplish this task.

Descriptive Statistics.   The first class of techniques is called **descriptive statistics** and is relevant in several different situations:

1. When a researcher needs to summarize or describe the distribution of a single variable. These statistics are called *univariate* ("one variable") descriptive statistics.
2. When the researcher wants to describe the relationship between two or more variables. These statistics are called *bivariate* ("two variables") or *multivariate* (more than two variables) descriptive statistics.

To describe a single variable, we would arrange the values or scores of that variable so that the relevant information can be quickly understood and appreciated. Many of the statistics that might be appropriate for this summarizing task are probably familiar to you. For example, we can use percentages, graphs, and charts to describe single variables.

To illustrate the usefulness of univariate descriptive statistics, consider the following problem: Suppose you wanted to summarize the distribution of the variable "family income" for a community of 10,000 families. How would you do it? Obviously, you couldn't simply list all incomes in the community and let it go at that. Imagine trying to make sense of a listing of 10,000 different incomes! Presumably, you would want to develop some summary measures of the overall income distributions—perhaps an arithmetic average or the proportions of incomes that fall in various ranges (such as low, middle, and high). Or perhaps a graph or a chart would be more useful. Whatever specific method you choose, its function is the same: to reduce these thousands of individual items of information into a few easily understood numbers. **Data reduction** allows a few numbers to summarize many numbers, and it is the basic goal of univariate descriptive statistical procedures. Part I of this text is devoted to these statistics; their primary goal is simply to report, clearly and concisely, essential information about a variable.

The second type of descriptive statistics is designed to help the investigator understand the relationship between two or more variables. These statistics, called **measures of association,** allow the researcher to quantify the strength and direction of a relationship. These statistics are very useful because they enable us to investigate two matters of central theoretical and practical importance to any science: causation and prediction. These techniques help us disentangle and uncover the connections between variables. They help us trace how some variables might have causal influences on others, and, depending on the strength of the relationship, they enable us to predict scores on one variable from the scores on another. Note that measures of association cannot, by themselves, prove that two variables are causally related. However, because these techniques can provide valuable clues about causation, they are extremely important for testing and constructing theories.

For example, suppose you were interested in the relationship between "time spent studying statistics" and "final grade in statistics" and had gathered data on these two variables from a group of college students. By calculating the appropriate measure of association, you could determine the strength of the bivariate relationship and its direction. Suppose you found a strong, positive relationship. This would indicate that "study time" and "grade" were closely related (strength of the relationship) and that as one increased in value, the other also increased (direction of the relationship). You could make predictions from one variable to the other ("the longer the study time, the higher the grade").

As a result of finding this strong, positive relationship, you might be tempted to make causal inferences. That is, you might jump to such conclusions as "longer study time leads to (causes) higher grades." Such a conclusion might make a good deal of common sense and would certainly be supported by your statistical analysis. However, the statistical analysis cannot prove the causal nature of the relationship. Measures of association can be important clues about causation, but the mere existence of a relationship can never be conclusive proof of causation: Causation and correlation are two different things and must not be confused.

In fact, other variables might have an effect on the relationship. In the preceding example, we probably would not find a perfect relationship between "study time" and "final grade." That is, we would likely find some individuals who spend a great deal of time studying but receive low grades and some individuals who fit the opposite pattern. We know intuitively that other variables besides study time affect grades (such as efficiency of study techniques, amount of background in mathematics, and even random chance). Fortunately, researchers can incorporate these other variables into the analysis and measure their effects. Part III of this text is devoted to bivariate (two-variable) and Part IV to multivariate (more than two variables) descriptive statistics.

Inferential Statistics. This second class of statistical techniques becomes relevant when we want to generalize our findings from a **sample** to a **population.** A population is the total collection of all cases in which the researcher is interested and wants to understand better. Examples of possible populations would be voters in the United States, all parliamentary democracies, unemployed Puerto Ricans in Atlanta, or sophomore college football players in the Midwest.

Populations can theoretically range from inconceivable in size ("all humanity") to quite small (all 35-year-old red-haired belly dancers currently residing in downtown Cleveland) but are usually fairly large. In fact, they are almost always too large to be measured. To put the problem another way, social scientists almost never have the resources or time to test every case in a population. This leads to the need for **inferential statistics,** which involve using information from a sample (a carefully chosen subset of the population) to make inferences about a population. Because they have fewer cases, samples are much cheaper to assemble, and—if the proper techniques are followed—generalizations based on these samples can be very accurate representations of the population.

Many of the concepts and procedures involved in inferential statistics may be unfamiliar. However, most of us are experienced consumers of inferential statistics—most familiarly, perhaps, in the form of public-opinion polls and election projections. When a public-opinion poll reports that 42% of the American electorate plans to vote for a certain presidential candidate, it is essentially

STATISTICS IN EVERYDAY LIFE: Introduction

In this age of information, statistical literacy is not just for academics or researchers. A critical attitude about statistics in everyday life—as well as in the social science research literature—can help us think more critically and carefully, assess the torrent of information, opinion, facts, and factoids that washes over us every day, and make better decisions on a broad range of issues.

To help develop these critical skills, I have included a series of boxed inserts in this text called "Statistics in Everyday Life" in which I examine how statistics can be applied to issues that you might encounter in your daily life: in the media, in conversations with family and friends, or in discussions with roommates or fellow workers. The topics range from traffic congestion to presidential politics to astrology and baseball. How can we evaluate information effectively? How can we assess the claims and counterclaims that blare at us from the media? The truth is elusive and multifaceted—how can we know it when we see it? Statistical literacy will not always lead you to the truth, of course, but it will enhance your ability to analyze and evaluate information.

reporting a generalization to a population ("the American electorate"—which numbers about 100 million people) from a carefully drawn sample (usually about 1,500 respondents). Matters of inferential statistics will occupy our attention in Part II of this book. *(For practice in describing different statistical applications, see problems 1.3 and 1.7.)*

1.5 LEVEL OF MEASUREMENT

In the next chapter, you will begin to encounter some of the broad array of statistics available to the social scientist. One aspect of using statistics that can be puzzling is deciding when to use which statistic. You will learn specific guidelines as you go along, but we will consider the most basic and important guideline at this point: the **level of measurement,** or the mathematical nature of the variables under consideration. Variables at the highest level of measurement have numerical scores and can be analyzed with a broad range of statistics. Variables at lower levels of measurement have "scores" that are really just labels, not numbers at all. Statistics that require numerical variables are inappropriate and, usually, completely meaningless when used with non-numerical variables. When selecting statistics, you must be sure that the level of measurement of the variable justifies the mathematical operations required to compute the statistic.

For example, consider the variables age (measured in years) and income (measured in dollars). Both of these variables have numerical scores and could be summarized with a statistic such as the mean or average (for example, "The average income in this city is $43,000" or "The average age of students on this campus is 19.7"). In contrast, the arithmetic average would be meaningless as a way of describing religious denomination or zip codes, variables with nonnumerical scores. Your personal zip code might *look* like a number but it is merely an arbitrary label that happens to be expressed in digits. The "numbers" in your zip code cannot be added or divided, and statistics like the average cannot be applied to this variable: The average zip code of a group of people is a meaningless statistic.

Determining the level at which a variable has been measured is one of the first steps in any statistical analysis, and we will consider this matter at some length. Throughout this text, I will introduce level-of-measurement considerations for each statistical technique.

There are three levels of measurement. In order of increasing sophistication, they are nominal, ordinal, and interval-ratio.

The Nominal Level of Measurement. Variables measured at the nominal level have non-numerical scores or categories. Examples of variables at the nominal level include gender, zip code, race, religious affiliation, and place of birth. At this lowest level of measurement, the only mathematical operation permitted is comparing the relative sizes of the categories of the variable (e.g., "there are more females than males in this dorm"). The categories or scores of nominal-level variables cannot be ranked with respect to each other and cannot be added, divided, or otherwise manipulated mathematically. Even when the scores or categories are expressed in digits (such as zip codes or street addresses), all we can do is compare relative sizes of categories (for example, "the most common zip code on this campus is 22033"). The scores of nominal-level variables do not form a mathematical scale: The scores are different from each other but not more or less or higher or lower than each other. Males and females differ in terms of gender but neither category has more or less gender than the other. In the same way, a zip code of 54398 is different from but not "more than" a zip code of 13427.

Nominal variables are rudimentary, but there are criteria and procedures that we need to observe to ensure adequate measurement. In fact, these criteria apply to variables measured at all levels, not just nominal variables. First, the categories of nominal-level variables must be mutually exclusive of each other so that no ambiguity exists concerning classification of any given case. There must be one and only one category for each case. Second, the categories must be exhaustive. In other words, there must be a category—at least an "other" or miscellaneous category—for every possible score that might be found.

Third, the categories of nominal variables should be relatively homogeneous. That is, our categories should include cases that are truly comparable, or to put it another way, we need to avoid categories that lump apples with oranges. There are no hard and fast guidelines for judging if a set of categories is appropriately homogeneous. The researcher must make that decision in terms of the specific purpose of the research; categories that are too broad for some purposes may be perfectly adequate for others.

Table 1.1 demonstrates some errors of measurement in four different schemes for measuring the nominal-level variable "religious preference." Scale A in the table violates the criterion of mutual exclusivity because of overlap between the categories Protestant and Episcopalian. Scale B is not exhaustive because it does not provide a category for people with no religious preference (None) or people who belong to religions other than the three listed. Scale C uses a category (Non-Protestant) that would be too broad for many research purposes. Scale D is the way religious preference is often measured in North America, but note that these categories may be too general for some research projects and not comprehensive enough for others. For example, an investigation of issues that have strong moral and religious content (assisted suicide, abortion, or capital punishment, for example) might need to distinguish between the various Protestant denominations, and an effort to document religious diversity would need to add categories for Buddhists, Muslims, and other religious preferences that are less common in North America.

TABLE 1.1 FOUR SCALES FOR MEASURING RELIGIOUS PREFERENCE

Scale A (not mutually exclusive)	Scale B (not exhaustive)	Scale C (not homogeneous)	Scale D (an adequate scale)
Protestant	Protestant	Protestant	Protestant
Episcopalian	Catholic	Non-Protestant	Catholic
Catholic	Jewish		Jewish
Jewish			None
None			Other
Other			

As is the case with zip codes, numerical labels are often used to identify the categories or scores of nominal-level variables, especially when the data are being prepared for computer analysis. For example, the various religions might be labeled with a 1 indicating Protestant, a 2 signifying Catholic, and so on. Remember that these numbers are merely labels or names and have no numerical quality to them. They cannot be ranked, added, subtracted, multiplied, or divided. The only mathematical operation permissible with nominal variables is counting and comparing the number of cases in each category of the variable.

The Ordinal Level of Measurement. Variables measured at the ordinal level are more sophisticated than nominal-level variables. They have scores or categories that can be ranked from high to low, so, in addition to classifying cases into categories, we can describe the categories in terms of "more or less" with respect to each other. Thus, with variables measured at this level, we can say not only that one case is different from another but also that one case is higher or lower, more or less than another.

For example, the variable socioeconomic status (SES) is usually measured at the ordinal level. The categories of the variable are often ordered according to the following scheme:

4. Upper class
3. Middle class
2. Working class
1. Lower class

Individuals can be compared in terms of the categories into which they are classified: a person classified as a 4 (upper class) would be ranked higher than someone classified as a 2 (working class) and a lower-class person (1) would rank lower than a middle-class person (3). Other variables that are usually measured at the ordinal level include attitude and opinion scales such as those that measure prejudice, alienation, or political conservatism.

The major limitation of the ordinal level of measurement is that a particular score represents only position with respect to some other score. We can distinguish between high and low scores, but the distance between the scores cannot be described in precise terms. Although we know that a score of 4 is more than a score of 2, we do not know if it is twice as much as 2.

Because we don't know what the exact distances are from score to score on an ordinal scale, our options for statistical analysis are limited. For example, addition assumes that the intervals between scores are exactly equal (as do most other mathematical operations). If the distances from score to score are not equal, 2 + 2 might equal 3 or 5 or even 15. Thus, strictly speaking, statistics such as the average or mean (which requires that the scores be added together and then divided by the number of scores) are not permitted with ordinal-level variables. The most sophisticated mathematical operation fully justified with an ordinal variable is the ranking of categories and cases (although, as we will see, it is common for social scientists to take liberties with this criterion).

STEP BY STEP	Determining the Level of Measurement of a Variable

To determine the level of measurement of a variable, focus on the scores or values of the variable and use the following questions, organized as steps. If the answer to the question in step 1 is yes, the variable is nominal and you do not need to go any farther. If the answer to the question in Step 2 is yes, the variable is ordinal. If the answers to the first two questions are no, the variable is interval-ratio.

Step 1: Change the order of the scores of the variable. If the reordered scores still make sense, the variable is nominal.

The key characteristic of nominal-level variables is that the scores or categories have no mathematical relationship with each other and can be stated in any order without violating the logic of the variable. To illustrate, the table below displays several different arrangements of the scores of marital status, a nominal-level variable. Each statement of the scores is just as sensible as every other statement. On a nominal-level variable, no score is higher or lower than any other score and the order of the scores is arbitrary.

Various Arrangements of Scores for Marital Status

1. Married	1. Divorced	1. Never Married
2. Divorced	2. Widowed	2. Divorced
3. Widowed	3. Separated	3. Married
4. Separated	4. Married	4. Widowed
5. Never Married	5. Never Married	5. Separated

If the rearranged scores do not make sense—if the scores have an ordered relationship and can be arrayed from high to low or more to less—go on to Step 2.

Step 2: Is the distance between the scores of the variable unequal or undefined? If the answer is yes, the variable is ordinal and there is no need to go on to Step 3.

To illustrate, consider a survey item that measures support for the death penalty by using a scale of five possible responses:

1. Strongly approve
2. Slightly approve
3. Neither approve nor disapprove
4. Slightly disapprove
5. Strongly disapprove

People who "strongly" approve of capital punishment are more in favor than people who "slightly" approve, but the distance from one level of endorsement to the next is undefined: We do not have enough information to ascertain "how much more or less" one score is than another.

Step 3: If you answered no to the questions presented in Steps 1 and 2, the variable is interval-ratio: It has scores that must be stated in a certain order and the scores have a defined, equal distance from score to score. Examples of interval-ratio variables include income, years of education, and age. Remember that interval-ratio variables also have a true zero point.

Source: This system for determining level of measurement was suggested Michael R. Bisciglia, Louisiana State University.

The Interval-Ratio Level of Measurement.[3]   The categories of nominal-level variables have no numerical quality to them. Ordinal-level variables have categories that can be arrayed along a scale from high to low, but the exact distances between categories or scores are undefined. Variables measured at the interval-ratio level not only permit classification and ranking but also allow the distance from category to category (or score to score) to be exactly defined.

---

[3]Many statisticians distinguish between the interval level (equal intervals) and the ratio level (true zero point). I find the distinction unnecessarily cumbersome in an introductory text and will treat these two levels as one.

TABLE 1.2 BASIC CHARACTERISTICS OF THE THREE LEVELS OF MEASUREMENT

Levels	Examples	Measurement Procedures	Mathematical Operations Permitted
Nominal	Sex, race, religion, marital status	Classification into categories	Counting number in each category, comparing sizes of categories
Ordinal	Social class, attitude and opinion scales	Classification into categories plus ranking of categories with respect to each other	All of the above plus statements of "greater than" and "less than"
Interval-ratio	Age, number of children, income	All of the above plus description of distances between scores in terms of equal units	All of the above plus all other mathematical operations (addition, subtraction, multiplication, division, square roots, and so on)

Interval-ratio variables have two characteristics. First, they are measured in units that have equal intervals. For example, asking people how old they are will produce an interval-ratio level variable (age) because the unit of measurement (years) has equal intervals (the distance from year to year is 365 days). Similarly, if we ask people how many siblings they have, we would produce a variable with equal intervals: 2 siblings are 1 more than 1 and 13 is 1 more than 12.

The second characteristic of interval-ratio variables is that they have a true zero point. That is, the score of zero for these variables is not arbitrary: It indicates the absence or complete lack of whatever is being measured. For example, the variable "number of siblings" has a true zero point because it is possible to have no siblings. Similarly, it is possible to have zero years of education, no income at all, a score of zero on a multiple-choice test, and to be zero years old (although not for very long). Other examples of interval-ratio variables would be number of children, life expectancy, and years married. All mathematical operations are permitted for data measured at the interval-ratio level.

Level of Measurement: A Summary. Table 1.2 presents the basic characteristics of the three levels of measurement. Note that the number of permitted mathematical operations increases as we move from nominal to ordinal to interval-ratio levels of measurement. Ordinal-level variables are more sophisticated and flexible than nominal-level variables, and interval-ratio-level variables permit the broadest range of mathematical operations.

Remember that level of measurement is important because different statistics require different mathematical operations. The level of measurement of a variable is the key characteristic that tells us which statistics are permissible and appropriate.

Also remember to examine the way in which the scores are *actually* stated. In particular, you should be aware that the scores of interval-ratio variables are often presented in ordinal format. For example, researchers usually do not ask respondents to report their income to an exact dollar value. Instead, respondents are asked to check one of a series of broad categories, as in this list:

___ Less than $24,999
___ $25,000 to $49,999
___ $50,000 to $99,999
___ More than $100,000

Because these categories as actually stated are unequal in size, the variable is ordinal in level of measurement.

Ideally, researchers would use only those statistics that were fully justified by the level-of-measurement criteria. In this imperfect world, however, the most powerful and useful statistics (such as the mean) require interval-ratio variables, while most of the variables of interest to the social sciences are only nominal (race, sex, marital status) or ordinal (attitude scales). Relatively few concepts of interest to the social sciences are so precisely defined that they can be measured at the interval-ratio level. This disparity creates some very real difficulties. On one hand, researchers generally should use the most sophisticated statistical procedures fully justified for a particular variable. Treating interval-ratio data as if they were only ordinal, for example, results in a significant loss of information and precision. Treated as an interval-ratio variable, the variable "age" can supply us with exact information regarding the differences between the cases (for example, "Individual A is exactly three years and two months older than Individual B"). Treated only as an ordinal variable, however, the precision of our comparisons would suffer and we could say only that "Individual A is older (or greater) than Individual B."

On the other hand, given the nature of the disparity, researchers are more likely to treat variables as if they were higher in level of measurement than they actually are. In particular, variables measured at the ordinal level, especially when they have many possible categories or scores, are often treated as if they were interval-ratio because the statistical procedures available at the higher level are more powerful, flexible, and interesting. This practice is common but researchers should be cautious in assessing statistical results and developing interpretations when the level-of-measurement criterion has been violated.

At any rate, level of measurement is a very basic characteristic of a variable, and we will always consider it when presenting statistical procedures. Level of measurement is also a major organizing principle for the material that follows, and you should make sure that you are familiar with these guidelines. *(For practice in determining the level of measurement of a variable, see problems 1.4 through 1.8.)*

SUMMARY

1. Within the context of social research, the purpose of statistics is to organize, manipulate, and analyze data so that researchers can test their theories and answer their questions. Along with theory and methodology, statistics are a basic tool by which social scientists attempt to enhance their understanding of the social world.

2. There are two general classes of statistics. Descriptive statistics are used to summarize the distribution of a single variable and the relationships between two or more variables. Inferential statistics provide us with techniques by which we can generalize to populations from random samples.

3. Variables may be measured at any of three different levels. At the nominal level, we can compare category sizes. At the ordinal level, categories and cases can be ranked with respect to each other. At the interval-ratio level, all mathematical operations are permitted.

GLOSSARY

Data. Any information collected as part of a research project and expressed as numbers.

Data reduction. Summarizing many scores with a few statistics. A major goal of descriptive statistics.

Dependent variable. A variable that is identified as an effect, result, or outcome variable. The dependent variable is thought to be caused by the independent variable.

Descriptive statistics The branch of statistics concerned with (1) summarizing the distribution of a single variable or (2) measuring the relationship between two or more variables.

Hypothesis. A statement about the relationship between variables that is derived from a theory. Hypotheses are more specific than theories, and all terms and concepts of a hypothesis are fully defined.

Independent variable. A variable that is identified as a causal variable. The independent variable is thought to cause the dependent variable.

Inferential statistics. The branch of statistics concerned with making generalizations from samples to populations.

Level of measurement. The mathematical characteristic of a variable and the major criterion for selecting statistical techniques. Variables can be measured at any of three levels, each permitting certain mathematical operations and statistical techniques. Table 1.2 summarizes the characteristics of the three levels.

Measure of association. Statistic that summarizes the strength and direction of the relationship between variables.

Population. The total collection of all cases in which the researcher is interested.

Quantitative research. Research focused on numerical information or data.

Research. Any process of gathering information systematically and carefully to answer questions or test theories. Statistics are useful for research projects in which the information is represented in numerical form or as data.

Sample. A carefully chosen subset of a population. In inferential statistics, information is gathered from a sample and then generalized to a population.

Statistics. A set of mathematical techniques for organizing and analyzing data.

Theory. A generalized explanation of the relationship between two or more variables.

Variable. Any trait that can change values from case to case.

PROBLEMS

1.1 In your own words, describe the role of statistics in the research process. Using the "wheel of science" as a framework, explain how statistics link theory with research.

1.2 Find a research article in any social science journal. Choose an article on a subject of interest to you and don't worry about being able to understand all of the statistics that are reported.

   a. How much of the article is devoted to statistics per se (as distinct from theory, ideas, discussion, and so on)?

   b. Is the research based on a sample from some population? How large is the sample? How were subjects or cases selected? Can the findings be generalized to some population?

   c. What variables are used? Which are independent and which are dependent? For each variable, determine the level of measurement.

   d. What statistical techniques are used? Try to follow the statistical analysis and see how much you can understand. Save the article and read it again after you finish this course and see if you do any better.

1.3 Distinguish between descriptive and inferential statistics. Describe a research situation that would use both types.

1.4 For each of the following items from a public-opinion survey, indicate the level of measurement:

a. What is your occupation? _____

b. How many years of school have you completed? _____

c. If you were asked to use one of these four names for your social class, which would you say you belonged in?

_____ Upper   _____ Middle
_____ Working   _____ Lower

d. What is your age? _____

e. In what country were you born? _____

f. What is your grade-point average? _____

g. What is your major? _____

h. The only way to deal with the drug problem is to legalize all drugs.

_____ Strongly agree
_____ Agree
_____ Undecided
_____ Disagree
_____ Strongly disagree

i. What is your astrological sign? _____

j. How many brothers and sisters do you have? _____

1.5 Below are brief descriptions of how researchers measured a variable. For each situation, determine the level of measurement of the variable.

a. Race. Respondents were asked to select a category from the following list:

_____ Black
_____ White
_____ Other

b. Honesty. Subjects were observed as they passed by a spot on campus where an apparently lost wallet was lying. The wallet contained money and complete identification. Subjects were classified into one of the following categories:

_____ Returned the wallet with money
_____ Returned the wallet but kept the money
_____ Did not return wallet

c. Social class. Subjects were asked about their family situation when they were 16 years old. Was their family:

_____ Very well off compared to other families?
_____ About average?
_____ Not so well off?

d. Education. Subjects were asked how many years of schooling they and each parent had completed.

e. Racial integration on campus. Students were observed during lunchtime at the cafeteria for a month. The number of students sitting with students of other races was counted for each meal.

f. Number of children. Subjects were asked: "How many children have you ever had? Please include any that may have passed away."

g. Student seating patterns in classrooms. On the first day of class, instructors noted where each student sat. Seating patterns were remeasured every two weeks until the end of the semester. Each student was classified as

_____ same seat as last measurement
_____ adjacent seat
_____ different seat, not adjacent
_____ absent

h. Physicians per capita. The number of practicing physicians was counted in each of 50 cities, and the researchers used population data to compute the number of physicians per capita.

i. Physical attractiveness. A panel of 10 judges rated each of 50 photos of a mixed-race sample of males and females for physical attractiveness on a scale from 0 to 20 with 20 being the highest score.

j. Number of accidents. The number of traffic accidents for each of 20 busy intersections in a city was recorded. Also, each accident was rated as

_____ minor damage, no injuries
_____ moderate damage, personal injury requiring hospitalization
_____ severe damage and injury

1.6. What is the level of measurement of each of the first 20 items in the General Social Survey (see Appendix G)?

1.7 For each research situation summarized below, identify the level of measurement of all variables. Also, decide which statistical applications are used: descriptive statistics with a single variable, descriptive statistics with two or more variables, or inferential statistics. Remember that it is common for a given situation to require more than one type of application.

a. The administration of your university is proposing a change in parking policy. You select a random sample of students and ask each one if he or she favors or opposes the change.

b. You ask everyone in your social research class to tell you the highest grade he or she ever received in a math course and the grade on a recent statistics test. You then compare the two sets of scores to see if there is any relationship.

c. Your aunt is running for mayor and hires you (for a huge fee, incidentally) to question a sample of voters about their concerns in local politics. Specifically, she wants a profile of the voters that will tell her what percent belong to each political party, what percent are male or female, and what percent favor or oppose the widening of the main street in town.

d. Several years ago, a state reinstituted the death penalty for first-degree homicide. Supporters of capital punishment argued that this change would reduce the homicide rate. To investigate this claim, a researcher has gathered information on the number of homicides in the state for the two-year periods before and after the change.

e. A local automobile dealer is concerned about customer satisfaction. He wants to mail a survey form to all customers during the past year and ask if they are satisfied, very satisfied, or not satisfied with their purchases.

1.8 For each research situation below, identify the independent and dependent variables. Classify each in terms of level of measurement.

a. A graduate student is studying sexual harassment on college campuses and asks 500 female students if they personally have experienced any such incidents. Each student is asked to estimate the frequency of these incidents as "often, sometimes, rarely, or never." The researcher also gathers data on age and major to see if there is any connection between these variables and frequency of sexual harassment.

b. A supervisor in the Solid Waste Management Division of a city government is attempting to assess two different methods of trash collection. One area of the city uses trucks with two-man crews who do "backyard" pickups, and "high-tech" single-person trucks with curbside pickup handle the rest of the city. The assessment measures include the number of complaints received from the two different areas during a six-month period, the amount of time per day required to service each area, and the cost per ton of trash collected.

c. Police have raided and closed the adult bookstore near campus. Your social research class has decided to poll the student body and get reactions and opinions. The class asks each student if he or she supports or opposes the closing of the store, how many times each one has visited the store, and if he or she agrees or disagrees that "pornography is a direct cause of sexual assaults on women." The class also collects information on the sex, age, religious and political philosophy, and major of each student to see if opinions are related to these characteristics.

d. For a research project in a political science course, a student has collected information about the quality of life and the degree of political democracy in 50 nations. Specifically, she used infant mortality rates to measure quality of life and the percentage of all adults who are permitted to vote in national elections as a measure of democratization. Her hypothesis is that quality of life is higher in more democratic nations.

e. A highway engineer wonders if a planned increase in the speed limit on a heavily traveled local avenue will result in any change in the number of accidents. He plans to collect information on traffic volume, number of accidents, and number of fatalities for the six-month periods before and after the change.

f. Students are planning a program to promote "safe sex" and awareness of a variety of other health concerns for college students. To measure the effectiveness of the program, they plan to give a survey measuring knowledge about these matters to a random sample of the student body before and after the program.

g. Several states have drastically cut their budgets for mental health care. Will this increase the number of homeless people in these states? A researcher contacts a number of agencies serving the homeless in each state and develops an estimate of the size of the homeless population before and after the cuts.

h. Does tolerance for diversity vary by race, ethnicity, or gender? Samples of white, black, Asian, Hispanic, and Native American adults have taken a survey that measures their interest in and appreciation of cultures and groups other than their own.

Introduction to SPSS and the General Social Survey

The problems at the end of chapters in this text have been written so that they can be solved with just a simple hand calculator. I've purposely kept the number of cases involved unrealistically low so that the tedium of mere calculation would not interfere unduly with the learning process. To provide a more realistic experience in the analysis of social science data, we will analyze a shortened version of the 2004 General Social Survey (GSS). This database can be downloaded from the companion website for this text (go to ***http://www.thomsonedu.com/sociology*** and search for this text's title). The GSS is a public-opinion poll that has been conducted on nationally representative samples of U.S. citizens since 1972. The full survey includes hundreds of questions covering a broad range of social and political issues. The version supplied with this text has a limited number of variables and cases but is still actual, real-life data, so you have the opportunity to practice your statistical skills in a more realistic context.

One of the problems with reality, of course, is that it is often cumbersome and confusing. It's hard enough to do your homework with simplified problems, and you should be a little leery, in terms of your own time and effort, of promises of relevance and realism. This brings us to the second purpose of this section: computers and statistical packages. A statistical package is a set of computer programs for the analysis of data. The advantage of these packages is that, because the programs are already written, you can capitalize on the power of the computer with minimal computer literacy and virtually no programming experience.

This text uses a set of statistical computer programs called the Statistical Package for the Social Sciences (SPSS). In these sections at the ends of chapters, I will explain how to use this package to manipulate and analyze the GSS data, and I will illustrate and interpret the results. Be sure to read Appendix F before attempting any data analysis.

Part I

Descriptive Statistics

Part I consists of four chapters, each devoted to a different application of univariate descriptive statistics. Chapter 2 covers "basic" descriptive statistics, including percentages, ratios, rates, frequency distributions, and graphs. It is a lengthy chapter but the material is relatively elementary and at least vaguely familiar to most students. Although the statistics covered in this chapter are "basic," they are not necessarily simple or obvious. You should consider the explanations and examples carefully before attempting the end-of-chapter problems or using them in actual research.

Chapters 3 and 4 cover measures of central tendency and dispersion, respectively. Measures of central tendency describe the typical case or average score (e.g., the mean) while measures of dispersion describe the amount of variety or diversity among the scores (e.g., the range or the distance from the high score to the low score). These two types of statistics appear in separate chapters to stress the point that centrality and dispersion are independent, separate characteristics of a variable. You should realize, however, that *both* measures are necessary and commonly reported together (along with some of the statistics presented in Chapter 2). To reinforce the idea that measures of centrality and dispersion are complementary descriptive statistics, many of the problems at the end of Chapter 4 require the computation of a measure of central tendency from Chapter 3.

Chapter 5 is a pivotal chapter in the flow of the text. It takes some of the statistics from Chapters 2 through 4 and applies them to the normal curve, a concept of great importance in statistics. The normal curve is a type of line chart or frequency polygon (see Chapter 2), which we can use to describe the position of scores using means (Chapter 3) and standard deviations (Chapter 4). Chapter 5 also uses proportions (Chapter 2) to introduce the concept of probability, a central component of social science research.

In addition to its role in descriptive statistics, the normal curve is a central concept in inferential statistics, the topic of Part II of this text. Thus, Chapter 5 serves a dual purpose: It ends the presentation of univariate descriptive statistics and lays essential groundwork for the material to come.

2

Basic Descriptive Statistics
Percentages, Ratios and Rates,
Tables, Charts, and Graphs

LEARNING OBJECTIVES

By the end of this chapter, you will be able to:

1. Explain the purpose of descriptive statistics in making data comprehensible.
2. Compute and interpret percentages, proportions, ratios, rates, and percentage change.
3. Construct and analyze frequency distributions for variables at each of the three levels of measurement.
4. Construct and analyze bar and pie charts, histograms, and line graphs.

Research results do not speak for themselves. They must be organized and manipulated so that the researcher and his or her readers can quickly and easily understand whatever meaning they have. Researchers use statistics to clarify their results and communicate effectively. In this chapter, we will consider some commonly used techniques for presenting research results: percentages and proportions, ratios and rates, percentage change, tables, charts, and graphs. Mathematically speaking, these univariate descriptive statistics are not very complex (although they are not as simple as they may appear at first glance), but they are extremely useful for presenting research results clearly and concisely.

2.1 PERCENTAGES AND PROPORTIONS

Consider the following statement: "Of the 269 cases handled by the court, 167 resulted in prison sentences of five years or more." While there is nothing wrong with this statement, the same fact could have been more clearly conveyed if it had been reported as a percentage: "About 62 percent of all cases resulted in prison sentences of five or more years."

Percentages and **proportions** supply a frame of reference for reporting research results in the sense that they standardize the raw data: percentages to the base 100 and proportions to the base 1.00. The mathematical definitions of proportions and percentages are:

FORMULA 2.1

$$\text{Proportion:} \quad p = \frac{f}{N}$$

FORMULA 2.2

$$\text{Percentage:} \quad \% = \left(\frac{f}{N}\right) \times 100$$

Where f = frequency, or the number of cases in any category
N = the number of cases in all categories

TABLE 2.1   DISPOSITION OF 269 CRIMINAL CASES (fictitious data)*

Sentence	Frequency (f)	Proportion (p)	Percentage (%)
Five Years or More	167	0.6208	62.08
Less Than Five Years	72	0.2677	26.77
Suspended	20	0.0744	7.44
Acquitted	10	0.0372	3.72
Totals =	269	1.0001	100.01%

*The slight discrepancies in the totals of the proportion and percentage columns are due to rounding error.

To illustrate the computation of percentages, consider the data presented in Table 2.1. Note that there are 167 cases in the category ($f = 167$) "Five years or more" and a total of 269 cases in all ($N = 269$). So:

$$\text{Percentage (\%)} = \left(\frac{f}{N}\right) \times 100 = \left(\frac{167}{269}\right) \times 100 = (0.6208) \times 100 = 62.08\%$$

Using the same procedures, we can also find the percentage of cases in the second category:

$$\text{Percentage (\%)} = \left(\frac{f}{N}\right) \times 100 = \left(\frac{72}{269}\right) \times 100 = (0.2677) \times 100 = 26.77\%$$

Both results can also be expressed as proportions. For example, the proportion of cases in the third category is 0.0744.

$$\text{Proportion } (p) = \frac{f}{N} = \frac{20}{269} = 0.0744$$

Percentages and proportions are easier to read and comprehend than frequencies. This advantage is particularly obvious when attempting to compare groups of different sizes. For example, based on the information presented in Table 2.2, which college has the higher relative number of social science majors?

Because the total enrollments are so different, comparisons are difficult to make from the raw frequencies. Computing percentages eliminates the difference in size of the two campuses by standardizing both distributions to the base of 100. Table 2.3 presents the same data in percentages.

The percentages in Table 2.3 make it easier to identify both differences and similarities between the two colleges. College A has a much higher percentage of social science majors (even though the absolute number of social science majors is less than at College B) and about the same percentage of humanities

TABLE 2.2   DECLARED MAJOR FIELDS ON TWO COLLEGE CAMPUSES (fictitious data)

Major	College A	College B
Business	103	312
Natural Sciences	82	279
Social Sciences	137	188
Humanities	93	217
	$N = 415$	996

TABLE 2.3   DECLARED MAJOR FIELDS ON TWO COLLEGE CAMPUSES IN PERCENTAGES (fictitious data)

Major	College A	College B
Business	24.82	31.33
Natural Sciences	19.76	28.01
Social Sciences	33.01	18.88
Humanities	22.41	21.79
	100.00%	100.01%
	(415)	(996)

Application 2.1

In Table 2.2, suppose that 237 of the 415 students enrolled in College A and 458 of the 996 students enrolled in College B are males. What percentage of each student body is male?

College A

$$\% = \left(\frac{237}{415}\right) \times 100 = (.5711) \times 100 = 57.11\%$$

College B

$$\% = \left(\frac{458}{996}\right) \times 100 = (.4598) \times 100 = 45.98\%$$

College B has the greater number of men but College A has the larger percentage.

---

majors. How would you describe the differences in the remaining two major fields? (*For practice in computing and interpreting percentages and proportions, see problems 2.1 and 2.2.*)

Some further guidelines on the use of percentages and proportions:

1. When working with a small number of cases (say, fewer than 20), it is usually preferable to report the actual frequencies rather than percentages or proportions. With a small number of cases, the percentages can change drastically with relatively minor changes in the data. For example, if you begin with a data set that includes 10 males and 10 females (that is, 50% of each gender) and then add another female, the percentage distributions will change noticeably to 52.38% female and 47.62% male. Of course, as the number of observations increases, each additional case will have a smaller impact. If we started with 500 males and 500 females and then added one more female, the percentage of females would change by only a tenth of a percent (from 50% to 50.10%).

2. Always report the number of observations along with proportions and percentages. This permits the reader to judge the adequacy of the sample size and, conversely, helps to prevent the researcher from lying with statistics. Statements such as "Two out of three people questioned prefer courses in statistics to any other course" might sound impressive, but the claim would lose its gloss if you learned that only three people were asked. *You should be extremely suspicious of reports that fail to report the number of cases that were tested.*

| STEP BY STEP | Computing Percentages and Proportions |

Step 1: Determine the values for f (number of cases in a category) and N (number of cases in all categories). Remember that f will be the number of cases in a *specific category* (e.g., males on your campus) and N will be the number of cases in *all* categories (e.g., all students, males and females, on your campus) and that f will be smaller than N, except when the category and the entire group are

the same (e.g., when all students are male). Proportions cannot exceed 1.00 and percentages cannot exceed 100.00%.

Step 2: For a proportion, divide f by N.

Step 3: For a percentage, multiply the value you calculated in step 2 by 100.

3. Percentages and proportions can be calculated for variables at the ordinal and nominal levels of measurement, in spite of the fact that these statistics require division. This is not a violation of the level-of-measurement guideline (see Table 1.2). Percentages and proportions do not require the division of the *scores* of the variable, as would be the case in computing the average score on a test, for example. Instead, we divide the *number of cases* in a particular category (f) of the variable by the *total number of cases* in the sample (N). When we state that "43% of the sample is female," we are merely expressing the relative size of a category (female) of the variable (gender) in a convenient way.

2.2 RATIOS, RATES, AND PERCENTAGE CHANGE

Ratios, rates, and percentage change provide some additional ways of summarizing results simply and clearly. Although they are similar to each other, each statistic has a specific application and purpose.

Ratios. **Ratios** are especially useful for comparing the number of cases in the categories of a variable. Instead of standardizing the distribution of the variable to the base 100 or 1.00, as we did in computing percentages and proportions, we determine ratios by dividing the frequency of one category by the frequency in another. Mathematically, a ratio can be defined as:

FORMULA 2.3
$$\text{Ratio} = \frac{f_1}{f_2}$$

Where f_1 = the number of cases in the first category
f_2 = the number of cases in the second category

To illustrate the use of ratios, suppose that you were interested in the relative sizes of the various religious denominations and found that a particular community included 1,370 Protestant families and 930 Catholic families. To find the ratio of Protestants (f_1) to Catholics (f_2), divide 1,370 by 930:

$$\text{Ratio} = \frac{f_1}{f_2} = \frac{1,370}{930} = 1.47$$

The resultant ratio is 1.47, which means that for every Catholic family, there are 1.47 Protestant families.

Ratios can be very economical ways of expressing the relative predominance of two categories. That Protestants outnumber Catholics in our example is obvious from the raw data. We could also use percentages or proportions to summarize the overall distribution (e.g., "59.56% of the families were Protestant, 40.44% were Catholic"). In contrast to these other methods, ratios express the relative size of the categories: They tell us exactly how much one category outnumbers the other.

Ratios are often multiplied by some power of 10 to eliminate decimal points. For example, the ratio computed above might be multiplied by 100 and reported as 147 instead of 1.47. This would mean that, for every 100 Catholic families, there are 147 Protestant families in the community. To ensure clarity, the comparison units for the ratio are often expressed as well. Based on a unit of ones, the ratio of Protestants to Catholics would be expressed as 1.47:1. Based on hundreds, the same statistic might be expressed as 147:100. (*For practice in computing and interpreting ratios, see problems 2.1 and 2.2.*)

Rates.   **Rates** provide still another way of summarizing the distribution of a single variable. Rates are the number of actual occurrences of some phenomenon divided by the number of possible occurrences per some unit of time. Rates are usually multiplied by some power of 10 to eliminate decimal points. For example, the crude death rate for a population is defined as the number of deaths in that population (actual occurrences) divided by the number of people in the population (possible occurrences) per year. Then we multiply this quantity by 1,000. The formula for the crude death rate can be expressed as:

$$\text{Crude death rate} = \frac{\text{Number of deaths}}{\text{Total population}} \times 1{,}000$$

If there were 100 deaths during a given year in a town of 7,000, the crude death rate for that year would be:

$$\text{Crude death rate} = \frac{100}{7{,}000} \times 1{,}000 = (0.01429) \times 1{,}000 = 14.29$$

Or, for every 1,000 people, there were 14.29 deaths during this particular year. In the same way, if a city of 237,000 people experienced 120 auto thefts during a certain year, the auto theft rate would be:

$$\text{Auto theft rate} = \frac{120}{237{,}000} \times 100{,}000 = (0.00005063) \times 100{,}000 = 50.63$$

Or, for every 100,000 people, there were 50.63 auto thefts during the year in question. (*For practice in computing and interpreting rates, see problems 2.3 and 2.4a.*)

Percentage Change.   Measuring social change, in all its variety, is an important task for all social sciences. One very useful statistic for this purpose is the **percentage change,** which tells us how much a variable has increased or decreased over a certain span of time.

To compute this statistic, we need the scores of a variable at two different times. The scores could be in the form of frequencies, rates, or percentages.

Application 2.2

How many natural science majors are there compared to social science majors at College B? We could answer this question with frequencies, but a more easily understood way of expressing the answer would be with a ratio. The ratio of natural science to social science majors would be:

$$\text{Ratio} = \frac{f_1}{f_2} = \frac{279}{188} = 1.48$$

For every social science major, there are 1.48 natural science majors at College B.

Application 2.3

In 2000, there were 2,500 births in a city of 167,000. In 1960, when the population of the city was only 133,000, there were 2,700 births. Is the birthrate rising or falling? Although this question can be answered from the preceding information, the trend in birthrates will be much more obvious if we compute birthrates for both years. As with crude death rates, crude birthrates are usually multiplied by 1,000 to eliminate decimal points. For 1960:

$$\text{Crude birthrate} = \frac{2,700}{133,000} \times 1,000 = 20.30$$

In 1960, there were 20.30 births for every 1,000 people in the city. For 2000:

$$\text{Crude birthrate} = \frac{2,500}{167,000} \times 1,000 = 14.97$$

In 2000, there were 14.97 births for every 1,000 people in the city. With the help of these statistics, the decline in the birthrate is clearly expressed.

The percentage change will tell us <u>how much the score has changed at the later time relative to the earlier time</u>. Using death rates as an example once again, imagine a society suffering from a devastating outbreak of disease in which the death rate rose from 16.00 per 1,000 population in 1995 to 24.00 per 1,000 in 2000. Clearly, the death rate is higher in 2000 but by how much relative to 1995?

The formula for the percent change is:

FORMULA 2.4
$$\text{Percent change} = \left(\frac{f_2 - f_1}{f_1}\right) \times 100 \quad \checkmark$$

Where f_1 = first score, frequency, or value
f_2 = second score, frequency, or value

In our example, f_1 is the death rate in 1995 (16.00) and f_2 is the death rate in 2000 (24.00). The formula tells us to subtract the earlier score from the later and then divide by the earlier score. The value that results expresses the size of the change in scores ($f_2 - f_1$) relative to the score at the earlier time (f_1). The value is then multiplied by 100 to express the change in the form of a percentage:

$$\text{Percent change} = \left(\frac{24 - 16}{16}\right) \times 100 = \left(\frac{8}{16}\right) \times 100 = (.50) \times 100 = 50\%$$

The death rate in 2000 is 50% higher than in 1995. This means that the 2000 rate was equal to the 1995 rate *plus* half of the earlier score. If the rate had risen to 32 per 1,000, the percent change would have been 100% (the rate would have doubled), and if the death rate had fallen to 8 per 1,000, the percent change would have been −50%. Note the negative sign: It means that the death rate has decreased by 50%. The 2000 rate would have been half the size of the 1995 rate.

An additional example should make the computation and interpretation of the percentage change clearer. Suppose we wanted to compare the projected population growth rates for various nations over the next 50 years. Table 2.4 presents the necessary information. Casual inspection will give us some information. For example, compare China and Nigeria. These societies are projected to add roughly similar numbers of people (about a million and a half for China, a little more for Nigeria) but, because China's 2000 population is more than 10 times the size of Nigeria's, its percentage change will be much lower (about 12% versus almost 150%).

TABLE 2.4   PROJECTED POPULATION GROWTH FOR SIX NATIONS, 2000–2050

Nation	Population, 2000 (f_1)	Population, 2050 (f_2)	Increase/Decrease ($f_2 - f_1$)	Percent Change $\left(\dfrac{f_2 - f_1}{f_1}\right) \times 100$
China	1,262,474,301	1,417,630,630	155,156,329	12.29%
U.S.	282,338,631	420,080,589	137,741,958	48.79%
Canada	31,278,097	41,429,579	10,151,482	32.46%
Mexico	100,349,766	153,162,145	52,812,379	52.63%
Italy	57,719,337	50,689,841	−7,029,496	−12.18%
Nigeria	123,749,589	307,420,055	183,670,466	148.42%

Source: U.S. Bureau of the Census: http://www.census.gov/cgi-bin/ipc/idbrank.pl.

Application 2.4

The American family has been changing rapidly over the past several decades. One major change has been an increase in the number of married women and mothers with jobs outside the home. For example, in 1975, 36.7% of women with children under the age of six worked outside the home. In 2001, this percentage had risen to 62.5%.[1] How large has this change been?

The 2001 percentage is obviously much higher, and calculating the percentage change will give us an exact idea of the magnitude of the change. The 1975 percentage is f_1 and the 2000 figure is f_2, so:

Percent change
$$= \left(\frac{62.5 - 36.7}{36.7}\right) \times 100 = \left(\frac{25.8}{36.7}\right) \times 100$$
$$= (.70299) \times 100 = 70.30\%$$

Between 1975 and 2001, the percentage of women with children younger than six who worked outside the home increased by 70.30%.

[1]U.S. Bureau of the Census. 2003. *Statistical Abstract of the United States, 2002.* Washington, DC: Government Printing Office. p. 373.

STATISTICS IN EVERYDAY LIFE: Road Rage

We have all witnessed acts of aggressive driving, and we all know that the nation's roadways can be dangerous places. Motorists can let their emotions get out of control and attack other drivers viciously, senselessly, and fatally. Even the safest and most even-tempered drivers can be caught up in "road rage."

These statements probably seem obvious and unremarkable but everyday statements like these are often incomplete, based on a misunderstanding, or simply false. In the case of road rage, for example, there is a difference between what everyone "knows" to be true and the actual facts.

First of all, it may surprise you to learn that "road rage" became part of everyday life in the United States only about 10 years ago. Angry drivers have been around since the invention of the automobile (and maybe since the invention of the wheel) but the term "road rage" entered the language in the mid-1990s, sparked by several violent incidents on the nation's highways and a frenzy of media coverage. Just how big is this problem?

Not very, according to sociologist Barry Glassner, in his popular book *The Culture of Fear*. Glassner argues that the problem of road rage has been hugely overblown by the media and in the popular imagination, partly because of incomplete reports and partly because of the failure to frame the information in a proper context. It's not that anyone lied with statistics. The errors were more subtle and can be identified only with a bit of statistical knowledge and some critical thinking.

Let's begin with what the media reported and then look at the realities. Glassner notes that, beginning in the mid-1990s, the media began to characterize road rage as a "growing American danger," "an exploding phenomenon," and a "plague" (Glassner, 1999, pp. 3–5). One widely cited statistic was that incidents of road rage rose almost 60% between 1990 and 1996. This percentage change was based on two numbers: there were 1,129 "road rage" incidents in 1990 and 1,800 in 1996. These values yield a percentage increase

of 59.43%:

Percent change

$$= \left(\frac{f_2 - f_1}{f_1} \right) \times 100 = \left(\frac{1,800 - 1,129}{1,129} \right) \times 100$$

$$= \left(\frac{671}{1,129} \right) = 59.43\%$$

The media reported the percentage increase—not the frequency of incidents—and a 60% increase certainly seems to justify the characterization of road rage as "an exploding phenomenon." However, these percentages are misleading and, in this case, the raw frequencies are actually the crucial pieces of information.

Note how the perception of these numbers changes when they are compared to other statistics and framed in a broader context:

- Between 1990 and 1996, there were 20 *million* injuries from traffic accidents, and about 250,000 fatalities on U.S. roadways.
- In this same period, there were a *total* of 11,000 acts of road rage.
- Alcohol is involved in about half of all traffic fatalities, road rage in about 1 in 1,000.

In the context of the total volume of traffic mayhem, injury, and death (and alcohol-related incidents), is it reasonable to label road rage a "plague"?

As Professor Glassner says, "Big percentages don't always have big numbers behind them" (1999, p. 5). Road rage might be growing but it accounts for a tiny percentage of traffic accidents and fatalities and represents a miniscule danger compared to drunk driving. In fact, concern about road rage actually may be harmful if it deflects attention from the more serious problem of drunk driving.

You should be suspicious of reports that don't include information about the numbers of cases or the size of the sample and don't provide a proper context for understanding. Considered in isolation, the increase in road rage seems very alarming. When viewed against the total volume of traffic injury and death, the problem fades to insignificance.

Barry Glassner, 1999. *The Culture of Fear: Why Americans Are Afraid of the Wrong Things.* New York: Basic Books.

STEP BY STEP	Computing Ratios, Rates, and Percent Change

Ratios

Step 1: Determine the values for f_1 and f_2. The value for f_1 will be the number of cases in the first category (e.g., the number of males on your campus) and the value for f_2 will be the number of cases in the second category (e.g., the number of females on your campus).

Step 2: Divide the value of f_1 by the value of f_2.

Rates

Step 1: Determine the number of actual occurrences (e.g., births, deaths, homicides, assaults). This value will be the numerator.

Step 2: Determine the number of possible occurrences. This value will usually be the total population for the area in question.

Step 3: Divide the number of actual occurrences by the number of possible occurrences.

Step 4: Multiply the value you calculated in step 3 by some power of 10. Conventionally, birthrates and death rates are multiplied by 1,000 and crime rates are multiplied by 100,000.

Percent Change

Step 1: Determine the values for f_1 and f_2. The former will be the score at time 1 (the earlier time) and the latter will be the score at time 2 (the later time).

Step 2: Subtract f_1 from f_2.

Step 3: Divide the quantity you found in step 2 by f_1.

Step 4: Multiply the quantity you found in step 3 by 100.

Calculating percent change will make these comparisons more precise. Table 2.4 shows the actual population for each nation in 2000 and the projected population for 2050. The "increase/decrease" column shows how many people will be added or lost. The right-hand column shows the percent change in projected population for each nation. These values were computed by subtracting the 2000 population (f_1) from the 2050 population (f_2), dividing by the 2000 population, and multiplying by 100.

Although China has the largest population of these six nations, it will grow at the slowest rate (12.29%). The United States and Mexico will increase by about 50% (in 2050, their populations will be half again larger than in 2000), and Canada will grow by about one-third. Italy's population will actually decline by more than 12%. Nigeria has by far the highest growth rate: It will add the most people and its population will increase in size by almost 150%. This means that, in 2050, the population of Nigeria will be almost 2 ½ times its 2000 size. (*For practice in computing and interpreting percent change, see problem 2.4b.*)

2.3 FREQUENCY DISTRIBUTIONS: INTRODUCTION

Frequency distributions are tables that summarize the distribution of a variable by reporting the number of cases contained in each of its categories. They are very helpful and commonly used ways of organizing and working with data. In fact, the construction of frequency distributions is almost always the first step in any statistical analysis.

To illustrate the usefulness of frequency distributions and to provide some data for examples, assume that the counseling center at a university is assessing the effectiveness of its services. Any realistic evaluation research would collect a variety of information from a large group of students but, for the sake of this example, we will confine our attention to just four variables and 20 students. Table 2.5 reports the data.

TABLE 2.5   DATA FROM COUNSELING CENTER SURVEY

Student	Sex	Marital Status	Satisfaction with Services	Age
A	Male	Single	4	18
B	Male	Married	2	19
C	Female	Single	4	18
D	Female	Single	2	19
E	Male	Married	1	20
F	Male	Single	3	20
G	Female	Married	4	18
H	Female	Single	3	21
I	Male	Single	3	19
J	Female	Divorced	3	23
K	Female	Single	3	24
L	Male	Married	3	18
M	Female	Single	1	22
N	Female	Married	3	26
O	Male	Single	3	18
P	Male	Married	4	19
Q	Female	Married	2	19
R	Male	Divorced	1	19
S	Female	Divorced	3	21
T	Male	Single	2	20

Note that, even though the data in Table 2.5 represent an unrealistically low number of cases, it is difficult to discern any patterns or trends. For example, try to ascertain the general level of satisfaction of the students from Table 2.5. You may be able to do so with just 20 cases, but it will take some time and effort. Imagine the difficulty with 50 cases or 100 cases presented in this fashion. Clearly the data need to be organized in a format that allows the researcher (and his or her audience) to understand easily the distribution of the variables.

One general rule that applies to all frequency distributions is that the categories of the frequency distribution must be exhaustive and mutually exclusive. In other words, the categories must be stated in a way that permits each case to be counted in one and only one category. This basic principle applies to the construction of frequency distributions for variables measured at all three levels of measurement.

Beyond this rule, there are only guidelines to help you construct useful frequency distributions. As you will see, the researcher has a fair amount of discretion in stating the categories of the frequency distribution (especially with variables measured at the interval-ratio level). I will identify the issues to consider as you make decisions about the nature of any particular frequency distribution. Ultimately, however, the guidelines I state for decision making are nothing more than helpful suggestions. As always, the researcher has the final responsibility for making sensible decisions and presenting his or her data in a meaningful way.

2.4 FREQUENCY DISTRIBUTIONS FOR VARIABLES MEASURED AT THE NOMINAL AND ORDINAL LEVELS

Nominal-Level Variables. For *nominal-level variables,* construction of the frequency distribution is typically very straightforward. For each category of the variable being displayed, the occurrences are counted and the subtotals, along with the total number of cases (*N*), are reported. Table 2.6 displays a frequency distribution for the variable "sex" from the counseling center survey. I have

TABLE 2.6 SEX OF RESPONDENTS, COUNSELING CENTER SURVEY

Sex	Tallies	Frequency (*f*)
Male	///// /////	10
Female	///// /////	10
		N = 20

included a column for tallies in this table to illustrate how the cases would be sorted into categories. (This column would not be included in the final form of the frequency distribution.) Notice several other features of the table. Specifically, the table has a descriptive title, clearly labeled categories (male and female), and a report of the total number of cases at the bottom of the frequency column. You must include these items in all tables regardless of the variable or level of measurement.

The meaning of the table is clear. There are 10 males and 10 females in the sample, a fact that is much easier to comprehend from the frequency distribution than from the unorganized data in Table 2.5.

For some nominal variables, the researcher might have to make some choices about the number of categories he or she wishes to report. For example, the distribution of the variable "marital status" could be reported using the categories listed in Table 2.5. Table 2.7 presents the resultant frequency distribution. Although this is a perfectly fine frequency distribution, it may be too detailed for some purposes. For example, the researcher might want to focus solely on "unmarried" as distinct from "married" students. That is, the researcher might not be concerned with the difference between single and divorced respondents but may want to treat both as simply "not married." In that case, these categories could be grouped together and treated as a single entity, as in Table 2.8. Notice that, when categories are collapsed like this, information and detail are lost. This latter version of the table would not allow the researcher to discriminate between these two unmarried states.

TABLE 2.7 MARITAL STATUS OF RESPONDENTS, COUNSELING CENTER SURVEY

Status	Frequency (*f*)
Single	10
Married	7
Divorced	3
	N = 20

TABLE 2.8 MARITAL STATUS OF RESPONDENTS WITH TWO CATEGORIES COLLAPSED, COUNSELING CENTER SURVEY

Status	Frequency (*f*)
Married	7
Not married	13
	N = 20

TABLE 2.9   SATISFACTION WITH SERVICES, COUNSELING CENTER SURVEY

Satisfaction	Frequency (f)	Percentage (%)
(4) Very Satisfied	4	20
(3) Satisfied	9	45
(2) Dissatisfied	4	20
(1) Very Dissatisfied	3	15
	N = 20	100%

TABLE 2.10   SATISFACTION WITH SERVICES WITH TWO CATEGORIES COLLAPSED, COUNSELING CENTER SURVEY

Satisfaction	Frequency (f)	Percentage (%)
Satisfied	13	65
Dissatisfied	7	35
	N = 20	100%

Ordinal-Level Variables.   Frequency distributions for *ordinal-level variables* are constructed following the same routines used for nominal-level variables. Table 2.9 reports the frequency distribution of the "satisfaction" variable from the counseling center survey. Note that a column of percentages by category has been added to this table. Such columns heighten the clarity of the table (especially with larger samples) and are common adjuncts to the basic frequency distribution for variables measured at all levels.

This table reports that most students were either satisfied or very satisfied with the services of the counseling center. The most common response (nearly half the sample) was "satisfied." If the researcher wanted to emphasize this major trend, the categories could be collapsed as in Table 2.10. Again, the price paid for this increased compactness is that some information (in this case, the exact breakdown of degrees of satisfaction and dissatisfaction) is lost. (*For practice in constructing and interpreting frequency distributions for nominal- and ordinal-level variables, see problem 2.5.*)

2.5 FREQUENCY DISTRIBUTIONS FOR VARIABLES MEASURED AT THE INTERVAL-RATIO LEVEL

Basic Considerations.   In general, the construction of frequency distributions for variables measured at the interval-ratio level is more complex than for nominal and ordinal variables. Interval-ratio variables usually have a large number of possible scores (that is, a wide range from the lowest to the highest score). The large number of scores requires some collapsing or grouping of categories to produce reasonably compact frequency distributions. To construct frequency distributions for interval-ratio-level variables, you must decide how many categories to use and how wide these categories should be.

For example, suppose you wanted to report the distribution of the variable "age" for a sample drawn from a community. Unlike the college data reported in Table 2.5, a community sample would have a very broad range of ages. If you simply reported the number of times that each year of age (or score) occurred, you could easily wind up with a frequency distribution that contained

70, 80, or even more categories. Such a large frequency distribution would not present a concise picture and would be difficult to understand. The scores (years of age) must be grouped into larger categories to heighten clarity and to ease comprehension. How large should these categories be? How many categories should the table include? Although there are no hard-and-fast rules for making these decisions, they always involve a trade-off between more detail (a greater number of narrow categories) or more compactness (a smaller number of wide categories).

Constructing the Frequency Distribution.   To introduce the mechanics and decision-making processes involved, we will construct a frequency distribution to display the ages of the students in the counseling center survey. Because of the narrow age range of a group of college students, we can use categories of only one year. (These categories are often called class intervals when working with interval-ratio data.) The frequency distribution is constructed by listing the ages from youngest to oldest, counting the number of times each score (year of age) occurs, and then totaling the number of scores for each category. Table 2.11 presents the information and reveals a concentration or clustering of scores in the 18 and 19 class intervals.

Even though the picture presented in this table is fairly clear, assume for the sake of illustration that you desire a more compact (less detailed) summary. To do this, you will have to group scores into wider class intervals. By increasing the interval width (say to two years), you can reduce the number of intervals and achieve a more compact expression. The grouping of scores in Table 2.12

TABLE 2.11   AGE OF RESPONDENTS, COUNSELING CENTER SURVEY (INTERVAL WIDTH = ONE YEAR OF AGE)

Class Intervals	Frequency (f)
18	5
19	6
20	3
21	2
22	1
23	1
24	1
25	0
26	1
	$N = 20$

TABLE 2.12   AGE OF RESPONDENTS, COUNSELING CENTER SURVEY (INTERVAL WIDTH = TWO YEARS OF AGE)

Class Intervals	Frequency (f)	Percentage (%)
18–19	11	55
20–21	5	25
22–23	2	10
24–25	1	5
26–27	1	5
	$N = 20$	100%

clearly emphasizes the relative predominance of younger respondents. You can stress this trend in the data even more by adding a column to display the percentage of cases in each category.

Note that the class intervals in Table 2.12 have been stated with an apparent gap between them (that is, the class intervals are separated by a distance of one unit). At first glance, these gaps may appear to violate the principle of exhaustiveness, but, because age has been measured in whole numbers, the gaps actually are not a problem. Given the level of precision of the measurement (in years, as opposed to 10ths or 100ths of a year), no case could have a score falling between these class intervals. In fact, for these data, the set of class intervals in Table 2.12 constitutes a scale that is exhaustive and mutually exclusive. Each of the 20 respondents in the sample can be sorted into one and only one age category.

However, consider the potential difficulties if age had been measured with greater precision. If age had been measured in 10ths of a year, into which class interval in Table 2.12 would a 19.4-year-old subject be placed? You can avoid this ambiguity by always stating the limits of the class intervals at the same level of precision as the data. Thus, if age were being measured in 10ths of a year, the limits of the class intervals in Table 2.12 would be stated in 10ths of a year. For example:

17.0–18.9
19.0–20.9
21.0–22.9
23.0–24.9
25.0–26.9

To maintain mutual exclusivity between categories, do not overlap the class intervals. If you state the limits of the class intervals at the same level of precision as the data (which might be in whole numbers, 10ths, 100ths, and so on) and maintain a "gap" between intervals, you will always produce a frequency distribution where each case can be assigned to one and only one category.

Midpoints.   On occasion, you will need to work with the **midpoints** of the class intervals, for example, when constructing or interpreting certain graphs. Midpoints are defined as the points exactly halfway between the upper and lower limits and can be found for any interval by dividing the sum of the upper and lower limits by two. Table 2.13 displays midpoints for two different sets of class intervals. (*For practice in finding midpoints, see problems 2.8b and 2.9b.*)

Cumulative Frequency and Cumulative Percentage.   Two commonly used adjuncts to the basic frequency distribution for interval-ratio data are the **cumulative frequency** and **cumulative percentage** columns. Their primary purpose is to allow researchers (and their audiences) to tell at a glance how many cases fall below a given score or class interval in the distribution.

To construct a cumulative frequency column, begin with the lowest class interval (that is, the class interval with the lowest scores) in the distribution. The entry in the cumulative frequency columns for that interval will be the same as the number of cases in the interval. For the next higher interval, the cumulative frequency will be all cases in the interval plus all the cases in the first interval.

TABLE 2.13 MIDPOINTS

Class interval width = three	
Class Intervals	Midpoints
0–2	1
3–5	4
6–8	7
9–11	10

Class interval width = six	
Class Intervals	Midpoints
100–105	102.5
106–111	108.5
112–117	114.5
118–123	120.5

STEP BY STEP	Computing Midpoints

Step 1: Find the upper and lower limits of the lowest interval in the frequency distribution. For any interval, the upper limit is the highest score included in the interval and the lower limit is the lowest score included in the interval. For example, for the top set of intervals in Table 2.13, the lowest interval (0–2) includes scores of 0, 1, and 2. The upper limit of this interval is 2 and the lower limit is 0.

Step 2: Add the upper and lower limits and divide by 2. For the interval 0–2: 0+2/2 = 1. The midpoint for this interval is 1.

Step 3: Midpoints for other intervals can be found by repeating steps 1 and 2 for each interval. As an alternative, you can find the midpoint for any interval by adding the value of the interval width to the midpoint of the next lower interval. For example, the lowest interval in Table 2.13 is 0–2 and the midpoint is 1. Intervals are three units wide (that is, they each include three scores), so the midpoint for the next higher interval (3–5) is 1 + 3 or 4. The midpoint for the interval 6–8 is 4 + 3 or 7, and so forth.

For the third interval, the cumulative frequency will be all cases in the interval plus all cases in the first two intervals. Continue adding (or accumulating) cases until you reach the highest class interval, which will have a cumulative frequency of all the cases in the interval plus all cases in all other intervals. For the highest interval, cumulative frequency equals the total number of cases. Table 2.14 shows a cumulative frequency column added to Table 2.12.

The cumulative percentage column is similar to the cumulative frequency column. Begin by adding a column to the basic frequency distribution for percentages as in Table 2.14. This column shows the percentage of all cases in each class interval. To find cumulative percentages, follow the same addition pattern explained earlier for cumulative frequency. That is, the cumulative

TABLE 2.14   AGE OF RESPONDENTS, COUNSELING CENTER SURVEY

Class Intervals	Frequency (f)	Cumulative Frequency
18–19	11	11
20–21	5	16
22–23	2	18
24–25	1	19
26–27	1	20
	N = 20	

TABLE 2.15   AGE OF 20 RESPONDENTS, COUNSELING CENTER SURVEY

Class Intervals	Frequency (f)	Cumulative Frequency	Percentage (%)	Cumulative Percentage
18–19	11	11	55	55
20–21	5	16	25	80
22–23	2	18	10	90
24–25	1	19	5	95
26–27	1	20	5	100
	N = 20		100%	

percentage for the lowest class interval will be the same as the percentage of cases in the interval. For the next higher interval, the cumulative percentage is the percentage of cases in the interval plus the percentage of cases in the first interval, and so on. Table 2.15 shows the age data with a cumulative percentage column added.

These cumulative columns are useful when the researcher wants to make a point about how cases are spread across the range of scores. For example, Tables 2.14 and 2.15 show clearly that most students in the counseling center survey are less than 21 years of age. If the researcher wants to impress this fact on his or her audience, then these cumulative columns are handy. Most realistic research situations will be concerned with many more than 20 cases and/or many more categories than our tables have. Because the cumulative percentage column is clearer and easier to interpret in such cases, it is normally preferred to the cumulative frequencies column.

Using Unequal Class Intervals.   As a general rule, for maximum clarity and ease of comprehension, the class intervals of frequency distributions should be equal in size. For example, note that all of the class intervals in Tables 2.14 and 2.15 are the same width (two years). There are several situations, however, in which the researcher may chose to use open-ended class intervals or intervals of unequal size. Open-ended intervals have an unspecified upper or lower limit and can be used when there are a few cases with extremely high or extremely low scores. Intervals of unequal size can be used to collapse a variable with a wide range of scores into more easily comprehended groupings. We will examine each situation separately.

TABLE 2.16   AGE OF 21 RESPONDENTS, COUNSELING CENTER SURVEY

Class Intervals	Frequency (f)	Cumulative Frequency
18–19	11	11
20–21	5	16
22–23	2	18
24–25	1	19
26–27	1	20
28 and older	1	21
	N = 21	

Open-Ended Intervals.   What would happen to the frequency distribution in Tables 2.14 and 2.15 if we added one more student who was 47 years of age? We would now have 21 cases and there would be a large gap between the oldest respondent (47) and the second oldest (age 26). If we simply added the older student to the frequency distribution, we would have to include nine new class intervals (28–30, 31–32, 32–33, and so on) with zero cases in them before we got to the 46–47 interval. This would waste space and probably be unclear and confusing. An alternative way to handle the situation would be to add an "open-ended" interval to the frequency distribution, as in Table 2.16.

The open-ended interval in Table 2.16 presents the distribution more compactly and efficiently than listing all of the empty intervals between 28–29 and 46–47. Note also that we could handle an extremely low score by adding an open-ended interval as the lowest class interval (for example, "17 and younger"). There is a small price to pay for this efficiency (Table 2.16 contains no information about the value of the scores included in the open-ended interval), so this technique should not be used indiscriminately.

Intervals of Unequal Size.   Another situation in which we might want to use unequal intervals occurs when the frequency distribution of a variable shows many cases clustered in a relatively narrow range of scores combined with some cases strewn across a wide range of scores. Consider, as an example, the distribution of income in the United States. Some households will have low incomes (for example, less than $20,000) and many will have moderate incomes (say, $20,000 to $60,000). Above this level, we will find some incomes spread between $60,000 and $100,000, a few incomes above $100,000, and very few in the high six-figure or even seven-figure range.

If we tried to summarize income with a frequency distribution with equal intervals of, say, $10,000, the table would have to have 20 or 30 (or more) intervals to include all the scores, and many of the intervals in the higher income ranges—those over $100,000—would have few or zero cases. In situations such as this, researchers sometimes use intervals of unequal size to summarize the distribution of the variable more efficiently. To illustrate, Table 2.17 uses unequal intervals to summarize the distribution of income in the United States.

TABLE 2.17   DISTRIBUTION OF INCOME BY HOUSEHOLD, UNITED STATES, 1999

Income	Households (Frequency)	Households (Percent)
Less than $20,000	23,325,275	22.1
$20,000 to $29,999	13,726,955	13.0
$30,000 to $39,999	12,954,424	12.3
$40,000 to $49,999	11,210,080	10.6
$50,000 to $74,999	20,570,604	19.5
$75,000 to $99,999	10,799,245	10.2
$100,000 to $149,999	8,147,826	7.7
$150,000 to $199,999	2,322,038	2.2
$200,000 or more	2,502,675	2.4
	105,559,122	100.00%

Source: U.S. Census Bureau, http://factfinder.census.gov/servlet/QTTable?_bm=y&-geo_id=01000US&-qr_name=DEC_2000_SF3_U_QTP32&-ds_name=DEC_2000_SF3_U&-_lang=en&-redoLog=false&-format=&-CONTEXT=qt.

Note that Table 2.17 uses open-ended intervals for both the lowest and highest scores. (*For practice in constructing and interpreting frequency distributions for interval-ratio-level variables, see problems 2.5 to 2.9.*)

STEP BY STEP   **Constructing Frequency Distributions for Interval-Ratio Variables**

Step 1: Decide how many class intervals (*k*) you want to use. One reasonable convention suggests that the number of intervals should be about 10. Many research situations may require fewer than 10 intervals, and it is common to find frequency distributions with as many as 15 intervals. Only rarely will more than 15 intervals be used, since the resultant frequency distribution would not be very concise.

Step 2: Find the range (*R*) of the scores by subtracting the low score from the high score.

Step 3: Find the size of the class intervals (*i*) by dividing *R* (from step 2) by *k* (from step 1). Round the value of *i* to a convenient whole number. This will be the interval size or width.

Step 4: State the lowest interval so that its lower limit is equal to or below the lowest score. By the same token, your highest interval will be the one that contains the highest score. Generally, intervals should be equal in size, but unequal and open-ended intervals may be used when convenient.

Step 5: State the limits of the class intervals at the same level of precision that you have used to measure the data. Do not overlap intervals. You will thereby define the class intervals so that each case can be sorted into one and only one category.

Step 6: Count the number of cases in each class interval and report these subtotals in a column labeled "frequency." Report the total number of cases (*N*) at the bottom of this column. The table may also include a column for percentages, cumulative frequencies, and cumulative percentages.

Step 7: Inspect the frequency distribution carefully. Have you lost too much detail? If so, reconstruct the table with a greater number of class intervals (or smaller interval size). Is the table too detailed? If so, reconstruct the table with fewer class intervals (or use wider intervals). Are there too many intervals with no cases in them? If so, consider using open-ended intervals or intervals of unequal size. Remember that the frequency distribution results from a number of somewhat arbitrary decisions. If the appearance of the table seems less than optimal given the purpose of the research, redo the table until you are satisfied that you have struck the best balance between detail and conciseness.

Step 8: Give your table a clear, concise title, and number the table if your report contains more than one. Clearly label all categories and columns.

2.6 CONSTRUCTING FREQUENCY DISTRIBUTIONS FOR INTERVAL-RATIO-LEVEL VARIABLES: A REVIEW

We covered a lot of ground in the preceding section, so let's pause and review this material by considering a specific research situation. Below are the numbers of visits received over the past year by 90 residents of a retirement community.

0	52	21	20	21	24	1	12	16	12
16	50	40	28	36	12	47	1	20	7
9	26	46	52	27	10	3	0	24	50
24	19	22	26	26	50	23	12	22	26
23	51	18	22	17	24	17	8	28	52
20	50	25	50	18	52	46	47	27	0
32	0	24	12	0	35	48	50	27	12
28	20	30	0	16	49	42	6	28	2
16	24	33	12	15	23	18	6	16	50

Listed in this format, the data are a hopeless jumble from which no one could derive much meaning. The function of the frequency distribution is to arrange and organize these data so that their meanings will be obvious.

First, we must decide how many class intervals to use in the frequency distribution. Following the guidelines presented in "Step by Step: Constructing Frequency Distributions for Interval-Ratio Variables," let's use about 10 intervals. By inspecting the data, we can see that the lowest score is 0 and the highest is 52. The range of these scores is 52 − 0, or 52. To find the approximate interval size, divide the range (52) by the number of intervals (10). Because 52/10 = 5.2, we can set the interval size at 5.

The lowest score is 0, so the lowest class interval will be 0–4. The highest class interval will be 50–54, which will include the high score of 52. All that remains is to state the intervals in table format, count the number of scores that fall into each interval, and report the totals in a frequency column. Table 2.18 shows these results and also includes columns for the percentages and cumulative percentages. Note that this table is the product of several relatively arbitrary decisions. The researcher should remain aware of this fact and inspect the frequency distribution carefully. If the table is unsatisfactory for any reason, it can be reconstructed with a different number of categories and interval sizes.

TABLE 2.18 NUMBER OF VISITS PER YEAR, 90 RETIREMENT COMMUNITY RESIDENTS

Class Interval	Frequency (f)	Cumulative Frequency	Percentage (%)	Cumulative Percentage
0–4	10	10	11.11	11.11
5–9	5	15	5.56	16.67
10–14	8	23	8.89	25.26
15–19	12	35	13.33	38.89
20–24	18	53	20.00	58.89
25–29	12	65	13.33	72.22
30-34	3	68	3.33	75.55
35–39	2	70	2.22	77.77
40–44	2	72	2.22	79.99
45–49	6	78	6.67	86.66
50–54	12	90	13.33	99.99
	N = 90		99.99%*	

*Percentage columns will occasionally fail to total to 100% because of rounding error. If the total is between 99.90% and 100.10%, ignore the discrepancy. Discrepancies of greater than ±.10% may indicate mathematical errors, and the entire column should be computed again.

Application 2.5

The following list shows the ages of 50 prisoners enrolled in a work-release program. Is this group young or old? A frequency distribution will provide an accurate picture of the overall age structure.

18	60	57	27	19
20	32	62	26	20
25	35	75	25	21
30	45	67	41	30
37	47	65	42	25
18	51	22	52	30
22	18	27	53	38
27	23	32	35	42
32	37	32	40	45
55	42	45	50	47

We will follow the steps listed in the step-by-step applications box at the end of Section 2.5 to construct the frequency distribution:

1. Set number of categories at 10 ($k = 10$).
2. By inspection we see that the youngest prisoner is 18 and the oldest is 75. The range is thus 57 ($R = 57$). Interval size will be 57/10, or 5.7, which we can round off to either 5 or 6. Let's use a six-year interval beginning at 18.

3. The limits of the lowest interval will be 18–23 and the highest will be 72–77.
4. The intervals, stated in whole numbers (like the scores), appear in the following table.
5. The table presents the number of cases in each interval and the total number of cases (N) and includes a percentage column

Ages	Frequency	Percentage
18–23	10	20
24–29	7	14
30–35	9	18
36–41	5	10
42–47	8	16
48–53	4	8
54–59	2	4
60–65	3	6
66–71	1	2
72–77	1	2
	$N = 50$	100

The prisoners seem to be fairly evenly spread across the age groups up to the 48–53 interval. There is a noticeable lack of prisoners in the oldest age groups and a concentration of prisoners in their 20s and 30s.

Now, with the aid of the frequency distribution, some patterns in the data can be discerned. There are three distinct clusterings of scores in the table. Ten residents were visited rarely, if at all (the 0–4 visits per year interval). The single largest interval, with 18 cases, is 20–24. Combined with the intervals immediately above and below, this represents a sizable grouping of cases (42 out of 90, or 46.67% of all cases) and suggests that the dominant visiting rate is about twice a month, or approximately 24 visits per year. The third grouping is in the 50–54 class interval with 12 cases, reflecting a visiting rate of about once a week. The cumulative percentage column indicates that the majority of the residents (58.89%) were visited 24 or fewer times a year.

2.7 CHARTS AND GRAPHS   Researchers frequently use charts and graphs to present their data in ways that are visually more dramatic than frequency distributions. These devices are particularly useful for conveying an impression of the overall shape of a distribution and for highlighting any clustering of cases in a particular range of scores. Many graphing techniques are available, but we will examine just four. The first two, pie and bar charts, are appropriate for variables at any level of measurement that have a limited number of categories. The last two, histograms and line charts or frequency polygons, are used with interval-ratio variables, particularly variables that have a large number of scores.

The sections that follow explain how to construct graphs and charts "by hand." These days, however, computer programs are almost always used to produce graphic displays. Graphing software is sophisticated and flexible but also relatively easy to use; if such programs are available to you, you should familiarize yourself with them. The effort required to learn these programs will be repaid in the quality of the final product. The section on computer applications at the end of this chapter includes a demonstration of how to produce bar charts and line charts.

Pie Charts.   To construct a **pie chart,** begin by computing the percentage of all cases that fall into each category of the variable. Then divide a circle (the pie) into segments (slices) proportional to the percentage distribution. Be sure to clearly label the chart and all segments.

Figure 2.1 is a pie chart that displays the distribution of "marital status" from the counseling center survey. The frequency distribution (Table 2.7) is reproduced as Table 2.19, with a column added for the percentage distribution. Because a circle's circumference is 360°, we will apportion 180° (or 50%) for the first category, 126° (35%) for the second, and 54° (15%) for the last category. The pie chart visually reinforces the relative preponderance of single respondents and the relative absence of divorced students in the counseling center survey.

Bar Charts.   Like pie charts, **bar charts** are relatively straightforward. Conventionally, the categories of the variable are arrayed along the horizontal axis (or abscissa) and frequencies, or percentages if you prefer, along the vertical axis (or ordinate). For each category of the variable, construct (or draw) a rectangle of constant width and with a height that corresponds to the number of cases in the category. The bar chart in Figure 2.2 reproduces the marital status data from Figure 2.1 and Table 2.19.

TABLE 2.19   MARITAL STATUS OF RESPONDENTS, COUNSELING CENTER SURVEY

Status	Frequency (f)	Percentage (%)
Single	10	50
Married	7	35
Divorced	3	15
	$N = 20$	100%

FIGURE 2.1   SAMPLE PIE CHART: MARITAL STATUS OF RESPONDENTS, COUNSELING CENTER SURVEY ($N = 20$)

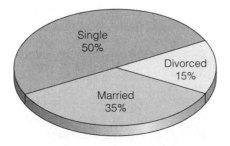

FIGURE 2.2   SAMPLE BAR CHART: MARITAL STATUS OF RESPONDENTS, COUNSELING CENTER SURVEY (*N* = 20)

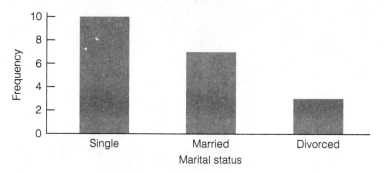

This chart would be interpreted in exactly the same way as the pie chart in Figure 2.1, and researchers are free to choose between these two methods of displaying data. However, if a variable has more than four or five categories, the bar chart would be preferred. With too many categories, the pie chart gets very crowded and loses its visual clarity. To illustrate, Figure 2.3 uses a bar chart to display the data on visiting rates for the retirement community presented in Table 2.18. A pie chart for this same data would have had 11 different "slices," a more complex or "busier" picture than that presented by the bar chart. In Figure 2.3, the clustering of scores in the "20 to 24" range (approximately two visits a month) is readily apparent, as are the groupings in the "0 to 4" and "50 to 54" ranges.

Bar charts are particularly effective ways to display the relative frequencies for two or more categories of a variable when you want to emphasize some comparisons. Suppose, for example, that you wanted to make a point about changing rates of homicide victimization for white males and females since 1955. Figure 2.4 displays the data in a dramatic and easily comprehended way. The bar chart shows that rates for males are higher than rates for females, that rates for both sexes were highest in 1975, and that rates have been declining since that time. (*For practice in constructing and interpreting pie and bar charts, see problems 2.5b and 2.10.*)

FIGURE 2.3   SAMPLE BAR CHART: VISITS PER YEAR, RETIREMENT COMMUNITY RESIDENTS (*N* = 90)

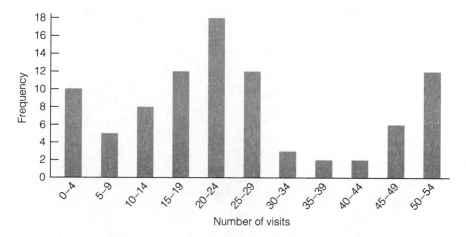

FIGURE 2.4   HOMICIDE VICTIMIZATION RATES, 1995–2000 (PER 100,000 POPULATION, WHITES ONLY)

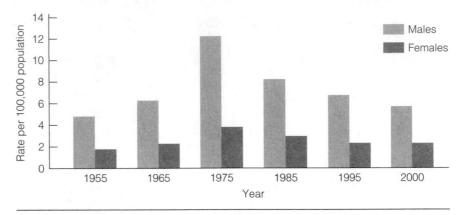

Source: U.S. Bureau of the Census, 2003. *Statistical Abstract of the United States 2002*. Washington DC: Government Printing Office.

Histograms.   **Histograms** look a lot like bar charts and, in fact, are constructed in much the same way. However, in histograms, the bars representing the frequency of each score border each other, as if they merged into each other in a continuous series from the lowest to highest scores. Histograms are particularly appropriate for interval-ratio variables (such as income or age) that have many scores covering a wide range. To create a histogram:

1. State the class intervals or scores along the horizontal axis (abscissa).

2. Array frequencies along the vertical axis (ordinate).

3. For each category in the frequency distribution, construct a bar with height corresponding to the number of cases in the category and with width corresponding to the real limits of the class intervals.

4. Label each axis of the graph.

5. Title the graph.

As an example, Figure 2.5 uses a histogram to display the distribution of ages for a sample of respondents to a national public-opinion poll. The graph

FIGURE 2.5   AGE OF RESPONDENTS TO 2004 GENERAL SOCIAL SURVEY

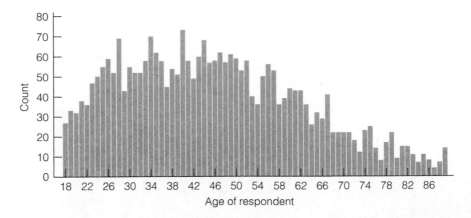

FIGURE 2.6   NUMBER OF VISITS PER YEAR, RETIREMENT COMMUNITY RESIDENTS (*N* = 90)

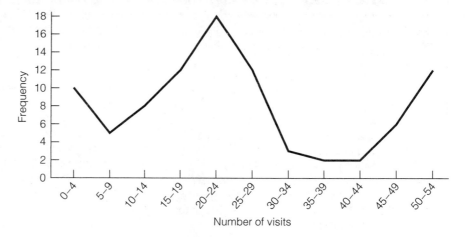

shows that the sample is concentrated in their late 20s, mid-30s, and early 40s and that the number of respondents declines with age. Note also that there are no people in the sample younger than age 18, the usual cutoff point for respondents to public-opinion polls.

Line Charts.   Construction of a **line chart** or **frequency polygon** is similar to construction of a histogram. Instead of using bars to represent the frequencies, however, use a dot at the midpoint of each interval. Straight lines then connect the dots. Figure 2.6 displays a line chart for the visiting data previously displayed in the bar chart in Figure 2.3.

Line charts are also useful to display trends across time. Figure 2.7 shows both marriage and divorce rates per 1,000 population for the United States since 1950. Note that both rates rose until the early 1980s and have been falling since with the marriage rate falling slightly faster.

Histograms and frequency polygons are alternative ways of displaying essentially the same message. Thus, the choice between the two techniques is left to the aesthetic pleasures of the researcher. (*For practice in constructing and interpreting histograms and line charts, see problems 2.7b, 2.8d, 2.9d, 2.11, and 2.12.*)

FIGURE 2.7   MARRIAGE AND DIVORCE RATES PER 1,000 POPULATION, UNITED STATES

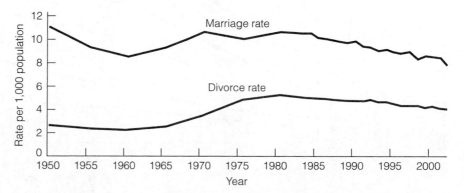

STATISTICS IN EVERYDAY LIFE: Increasing Congestion and Lost Time on the Road

Earlier, we discussed road rage. Another everyday (and possibly related) driving concern is traffic congestion. Americans are spending more and more time stopped in traffic on our freeways, wasting time and energy and, perhaps, growing increasingly frustrated and hostile as the minutes tick by. How serious is this problem today and how rapidly is it growing?

Graphs and charts can be very useful devices for expressing information about rates and changes over time, especially when attempting to communicate with the general public. To illustrate, let's consider some results of a study of traffic congestion conducted by the Texas Traffic Institute. The graph below presents "average annual hours of delay per driver," a measure of how much time is wasted for the average driver because of traffic congestion. Results are presented for all cities in the study (*N* = 85) and four groups of cities based on size ranging from "very large" (for example, Los Angeles) to "small" (for example, Bakersfield, California). Data are presented for each of three years: 1982, 1992, and 2002.

The information in this line chart is fairly straightforward: Traffic congestion is growing worse in cities of all sizes and is especially serious in the largest cities, where average annual hours of delay rose from 24 in 1982 to 62 in 2002. The latter figure

means that the average motorist in the largest cities spent the equivalent of a one and a half work weeks (more than 60 hours) sitting in traffic over the course of the year.

The chart makes it obvious that traffic congestion is a serious problem and that increases are related to the size of the city (the largest cities had the most congestion in all three years and the smallest cities had the least). What may not be so obvious from the chart is how the magnitude of change over the years varies by size of city. We can clarify this by using percentage change over the period:

City Size	Percent Increase, 1982–2002
Very large	158
Large	280
Medium	317
Small	200

The largest cities have the most congestion but the largest *increase* in congestion was in medium-size cities, followed by large cities and then small cities. Very large cities experienced the smallest increase in congestion, perhaps because they already had so much.

Source: Shrank, David, and Lomax, Tim. 2004. *The 2004 Urban Mobility Report*. Downloaded from the Texas Transportation Institute at the Texas A&M University: http://mobility.tamu.edu.

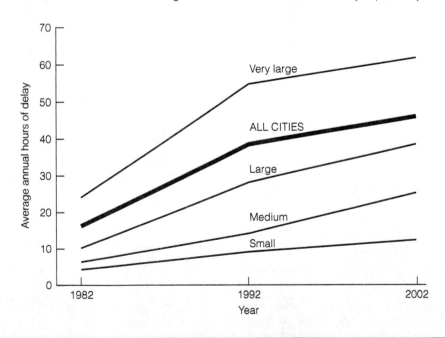

SUMMARY

1. We considered several different ways of summarizing the distribution of a single variable and, more generally, reporting the results of our research. Our emphasis throughout was on the need to communicate results clearly and concisely. You will often find that, as you strive to communicate statistical information to others, the meanings of the information will become clearer to you as well.

2. Percentages and proportions, ratios, rates, and percentage change represent several different techniques for enhancing clarity by expressing our results in terms of relative frequency. Percentages and proportions report the relative occurrence of some category of a variable compared with the distribution as a whole. Ratios compare two categories with each other, and rates report the actual occurrences of some phenomenon compared with the number of possible occurrences per some unit of time. Percentage change shows the relative increase or decrease in a variable over time.

3. Frequency distributions are tables that summarize the entire distribution of some variable. It is very common to construct these tables for each variable of interest as the first step in a statistical analysis. Columns for percentages, cumulative frequency, and/or cumulative percentages often enhance the readability of frequency distributions.

4. Pie and bar charts, histograms, and line charts or frequency polygons are graphic devices used to express the basic information contained in the frequency distribution in a compact and visually dramatic way.

SUMMARY OF FORMULAS

Proportion	2.1	$p = \dfrac{f}{N}$
Percentage	2.2	$\% = \left(\dfrac{f}{N}\right) \times 100$
Ratio	2.3	$\text{Ratio} = \dfrac{f_1}{f_2}$
Percent change	2.4	$\text{Percent change} = \left(\dfrac{f_2 - f_1}{f_1}\right) \times 100$

GLOSSARY

Bar chart. A graphic display device for nominal or ordinal variables with few categories. Categories are represented by bars of equal width, the height of each corresponding to the number (or percentage) of cases in the category.

Cumulative frequency. An optional column in a frequency distribution that displays the number of cases within an interval and all preceding intervals.

Cumulative percentage. An optional column in a frequency distribution that displays the percentage of cases within an interval and all preceding intervals.

Frequency distribution. A table that displays the number of cases in each category of a variable.

Frequency polygon. A graphic display device for interval-ratio variables. Class intervals are represented by dots placed over the midpoints, the height of each corresponding to the number (or percentage) of cases in the interval. All dots are connected by straight lines. Same as a line chart.

Histogram. A graphic display device for interval-ratio variables. Class intervals are represented by contiguous bars of equal width (equal to the class limits), the height of each corresponding to the number (or percentage) of cases in the interval.

Line chart. See Frequency polygon.

Midpoint. The point exactly halfway between the upper and lower limits of a class interval.

Percentage. The number of cases in a category of a variable divided by the number of cases in all categories of the variable, the entire quantity multiplied by 100.

Percent change. A statistic that expresses the magnitude of change in a variable from time 1 to time 2.

Pie chart. A graphic display device especially for nominal or ordinal variables with few categories. A circle (the pie) is divided into segments proportional in size to the percentage of cases in each category of the variable.

Proportion. The number of cases in one category of a variable divided by the number of cases in all categories of the variable.

Rate. The number of actual occurrences of some phenomenon or trait divided by the number of possible occurrences per some unit of time.

Ratio. The number of cases in one category divided by the number of cases in some other category.

MULTIMEDIA RESOURCES

The Wadsworth Sociology Resource Center: Virtual Society at
http://www.thomsonedu.com/sociology

Visit the companion website for the seventh edition of *Statistics: A Tool for Social Research* to access a wide range of student resources. Begin by clicking on the Student Resources section of the book's website to access the following study tools:

- Basic math review
- Flash cards

- Additional chapter problems
- Internet links
- Table of random numbers
- SPSS examples and exercises
- Hypothesis testing for variables measured at the ordinal level

PROBLEMS

2.1 SOC The table that follows report the marital status of 20 respondents in two different apartment complexes. (*HINT: Make sure that you have the correct numbers in the numerator and denominator before solving the following problems. For example, problem 2.1a asks for "the percentage of married respondents in each complex," and the denominators will be 20 for these two fractions. Problem 2.1d, on the other hand, asks for the "percentage of the single respondents living in Complex B," and the denominator for this fraction will be 4 + 6, or 10.*)

Status	Complex A	Complex B
Married	5	10
Unmarried ("Living Together")	8	2
Single	4	6
Separated	2	1
Widowed	0	1
Divorced	1	0
	20	20

a. What is the percentage of married respondents in each complex?

b. What is the ratio of single to married respondents at each complex?

c. What is the proportion of widowed respondents in each complex?

d. What is the percentage of single respondents living in Complex B?

e. What is the ratio of the "unmarried/living together" to the married at each complex?

2.2 At St. Algebra College, the numbers of males and females in the various major fields of study are as follows:

Major	Males	Females	Totals
Humanities	117	83	200
Social sciences	97	132	229
Natural sciences	72	20	92
Business	156	139	295
Nursing	3	35	38
Education	30	15	45
Totals	475	424	899

Read each of the following problems carefully before constructing the fraction and solving for the answer. (*HINT: Be sure you place the proper number in the denominator of the fractions. For example, some problems use the total number of males or females as the denominator but others use the total number of majors.*)

a. What is the percentage of male social science majors?

b. What is the proportion of female business majors?

c. For the humanities, what is the ratio of males to females?

d. What is the percentage of males in the total student body?

e. What is the ratio of males to females for the entire sample?

f. What is the proportion of male nursing majors?

g. What is the percentage of social science majors in the entire sample?

h. What is the ratio of humanities majors to business majors?

i. What is the ratio of female business majors to female nursing majors?

j. What is the proportion of male education majors?

2.3 CJ The town of Shinbone, Kansas, has a population of 211,732 and experienced 47 bank robberies, 13 murders, and 23 auto thefts during the past year. Compute a rate for each type of crime per 100,000 population. (*HINT: Make sure that you set up the fraction with size of population in the denominator.*)

2.4 CJ The numbers of homicides in five states and five Canadian provinces for the years 1997 and 2001 are reported below:

	1997		2001	
State	Homicides	Population	Homicides	Population
New Jersey	338	8,053,000	336	8,484,431
Iowa	52	2,852,000	50	2,923,179
Alabama	426	4,139,000	379	4,464,356
Texas	1,327	19,439,000	1,332	21,325,018
California	2,579	32,268,000	2,206	34,501,130

	1997		2001	
Province	Homicides	Population	Homicides	Population
Nova Scotia	24	936,100	9	942,900
Quebec	132	7,323,600	149	7,417,700
Ontario	178	11,387,400	170	12,068,300
Manitoba	31	1,137,900	34	1,150,800
British Columbia	116	3,997,100	85	4,141,300

Source: Statistics Canada: http://www.statcan.ca.

a. Calculate the homicide rate per 100,000 population for each state and each province for each year. Relatively speaking, which state and which province had the highest homicide rates in each year? Which society seems to have the higher homicide rate? Write a paragraph describing these results.

b. Using the rates you calculated in part a, calculate the percent change between 1997 and 2001 for each state and each province. Which states and provinces had the largest increase and decrease? Which society seems to have the largest change in homicide rates? Summarize your results in a paragraph.

2.5 SOC The scores of 15 respondents on four variables are reported below. These scores were taken from a public-opinion survey called the General Social Survey, or the GSS. This data set, which is described in Appendix G, is used for the computer exercises in this text. Small subsamples from the GSS appear throughout the text to provide "real" data for problems. For the actual questions and other details, see Appendix G. The numerical codes for the variables are as follows:

Sex	Support for Gun Control	Level of Education	Age
1 = Male	1 = In favor	0 = Less than HS	Years
2 = Female	2 = Opposed	1 = HS	
		2 = Jr. college	
		3 = Bachelor's degree	
		4 = Graduate school	

Case Number	Sex	Support for Gun Control	Level of Education	Age
1	2	1	1	45
2	1	2	1	48
3	2	1	3	55
4	1	1	2	32
5	2	1	3	33
6	1	1	1	28
7	2	2	0	77
8	1	1	1	50
9	1	2	0	43
10	2	1	1	48
11	1	1	4	33
12	1	1	4	35
13	1	1	0	39
14	2	1	1	25
15	1	1	1	23

a. Construct a frequency distribution for each variable. Include a column for percentages.
b. Construct pie and bar charts to display the distributions of sex, support for gun control, and level of education.

2.6 |SW| A local youth service agency has begun a sex education program for teenage girls who have been referred by the juvenile courts. The girls were given a 20-item test for general knowledge about sex, contraception, and anatomy and physiology upon admission to the program and again after completing the program. The scores of the first 15 girls to complete the program are listed below.

Case	Pretest	Posttest
A	8	12
B	7	13
C	10	12
D	15	19
E	10	8
F	10	17
G	3	12
H	10	11
I	5	7
J	15	12
K	13	20
L	4	5
M	10	15
N	8	11
O	12	20

Construct frequency distributions for the pretest and posttest scores. Include a column for percentages. (*HINT: The test had 20 items so the maximum range for these scores is 20. If you use 10 class intervals to display these scores, the interval size will be 2. Because there are no scores of 0 or 1 for either test, you may state the first interval as 2–3. To make comparisons easier, both frequency distributions should have the same intervals.*)

2.7 |SOC| Sixteen high school students completed a class to prepare them for the College Boards. Their scores are reported below.

420	345	560	650
459	499	500	657
467	480	505	555
480	520	530 ，	589

These same 16 students took a test of math and verbal ability to measure their readiness for college-level work. Scores are reported below in terms of the percentage of correct answers for each test:

Math Test

67	45	68	70
72	85	90	99
50	73	77	78
52	66	89	75

Verbal Test

89	90	78	77
75	70	56	60
77	78	80	92
98	72	77	82

a. Display each of these variables in a frequency distribution with columns for percentages and cumulative percentages.
b. Construct a histogram and frequency polygon for these data.

2.8 |GER| Here are the number of times 25 residents of a community for senior citizens left their homes for any reason during one week:

0	2	1	7	3
7	0	2	3	17
14	15	5	0	7
5	21	4	7	6
2	0	10	5	7

a. Construct a frequency distribution to display these data.
b. What are the midpoints of the class intervals?
c. Add columns to the table to display the percentage distribution, cumulative frequency, and cumulative percentages.
d. Construct a histogram and a frequency polygon to display this distribution.
e. Write a paragraph summarizing this distribution of scores.

2.9 |SOC| Twenty-five students completed a questionnaire that measured their attitudes toward interpersonal violence. Respondents who scored high believed that in many situations a person could legitimately use physical force against another person. Respondents who scored low believed that no situation (or very few situations) justified the use of violence.

52	47	17	8	92
53	23	28	9	90
17	63	17	17	23
19	66	10	20	47
20	66	5	25	17

a. Construct a frequency distribution to display these data.
b. What are the midpoints of the class intervals?
c. Add columns to the table to display the percentage distribution, cumulative frequency, and cumulative percentage.
d. Construct a histogram and a frequency polygon to display these data.
e. Write a paragraph summarizing this distribution of scores.

2.10 PA/CJ As part of an evaluation of the efficiency of your local police force, you have gathered the following data on police response time to calls for assistance during two different years. (Response times were rounded off to whole minutes.) Convert both frequency distributions into percentages and construct pie charts and bar charts to display the data. Write a paragraph comparing the changes in response time between the two years.

Response Time, 1990	Frequency (f)
21 minutes or more	35
16–20 minutes	75
11–15 minutes	180
6–10 minutes	375
Less than 6 minutes	210
	875

Response Time, 2000	Frequency (f)
21 minutes or more	45
16–20 minutes	95
11–15 minutes	155
6–10 minutes	350
Less than 6 minutes	250
	895

SPSS or freehand.

2.11 SOC Figures 2.8 through 2.10 display trends in crime in the United States. Write a paragraph describing each of these graphs. What similarities and differences can you observe among the three graphs? (For example, do crime rates always change in the same direction?) Note the differences in the vertical axes from chart to chart—for homicide the axis ranges from 0 to 12, while for burglary and auto theft the range is from 0 to 1,600. The latter crimes are far more common, and a scale with smaller intervals is needed to display the rates.

2.12 PA The city's Department of Transportation has been keeping track of accidents on a particularly dangerous stretch of highway. Early in the year, the city lowered the speed limit on this highway and increased police patrols. Data on the number of accidents before and after the changes are presented below. Did the changes work? Is the highway safer? Construct a line chart to display these two sets of data (use graphics software if available), and write a paragraph describing the changes.

Month	12 Months Before	12 Months After
January	23	25
February	25	21
March	20	18
April	19	12
May	15	9
June	17	10
July	24	11
August	28	15
September	23	17
October	20	14
November	21	18
December	22	20

FIGURE 2.8 U.S. HOMICIDE RATES

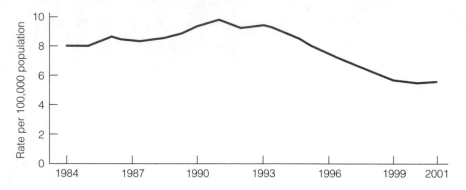

FIGURE 2.9 ROBBERY AND ASSAULT

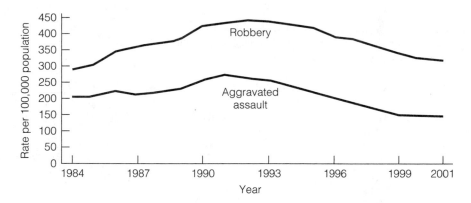

FIGURE 2.10 BURGLARY AND AUTO THEFT

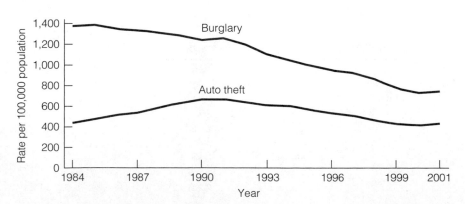

Using *SPSS for Windows* to Produce Frequency Distributions and Graphs

Click the SPSS icon on your monitor screen to start *SPSS for Windows*. Load the 2004 GSS by clicking the file name on the first screen or by clicking **File, Open,** and **Data** on the **SPSS Data Editor** screen. You may have to change the drive specification to locate the 2004 GSS data supplied with this text (probably named **GSS2004.sav**). Double-click the file name to open the data set. When you see the message "SPSS Processor is Ready" on the bottom of the screen, you are ready to proceed.

SPSS DEMONSTRATION 2.1 Frequency Distributions

We produced and examined a frequency distribution for the variable *sex* in Appendix F. Use the same procedures to produce frequency distributions for the variables *age* and *marital* (marital status). From the menu bar, click **Analyze.** From the menu that drops down, click **Descriptive Statistics** and **Frequencies.** The **Frequencies** window appears with the variables listed in alphabetical order in the left-hand box. The window may display variables by name (for example, *abany*, *abhlth*) or by label (for example, ABORTION IF WOMAN WANTS FOR ANY REASON). If labels are displayed, you may switch to variable names by clicking **Edit, Options,** and then making the appropriate selections on the "General" tab. See Appendix F and Table F.2 for further information. The variable *age* (AGE OF RESPONDENT) will be visible. Click on it to highlight it and then click the arrow button in the middle of the screen to move *age* to the right-hand window.

Find *marital* in the left-hand box by using the slider button or the arrow keys on the right-hand border to scroll through the variable list. As an alternative, type "m," and the cursor will move to the first variable name in the list that begins with that letter. Scroll down the list until you find *marital*, the variable in which we are interested. Once *marital* is highlighted, click the arrow button in the center of the screen to move the variable name to the **Variables** box. There should now be two variable names in the box, *age* and *marital*. SPSS will process all variables listed in the right-hand box together. Click **OK** in the upper-right-hand corner, and SPSS will create the frequency distributions you requested.

The table will be in the **Output** window that will now be "closest" to you on the screen. The tables, along with other information, will be in the right-hand box of the **Output** window. To change the size of the output window, click the middle symbol (either a square or two intersecting squares) in the upper-right-hand corner of the **Output** window.

The frequency distribution for *marital* will look like this:

MARITAL STATUS[*]

		Frequency	Percent	Valid Percent	Cumulative Percent
Valid	MARRIED	735	51.9	51.9	51.9
	WIDOWED	110	7.8	7.8	59.7
	DIVORCED	216	15.3	15.3	75.0
	SEPARATED	47	3.3	3.3	78.3
	NEVER MARRIED	307	21.7	21.7	100.0
	Total	1415	100.0	100.0	

[*]The format of this table has been slightly modified to improve clarity.

Let's examine the elements of this table. The variable label is printed at the top of the output ("MARITAL STATUS"). The various categories are printed on the left. Moving one column to the right, we find the actual frequencies or the number of times each score of the variable occurred. We see that 735 of the respondents were married, 110 were widowed, and so forth. Next are two columns that report percentages. The entries in the "Percent" column are based on all respondents who were asked this question and include the scores "NA" (No Answer), "DK" (Don't Know), or "NAP" (Not Applicable). The "Valid Percent" column eliminates all cases with missing values. Because we almost always ignore missing values, we will pay attention only to the "Valid Percent" column (even though, in this case, all respondents supplied this information and there is no difference between the two columns). The final column is a cumulative percentage column (see Table 2.14). For nominal-level variables such as *marital*, this information is not meaningful because the order in which the categories are stated is arbitrary.

Turning to the frequency distribution for *age*, note that the table follows the same format as the table for *marital* but is much longer. This reflects the fact that *age* has many more possible scores than *marital*—so many scores, in fact, that the table is not easy to read or understand. The table needs to be made more compact by collapsing scores into fewer categories. We could collapse the categories by hand (see Section 2.5), or we could have SPSS do the work. Demonstration 2.2 explains the procedures for recoding or collapsing variables.

SPSS DEMONSTRATION 2.2 Collapsing Categories with the Recode Command

The scores of interval-ratio level variables will often have to be collapsed to produce readable frequency distributions. *SPSS for Windows* provides a number of ways to change the scores of a variable, and one of the most useful of these is the **Recode** command. We will use this command to create a new version of *age* that has fewer categories and is more suitable for display in a frequency distribution. When we are finished, we will have two different versions of the same variable in the data set: the original interval-ratio version with age measured in years and a new ordinal-level version with collapsed categories. If we wish, the new version of age can be added to the permanent data file and used in the future.

We will collapse the values of *age* into three categories of unequal size. Because the youngest respondents are 18, let's (arbitrarily) begin with an interval of 18–29. The next interval will be 30–59, and the third will be 60 and older. As you will see, recoding requires many small steps, so please be patient and execute the commands as they are discussed.

1. In the **SPSS Data Editor** window, click **Transform** from the menu bar and then click **Recode**. A window will open that gives us two choices: **into same variable** or **into different variable.** If we choose the former (**into same variable**), the new version of the variable will replace the old version—the original version of *age* (with actual years) would disappear. We definitely do not want this to happen, so we will choose (click on) **into different variable.** This option will allow us to keep both the old and new versions of the variable.

2. The **Recode into Different Variable** window will open. A box containing an alphabetical list of variables will appear on the left. Use the cursor to highlight *age* and then click on the arrow button to move the variable to the **Input Variable → Output Variable** box. The input variable is the old version of age, and the output variable is the new, recoded version we will soon create.

3. In the **Output Variable** box on the right, click in the **Name** box and type a name for the new (output) variable. I suggest *ager* (age recoded) for the new variable, but you can assign any name as long as it does not duplicate the name of some other variable in the data set and is no longer than eight characters. Click the **Change** button and the expression *age → ager* will appear in the **Input Variable → Output Variable** box.

4. Click on the **Old and New Values** button in the middle of the screen, and a new dialog box will open. Read down the left-hand column until you find the **Range** button. Click on the button, and the cursor will move to the small box immediately below. In these boxes we will specify the low and high points of each interval of the new variable *ager*.

5. Type 18 into the left-hand **Range** dialog box and then click on the right-hand box and type 29. In the **New Value** box in the upper-right-hand corner of the screen, click the **Value** button. Type 1 in the **Value** dialog box and then click the **Add** button directly below. The expression 18–29 → 1 will appear in the **Old → New** dialog box. This completes the first recode instruction to SPSS.

6. Continue recoding by returning to the **Range** dialog boxes on the left. Type 30 in the left-hand box and 59 in the right-hand box and then click the **Value** button in the **New Value** box. Type 2 in the **Value** dialog box and then click the **Add** button. The expression 30–59 → 2 appears in the **Old → New** dialog box.

7. Finish the recoding by typing 60 and 89 (the age of the oldest respondent) in the left- and right-hand boxes under **Range,** clicking the **Value** button in the **New Value** box, and typing 3 in the **Value** dialog box. Don't forget to click the **Add** button.

8. Now click the **Continue** button at the bottom of the screen, and you will return to the **Recode into Different Variable** dialog box. Click **OK,** and SPSS will execute the transformation.

You now have a data set with one more variable named *ager* (or whatever name you gave the recoded variable). SPSS adds the new variable to the data set and you can find it in the last column to the right in the data window. You can make the new variable a permanent part of the data set by saving the data file. If you do not want to save the new, expanded data file, click **No** when you are asked if you want to save the data file. If you are using the student version of *SPSS for Windows*, remember that you are limited to a maximum of 50 variables and you may not be able to save the new variable.

To produce a frequency distribution for the new variable, click **Analyze, Descriptive Statistics,** and **Frequencies.** Find *ager* in the left-hand box and click the arrow key to move it to the right-hand box. Click **OK,** and you will soon be looking at the following table:

AGER*

		Frequency	Percent	Valid Percent	Cumulative Percent
Valid	1	261	18.4	18.5	18.5
	2	828	58.5	58.7	77.2
	3	322	22.8	22.8	100.0
	Total	1411	99.7	100.0	
Missing	System	4	.3		
	Total	1415	100.0		

*The format of this table has been slightly modified to improve clarity.

The youngest age category (1 or 18–29) has the fewest respondents (18.5%) and the sample is clustered (58.7%) in the middle age range (2 or 30–59). Note that we are missing data on four respondents, so be sure to use the "valid percent" column when reading this table.

SPSS DEMONSTRATION 2.3 Graphs and Charts

SPSS for Windows can produce a variety of graphs and charts and we will use the program to produce a bar chart and a line chart (frequency polygon) in this demonstration. To conserve space, I will keep the choices as simple as possible, but you should explore the options for yourself. For any questions you might have that are not answered in this demonstration, click on **Help** on the main menu bar.

To produce a bar chart, first click on **Graphs** on the main menu bar and then click **Bar.** The **Bar Chart** dialog box appears with three choices for the type of graph we want. The **Simple** option is already highlighted, and this is the one we want, so just click **Define.** The **Define Simple Bar** dialog box appears with variable names listed on the left. Choose *degree* from the variable list by moving the cursor to highlight this variable name. Click the arrow button in the middle of the screen to move degree to the **Category Axis** box.

Note the **Bars Represent** box above the **Category Axis** box. This dialog box gives you control over the vertical axis of the graph, which can be calibrated in frequencies, percentages, or cumulative frequencies or percentages. Let's choose *N* of cases or frequencies, the option that is already selected.

Click the **Options** button in the lower-right-hand corner. In the **Options** dialog box, make sure that the button next to the **Display groups defined by missing values** option is not selected (or checked). Click **Continue** and then click **OK** in the **Define Simple Bar** dialog box, and the following bar chart will be produced:

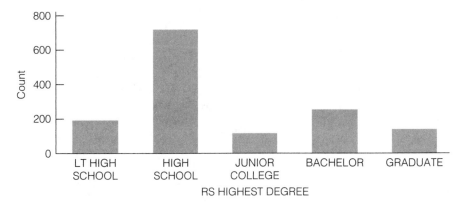

Click on any part of the graph and the **SPSS Chart Editor** window will appear. The menu bar across the top of the window gives you a wide array of options for the final appearance of the chart, including—under the **Format** command—the color of the bars and their borders. Explore these options at your leisure using the **Help** function as necessary.

Turning to the chart itself, we see that by far the most common level of education for this sample was high school, followed by "bachelor" and "less than high school." When you are ready, you can print or save the bar chart by selecting the appropriate command from the **File** menu.

The procedures for producing a line chart are very similar to those for a bar chart. Click **Graphs** and then **Line.** The **Line Charts** dialog box will appear with the

Simple option already chosen. This is the one we want, so click **Define,** and the **Define Simple Line** dialog box will appear. Your choices here are the same as in the **Define Simple Bar** dialog box. Line charts are appropriate for interval-ratio data, so let's use *educ* for this demonstration. You might think of this variable as a non-collapsed form of *degree*, and we should be able to compare the two variables quickly and easily.

Select *educ* from the variable list on the left and click the arrow button in the middle of the screen to move *educ* to the **Category Axis** box. Note that the **Lines Represent** box is the same as in the **Simple Bar** dialog box. We want the lines to represent frequencies, and this option is already selected. Click the **Options** button and make sure that the **Display groups defined by missing values** option is not selected (or checked). Click **Continue** and then click **OK** in the **Define Simple Line** dialog box, and the following line chart will be produced:

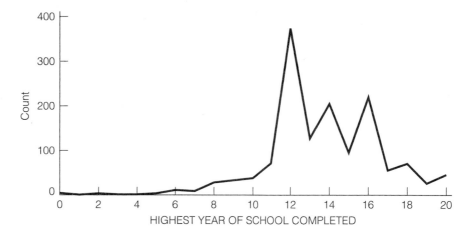

You can edit the line chart by clicking on any part of the chart and activating the **Chart Editor** window. For example, to change the color of the line, click anywhere on the line in the **Chart Editor** window, then click **Format** and **Color.** Click on the new color and then click **Apply.** Don't forget to **Save** or **Print** the chart if you wish.

The chart has a very high peak at 12 years of education (high school) and other peaks at 14 (junior college) and 16 (college). This essentially duplicates the information in the bar chart for degree.

Exercises

2.1 Get frequency distributions for 5 or 10 nominal or ordinal variables, including race (*raccen1*) and religion (*relig*). Get bar charts for each variable and write a sentence or two summarizing each frequency distribution. Your description should clearly identify the most and least common scores and any other noteworthy patterns you observe.

2.2 Get frequency distributions for the two variables (*degree* and *educ*) that measure education. Use the **Recode** command to collapse *educ* into categories comparable to degree. That is, use the categories 0–11 for less than high school, 12 for high school graduates, 13–15 to approximate the number of junior college graduates, 16 for college graduates, and 17–20 for graduate degrees. Get a frequency distribution for recoded *educ* and compare it to the frequency distribution for *degree*. Describe the three frequency distributions in a few sentences.

2.3 Get a frequency distribution for *prestg80* (respondent's occupational prestige). Find the range of this variable and use this information to collapse the variable into four categories of roughly equal number of cases. Use the **Recode** command to produce a final frequency distribution for the recoded variable and write a sentence or two of description and interpretation.

2.4 Choose five or six more variables of interest to you and produce bar charts or line charts for each. Remember that bar charts are more appropriate for nominal and ordinal variables and line charts for interval-ratio variables. Write a sentence or two of interpretation for each chart.

3

Measures of Central Tendency

LEARNING OBJECTIVES   By the time you finish this chapter, you will be able to:

1. Explain the purposes of measures of central tendency and interpret the information they convey.
2. Calculate, explain, and compare and contrast the mode, median, and mean.
3. Explain the mathematical characteristics of the mean.
4. Select an appropriate measure of central tendency according to level of measurement.

3.1 INTRODUCTION

One clear benefit of frequency distributions, graphs, and charts is that they summarize the overall shape of a distribution of scores in a way that can be quickly comprehended. Often, however, you will need to report more detailed information about the distribution. Specifically, two additional kinds of statistics are almost always useful. First, we will want to have some idea of the typical or average case in the distribution (for example, "the average starting salary for social workers is $35,000 per year") and some idea of how much variety or heterogeneity there is in the distribution ("In this state, starting salaries for social workers range from $30,000 per year to $40,000 per year"). Statistics that give us the first type of information are called **measures of central tendency,** and this chapter covers them. Statistics that give us the second type of information are called **measures of dispersion,** and Chapter 4 will discuss them.

The three commonly used measures of central tendency—the mode, median, and mean—are probably familiar to you. All three summarize an entire distribution of scores by describing the most common score (the mode), the score of the middle case (the median), or the average score (the mean). These statistics are powerful because they can reduce huge arrays of data to a single, easily understood number. Remember that the central purpose of descriptive statistics is to summarize or "reduce" data.

Even though they share a common purpose, the three measures of central tendency are quite different from each other. In fact, they will have the same value only under specific and limited conditions. As we shall see, they vary in terms of level-of-measurement considerations, and perhaps more importantly, they also vary in terms of how they define central tendency—they will not necessarily identify the same score or case as "typical." Thus, your choice of an appropriate measure of central tendency will depend in part on the way you measure the variable and in part on the purpose of the research.

3.2 THE MODE

The **mode** of any distribution is the value that occurs most frequently. For example, in the set of scores 58, 82, 82, 90, 98, the mode is 82 because it occurs twice and the other scores occur only once.

TABLE 3.1   RELIGIOUS PREFERENCE (fictitious data)

Protestant	128
Catholic	57
Jewish	10
None	32
Other	15
	N = 242

TABLE 3.2   A DISTRIBUTION OF 100 TEST SCORES

Scores (% correct)	Frequency
97	10
91	7
90	10
86	8
82	10
80	3
78	10
77	6
75	9
70	7
66	10
55	10
	N = 100

The mode is a simple statistic, most useful when you want a quick and easy indicator of central tendency and when you are working with nominal-level variables. In fact, the mode is the only measure of central tendency that you can use with nominal-level variables. Such variables do not, of course, have numerical "scores" per se, and the mode of a nominally measured variable would be its largest category. For example, Table 3.1 reports the religious affiliations of a fictitious sample of 242 respondents. The mode of this distribution, the single largest category, is Protestant.

If a researcher wants to report only the most popular or common value of a distribution, or if the variable under consideration is nominal, then the mode is the appropriate measure of central tendency. However, keep in mind that the mode does have several limitations. First, some distributions have no mode at all (see Table 2.6). Other distributions may have so many modes that the statistic loses its usefulness. For example, consider the distribution of test scores for a class of 100 students presented in Table 3.2. The distribution has modes at 55, 66, 78, 82, 90, and 97. Would reporting all six of these modes actually convey any useful information about central tendency in the distribution?

Second, with ordinal and interval-ratio data, the modal score may not be central to the distribution as a whole. That is, *most common* does not necessarily mean "typical" in the sense of identifying the center of the distribution. For example, consider the rather unusual (but not impossible) distribution of scores on a statistics test given in Table 3.3. The mode of the distribution is 93. Is this score very close to the majority of the scores? If the instructor summarized this

TABLE 3.3   A DISTRIBUTION OF 24 TEST SCORES

Scores (% correct)	Frequency
93	5
70	2
68	1
67	4
66	3
64	4
62	3
58	2
	$N = 24$

distribution by reporting only the modal score, would he or she be conveying an accurate picture of the distribution as a whole?

Remember that the mode is the most common *category* or *score* (e.g., "Protestant" or "93") of the variable, not the *frequency* in the modal category (e.g., 128 or 5). In a frequency distribution, we find the mode by looking in the frequency column for the largest number of cases, but the mode itself is the score or the name of the category. *(For practice in finding and interpreting the mode, see problems 3.1 to 3.7.)*

3.3 THE MEDIAN

Unlike the mode, the **median (Md)** is always at the exact center of a distribution of scores. The median is the score of the case that is in the middle: Half the cases have scores higher and half the cases have scores lower than the case with the median score. Thus, if the median family income for a community is $35,000, half the families earn more than $35,000 and half earn less.

To find the middle case, you must place the cases in order from the highest to the lowest (or lowest to highest). Then you can determine the median by locating the case that divides the sample into two equal halves. For example, if five students received grades of 93, 87, 80, 75, and 61 on a test, the median would be 80, the score that splits the distribution into two equal halves.

When the number of cases (N) is odd, the value of the median is unambiguous because there will always be a middle case. With an even number of cases, however, there will be two middle cases. In this situation, the median is the score exactly halfway between the scores of the two middle cases.

To illustrate, assume that seven students were asked to indicate their level of support for the intercollegiate athletic program at their universities on a scale ranging from 10 (indicating great support) to 0 (no support). After arranging their responses from high to low, you can find the median by locating the case that divides the distribution into two equal halves. With a total of seven cases, the middle case would be the fourth case because there will be three cases above and three cases below it. Table 3.4 lists the cases in order and identifies the median. With seven cases, the median is the score of the fourth case.

To summarize: When N is odd, find the middle case by adding 1 to N and then dividing that sum by 2. With an N of 7, the median is the score associated with the $(7 + 1)/2$, or fourth, case. If N had been 25, the median would be the score associated with the $(25 + 1)/2$, or 13th, case.

TABLE 3.4   FINDING THE MEDIAN: ODD NUMBER OF SCORES

Case #	Score	
1	10	
2	10	
3	8	
4	7	← Md
5	5	
6	4	
7	2	

TABLE 3.5   FINDING THE MEDIAN: EVEN NUMBER OF SCORES

Case #	Score	
1	10	
2	10	
3	8	
4	7	
		Md = 6
5	5	
6	4	
7	2	
8	1	

Now, if we make N an even number (8) by adding a student to the sample whose support for athletics was measured as a 1, we would no longer have a single middle case. Table 3.5 presents the new scores and, as you can see, any value between 7 and 5 would technically satisfy the definition of a median (that is, would split the distribution into two equal halves of four cases each). This ambiguity is resolved by defining the median as the average of the scores of the two middle cases. In this example, the median would be defined as (7 + 5)/2, or 6.

To summarize: To identify the two middle cases when N is an even number, divide N by 2 to find the first middle case and then increase that number by 1 to find the second middle case. In the example above with eight cases, the first middle case would be the fourth case ($N/2 = 4$) and the second middle case would be the ($N/2$) + 1, or fifth, case. If N had been 142, the first middle case would have been the 71st case and the second the 72nd case. Remember that the median is the average of the scores associated with the two middle cases.[1]

Because the median requires that scores be ranked from high to low, it cannot be calculated for variables measured at the nominal level. Remember that the scores of nominal-level variables cannot be ordered or ranked: The scores differ from each other but do not form a mathematical scale of any sort. We can find the median for either ordinal or interval-ratio data, but generally it is more appropriate for ordinal data. *(You can try finding the median for any problem at the end of this chapter.)*

---

[1]If the middle cases have the same score, that score is defined as the median. In the distribution 10, 10, 8, 6, 6, 4, 2, 1 the middle cases both have scores of 6 and, thus, the median is 6.

STEP BY STEP	**Computing the Median**

Step 1. Array the scores in order from high score to low score.

Step 2. Determine if N is even or odd.

If N is odd:

Step 3. The median will be the score of the middle case.

Step 4. To find the middle case, add 1 to N and divide by 2.

Step 5. The value you calculated in step 4 is the number of the middle case. For example, if $N = 13$, the median will be the score of $(13+1)/2$, or the seventh case. Remember that the median is the score of this case.

If N is even:

Step 3. The median is halfway between the scores of the two middle cases.

Step 4. To find the first middle case, divide N by 2.

Step 5. To find the second middle case, increase the value you computed in step 4 by 1.

Step 6. Add the scores of the two middle cases and divide by 2. The result is the median. If $N = 14$, the median is the score halfway between the scores of the seventh and eighth cases.

3.4 THE MEAN

The **mean** ($\overline{X}$, read this as "ex-bar"),[2] or arithmetic average, is by far the most commonly used measure of central tendency. It reports the average score of a distribution, and its calculation is straightforward: To compute the mean, add the scores and then divide by the number of scores (N). To illustrate: A birth control clinic administered a 20-item test of general knowledge about contraception to 10 clients. The number of correct responses was 2, 10, 15, 11, 9, 16, 18, 10, 11, 7. To find the mean of this distribution, add the scores (total = 109) and divide by the number of scores (10). The result (10.9) is the average score on the test.

The mathematical formula for the mean is:

FORMULA 3.1

$$\overline{X} = \frac{\Sigma(X_i)}{N}$$

Where: $\overline{X}$ = the mean
$\Sigma(X_i)$ = the summation of the scores
N = the number of cases

This formula introduces some new symbols, so let's take a moment to consider it. First, the symbol Σ (uppercase Greek letter sigma) is a mathematical operator just like the plus sign ($+$) or divide sign ($\div$). It stands for "the summation of" and directs us to add whatever quantities are stated immediately following it. The second new symbol is X_i (**"X sub i"**), which refers to any single score—the "ith" score. If we wanted to refer to a particular score in the distribution, the specific number of the score could replace the subscript. Thus, X_1 would refer to the first score, X_2 to the second, X_{26} to the 26th, and so forth. The operation of adding all the scores is symbolized as $\Sigma(X_i)$. This combination of

---

[2]This is the symbol for the mean of a sample. The mean of a population is symbolized with the lowercase Greek letter mu (μ—read this symbol as "mew").

Application 3.1

Ten students have been asked how many hours they spent in the college library during the past week. What is the average "library time" for these students? The hours appear in the following list, and we will find the mode, the median, and the mean for these data.

Student:	Number of Hours Spent in the Library Last Week (X_i)
1	0
2	2
3	5
4	5
5	7
6	10
7	14
8	14
9	20
10	30
	$\Sigma(X_i)$ = 107

By scanning the scores, we can see that two scores, 5 and 14, occurred twice, and no other score occurred more than once. This distribution has two modes, 5 and 14.

Because the number of cases is even, the median will be the average of the two middle cases. Note that the scores have been ordered from low to high. With 10 cases, the first middle case will be (N/2), or (10/2), or the fifth case. The second middle case is (N/2) + 1, or (10/2) + 1, or the sixth case. The median will be the score halfway between the scores of the fifth and sixth cases. The score of the fifth case is 7 and the score of the sixth case is 10. The median is (7 + 10)/2, or (17/2), or 8.5.

We find the mean by first adding all the scores (X_i) and then dividing by the number of scores. The sum of the scores is 107, so the mean is:

$$\overline{X} = \frac{\Sigma(X_i)}{N} = \frac{107}{10} = 10.7$$

These 10 students spent an average of 10.7 hours in the library during the week in question.

Note that the mean is a higher value than the median. This indicates a positive skew in the distribution (a few extremely high scores). By inspection we can see that the positive skew is caused by the two students who spent many more hours (20 hours and 30 hours) in the library than the other eight students.

symbols directs us to sum the scores, beginning with the first score and ending with the last score in the distribution. Formula 3.1 states in symbols what has already been stated in words (to calculate the mean, add the scores and divide by the number of scores), but in a very succinct and precise way. *(For practice in computing the mean, see any problem at the end of this chapter.)*

Because computation of the mean requires addition and division, it should be used with variables measured at the interval-ratio level. However, researchers often calculate the mean for variables measured at the ordinal level because the mean is much more flexible than the median and is a central feature of many interesting and powerful advanced statistical techniques. Thus, if the researcher plans to do more than merely describe his or her data, the mean will probably be the preferable measure of central tendency even for ordinal-level variables.

3.5 SOME CHARACTERISTICS OF THE MEAN

The mean is the most commonly used measure of central tendency, and we will consider its mathematical and statistical characteristics in some detail. First, the mean is always the center of any distribution of scores in the sense that it is the

STEP BY STEP	Computing the Mean

Step 1. Add up the scores (X_i).

Step 2. Divide the quantity you found in step 1, $\Sigma(X_i)$, by N.

TABLE 3.6   A DEMONSTRATION: ALL SCORES CANCEL OUT
AROUND THE MEAN

X_i	$(X_i - \bar{X})$
65	$65 - 78 = -13$
73	$73 - 78 = -5$
77	$77 - 78 = -1$
85	$85 - 78 = 7$
90	$90 - 78 = 12$
$\Sigma(X_i) = 390$	$\Sigma(X_i - \bar{X}) = 0$
$\bar{X} = 390/5 = 78$	

point around which all of the scores cancel out. Symbolically:

$$\Sigma(X_i - \bar{X}) = 0$$

Or, if we take each score in a distribution, subtract the mean from it, and add all of the differences, the resultant sum will always be zero.

To illustrate, consider the test scores in Table 3.6. The mean of these five scores is 78, and the sum of the differences between the scores and the mean is zero. The total of the negative differences (-19) is exactly equal to the total of the positive differences ($+19$), as will always be the case. This algebraic relationship between the scores and the mean indicates that the mean is a good descriptive measure of the centrality of scores. Think of the mean as a fulcrum that exactly balances all of the scores.

A second characteristic of the mean is called the "least squares" principle, a characteristic that is expressed in the statement:

$$\Sigma(X_i - \bar{X})^2 = \text{minimum}$$

In other words, the mean is the point in a distribution around which the variation of the scores (as indicated by the squared differences) is minimized. If the differences between the scores and the mean are squared and then added, the resultant sum will be less than the sum of the squared differences between the scores and any other point in the distribution.

To illustrate this principle, consider the distribution of five scores mentioned above: 65, 73, 77, 85, and 90. The differences between the scores and the mean have already been found. As Table 3.7 illustrates, if we square and sum these differences, we would get a total of 388. If we performed that same mathematical operations with any number other than the mean—say the value 77—the resultant sum would be greater than 388. Illustrating this point, Table 3.7 shows that the sum of the squared differences around 77 is 393, a value greater than 388.

TABLE 3.7   A DEMONSTRATION: THE MEAN IS THE POINT OF MINIMIZED VARIATION

X_i	$(X_i - \bar{X})$	$(X_i - \bar{X})^2$	$(X_i - 77)^2$
65	$65 - 78 = -13$	$(-13)^2 = 169$	$(65 - 77)^2 = (-12)^2 = 144$
73	$73 - 78 = -5$	$(-5)^2 = 25$	$(73 - 77)^2 = (-4)^2 = 16$
77	$77 - 78 = -1$	$(-1)^2 = 1$	$(77 - 77)^2 = (0)^2 = 0$
85	$85 - 78 = 7$	$(7)^2 = 49$	$(85 - 77)^2 = (8)^2 = 64$
90	$90 - 78 = 12$	$(12)^2 = 144$	$(90 - 77)^2 = (13)^2 = 169$
$\Sigma(X_i) = 390$	$\Sigma(X_i - \bar{X}) = 0$	$\Sigma(X_i - \bar{X})^2 = 388$	$\Sigma(X_i - 77)^2 = 393$

TABLE 3.8   A DISTRIBUTION OF FIVE SCORES

15	
20	
25	← Md = 25
30	
35	

$\Sigma(X_i) = 125$

$\bar{X} = 125/5 = 25$

TABLE 3.9   A DISTRIBUTION OF FIVE SCORES WITH ONE EXTREME SCORE

15	
20	
25	← Md = 25
30	
3,500	

$\Sigma(X_i) = 3,590$

$\bar{X} = 3,590/5 = 718$

In a sense, the least-squares principle merely underlines the fact that the mean is closer to all of the scores than the other measures of central tendency. However, this characteristic of the mean is also important for the statistical techniques of correlation and regression, topics we will take up toward the end of this text.

The final important characteristic of the mean is that every single score in the distribution affects it. The mode (the single most common score) and the median (the score of the middle case or cases) take account of only one or (in the case of the median when N is even) two scores. This quality of the mean is both an advantage and a disadvantage. On one hand, the mean uses all the available information—every score in the distribution affects the mean. On the other hand, when a distribution has a few very high or very low scores, the mean may become a very misleading measure of centrality.

To illustrate, consider the distribution of five scores in Table 3.8. Both the mean and median of this distribution are 25.

What will happen if we change the last score from 35 to 3,500, as in Table 3.9? This change would not affect the median at all; it would remain 25. This is because the median is based *only* on the score of the middle case and is not affected by changes in the scores of other cases in the distribution.

The mean, in contrast, would be very much affected because it takes *all* scores into account. The mean would become 3,590/5, or 718. Clearly, the one extreme score in the data set (3,500) disproportionately affects the mean. For a distribution that has a few scores much higher or lower than the other scores, the mean may present a very misleading picture of the typical or central score. *(For practice in dealing with the effects of extreme scores on means and medians, see problems 3.7, 3.10, 3.12, 3.13, and 3.14.)*

The general principle to remember is that, relative to the median, the mean is always pulled in the direction of extreme scores, those that are much higher or lower than other scores. The mean and median will have the same value when and only when a distribution is symmetrical. Unsymmetrical distributions can have either a positive or negative **skew.** When a distribution has a positive skew, or some extremely high scores, the mean will always have a greater numerical value than the median. If the distribution has a negative skew, or some very low scores, the mean will be lower in value than the median. Figures 3.1 to 3.3 depict three different frequency polygons that demonstrate these relationships.

FIGURE 3.1   A POSITIVELY SKEWED DISTRIBUTION (TI IE MEAN IS GREATER IN VALUE THAN THE MEDIAN)

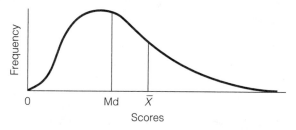

FIGURE 3.2   A NEGATIVELY SKEWED DISTRIBUTION (THE MEAN IS LESS THAN THE MEDIAN)

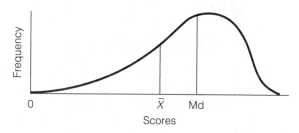

FIGURE 3.3   AN UNSKEWED, SYMMETRICAL DISTRIBUTION (THE MEAN AND THE MEDIAN ARE EQUAL)

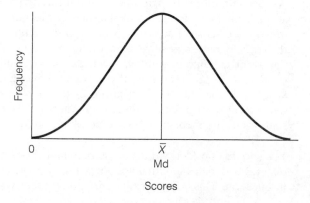

STATISTICS IN EVERYDAY LIFE: Housing Costs, Football Salaries, and Skew

As we noted in this chapter, the mean is always pulled in the direction of extreme scores relative to the median. It is likely that when variables are skewed, the mean will present a misleading picture of the actual typical case or central score. In everyday life, it is most likely that you will encounter problems with skew when dealing with variables that measure wealth, income, or prices—generally, anything to do with money. For example, consider the average sale price of houses, a variable that can be very meaningful to home owners and potential home buyers. Because there will be only a few mansions in any given community, this variable is almost always positively skewed. That is, the few very expensive houses will create an artificially high value for the mean and exaggerate the price of the "typical" house (and possibly discourage potential home buyers or encourage a false sense of affluence in home owners).

Medians rather than means are almost always used to report housing prices because they will give the more useful view of what the "typical" house costs. The following table presents some of the highest and lowest median sales prices from around the nation:

Median Sales Prices of Houses for Various Areas, 2003

Area	Median Sales Price
Albany, NY	$141,900
Salt Lake City, UT	$148,000
Charlotte, NC	$151,500
Portland, OR	$192,000
Boston, MA	$354,000
Los Angeles, CA	$354,700
San Francisco, CA	$558,100

Source: U.S. Bureau of the Census. 2004. *Statistical Abstracts of the United States,* 2003. Washington, DC: Government Printing Office, p. 603.

Each of the seven communities listed in the table has a full range of housing options—from mansions costing millions to tumble-down shacks—but the median gives a more realistic picture of the affordability of housing for most citizens. The table also suggests the range of housing prices across the nation: San Francisco is one of the most expensive areas, and the average house in the Bay area will cost almost four times as much as one in Albany, NY.

(continued next page)

These relationships between medians and means also have a practical value. For one thing, a quick comparison of the median and mean will always tell you if a distribution is skewed and the direction of the skew. If the mean is less than the median, the distribution has a negative skew. If the mean is greater than the median, the distribution has a positive skew.

Second, these characteristics of the mean and median also provide a simple and effective way to "lie" with statistics. For example, if you want to maximize the average score of a positively skewed distribution, report the mean. Income data usually have a positive skew (there are only a few very wealthy people). If you want to impress someone with the general affluence of a mixed-income community, report the mean. If you want a lower figure, report the median.

Which measure is most appropriate for skewed distributions? This will depend on what point the researcher wants to make, but as a rule, either both statistics or the median alone should be reported.

3.6 CHOOSING A MEASURE OF CENTRAL TENDENCY

You should consider two main criteria when choosing a measure of central tendency. First, make sure that you know the level of measurement of the variable in question: This will generally tell you whether you should report the mode,

---

STATISTICS IN EVERYDAY LIFE *(continued)*

Measures of salary, income, or wealth are also positively skewed because there will be only a few fabulously well-to-do individuals in any community or group. To illustrate, consider professional football players, some of our best-paid entertainers. I collected the salaries of about one-third of the more than 1,800 players in the National Football League for 2004 and found that yearly salaries ranged from a low of $152,000 to more than $19 million, with the huge majority of players closer to the former than the latter. For example, nearly 40% of the players earned less than $500,000, almost two-thirds earned less than $1,000,000, and 90% earned less than $3,500,000. Only 5 of the 643 players had incomes greater than $10 million. The histogram below clarifies the clustering of scores in the lower salaries.

The extreme skew of this distribution is evident in the difference between the mean and median. The average yearly salary (the mean) of these 643 players was almost $1,400,000 but the average player earned $632,300 (the median). Although both measures are much closer to the low end of the salary scale (reflecting the clustering of players in that range), the mean is strongly affected by the few players with extremely high salaries.

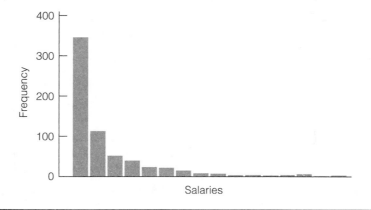

---

median, or mean. Table 3.10 shows the relationship between the level of measurement and measure of central tendency. The capitalized, boldface **"YES"** entries identify the most appropriate measure of central tendency for each level of measurement, and the lowercase "Yes" indicates the levels of measurement for which the measure is permitted. An entry of "No" in the table means that the statistic cannot be computed for that level of measurement. Finally, the "Yes (?)" entry in the bottom row indicates that the mean is often used with ordinal-level variables even though, strictly speaking, this practice violates level-of-measurement guidelines.

Second, consider the definitions of the three measures of central tendency and remember that they provide different types of information. They will be the same value only under certain, specific conditions (that is, for symmetrical distributions with one mode) and each has its own message to report. In many circumstances, you might want to report all three.

The guidelines in Table 3.11 stress both of the selection criteria and may be helpful when choosing a specific measure of central tendency.

TABLE 3.10   THE RELATIONSHIP BETWEEN LEVEL OF MEASUREMENT AND MEASURE OF CENTRAL TENDENCY

	Level of Measurement		
Measure of Central Tendency	Nominal	Ordinal	Interval-Ratio
Mode	**YES**	Yes	Yes
Median	No	**YES**	Yes
Mean	No	Yes (?)	**YES**

TABLE 3.11   CHOOSING A MEASURE OF CENTRAL TENDENCY

Use the Mode When:	1. The variable is measured at the nominal level.
	2. You want a quick and easy measure for ordinal and interval-ratio variables.
	3. You want to report the most common score.
Use the Median When:	1. The variable is measured at the ordinal level.
	2. A variables measured at the interval-ratio level has a highly skewed distribution.
	3. You want to report the central score. The median always lies at the exact center of a distribution.
Use the Mean When:	1. The variable is measured at the interval-ratio level (except when the variable is highly skewed).
	2. You want to report the typical score. The mean is "the fulcrum that exactly balances all of the scores."
	3. You anticipate additional statistical analysis.

SUMMARY

1. Measures of central tendency provide information about the most typical or representative value in a distribution. These statistics permit the researcher to report important information about an entire distribution of scores in a single, easily understood number.

2. The mode reports the most common score and is most appropriate to use with nominally measured variables.

3. The median (Md) reports the score that is the exact center of the distribution. It is most appropriately used with variables measured at the ordinal level and with variables measured at the interval-ratio level when the distribution is skewed.

4. The mean ($\overline{X}$), the most frequently used of the three measures, reports the most typical score. It is most appropriate to use with variables measured at the interval-ratio level (except when the distribution is highly skewed).

5. The mean has several mathematical characteristics that are significant for statisticians. First, it is the point in a distribution of scores around which all other scores cancel out. Second, the mean is the point of minimized variation. Last, as distinct from the mode or median, the mean is affected by every score in the distribution and is therefore pulled in the direction of extreme scores.

SUMMARY OF FORMULAS

Mean   3.1 $\overline{X} = \dfrac{\Sigma(X_i)}{N}$

GLOSSARY

Mean. The arithmetic average of the scores. $\overline{X}$ represents the mean of a sample, and μ, the mean of a population.

Measure of central tendency. Statistic that summarizes a distribution of scores by reporting the most typical or representative value of the distribution.

Measure of dispersion. Statistic that summarizes the amount of variability or heterogeneity in a set of scores. See Chapter 4.

Median (Md). The point in a distribution of scores above and below which exactly half of the cases fall.

Mode. The most common value in a distribution or the largest category of a variable.

Σ. Uppercase Greek letter sigma. A mathematical operator that stands for "the summation of."

Skew. The extent to which a distribution of scores has a few scores that are extremely high (positive skew) or extremely low (negative skew).

X_i ("X sub i"). Any score in a distribution.

MULTIMEDIA RESOURCES

The Wadsworth Sociology Resource Center: Virtual Society at
http://www.thomsonedu.com/sociology

Visit the companion website for the seventh edition of *Statistics: A Tool for Social Research* to access a wide range of student resources. Begin by clicking on the Student Resources section of the book's website to access the following study tools:

- Basic math review
- Flash cards

- Additional chapter problems
- Internet links
- Table of random numbers
- MicroCase and SPSS examples and exercises
- Hypothesis testing for variables measured at the ordinal level

PROBLEMS

3.1 $\boxed{\text{SOC}}$ A variety of information has been gathered from a sample of college freshmen and seniors, including their region of birth, the extent to which they support legalization of marijuana (measured by a scale on which 7 = strong support, 4 = neutral, and 1 = strong opposition), the amount of money they spend each week out of pocket for food, drinks, and entertainment, how many movies they watched in their dorm rooms last week, their opinion of cafeteria food (10 = excellent, 0 = very bad), and their religious affili-

ation. Some results follow. Find the *most appropriate* measure of central tendency for each variable for freshmen and then for seniors. Report both the measure you selected as well as its value for each variable (e.g., "mode = 3" or "median = 3.5"). (*HINT: Determine the level of measurement for each variable first. In general, this will tell you which statistic is appropriate. See Section 3.6 to review the relationship between measure of central tendency and level of measurement.*)

FRESHMEN

Student	Region of Birth	Legalization	Out-of-Pocket Expenses	Movies	Cafeteria Food	Religion
A	North	7	33	0	10	Protestant
B	North	4	39	14	7	Protestant
C	South	3	45	10	2	Catholic
D	Midwest	2	47	7	1	None
E	North	3	62	5	8	Protestant
F	North	5	48	1	6	Jewish
G	South	1	52	0	10	Protestant
H	South	4	65	14	0	Other
I	Midwest	1	51	3	5	Other
J	West	2	43	4	6	Catholic

SENIORS

Student	Region of Birth	Legalization	Out-of-Pocket Expenses	Movies	Cafeteria Food	Religion
K	North	7	65	0	1	None
L	Midwest	6	62	5	2	Protestant
M	North	7	60	11	8	Protestant
N	North	5	90	3	4	Catholic
O	South	1	62	4	3	Protestant
P	South	5	57	14	6	Protestant
Q	West	6	40	0	2	Catholic
R	West	7	49	7	9	None
S	North	3	45	5	4	None
T	West	5	85	3	7	Other
U	North	4	78	5	4	None

3.2 A variety of information has been collected for each of the nine high schools in a district. Find the most appropriate measure of central tendency for each variable and summarize this information in a paragraph. *(HINT: The level of measurement of the variable will generally tell you which statistic is appropriate. Remember to organize the scores from high to low before finding the median.)*

High School	Enrollment	Largest Racial/Ethnic Group	Percent College Bound	Most Popular Sport	Condition of Physical Plant (scale of 1–10, 10 = high)
1	1,400	White	25	Football	10
2	1,223	White	77	Baseball	7
3	876	Black	52	Football	5
4	1,567	Hispanic	29	Football	8
5	778	White	43	Basketball	4
6	1,690	Black	35	Basketball	5
7	1,250	White	66	Soccer	6
8	970	White	54	Football	9
9	1,109	Hispanic	64	Soccer	3

3.3 PS You have been observing the local Democratic Party in a large city and have compiled some information about a small sample of party regulars. Find the appropriate measure of central tendency for each variable.

Respondent	Sex	Social Class	No. of Years in Party	Education	Marital Status	No. of Children
A	M	High	32	High school	Married	5
B	M	Medium	17	High school	Married	0
C	M	Low	32	High school	Single	0
D	M	Low	50	Eighth grade	Widowed	7
E	M	Low	25	Fourth grade	Married	4
F	M	Medium	25	High school	Divorced	3
G	F	High	12	College	Divorced	3
H	F	High	10	College	Separated	2
I	F	Medium	21	College	Married	1
J	F	Medium	33	College	Married	5
K	M	Low	37	High school	Single	0
L	F	Low	15	High school	Divorced	0
M	F	Low	31	Eighth grade	Widowed	1

3.4 SOC You have compiled the information below on each of the graduates voted "most likely to succeed" by a local high school for a 10-year period. For each variable, find the appropriate measures of central tendency.

Case	Present Income	Marital Status	Owns a BMW?	Years of Education After High School
A	24,000	Single	No	8
B	48,000	Divorced	No	4
C	54,000	Married	Yes	4
D	45,000	Married	No	4
E	30,000	Single	No	4
F	35,000	Separated	Yes	8
G	30,000	Married	No	3
H	17,000	Married	No	1
I	33,000	Married	Yes	6
J	48,000	Single	Yes	4

3.5 SOC For 15 respondents, data have been gathered on four variables (table below). Find and report the appropriate measure of central tendency for each variable.

Respondent	Marital Status	Racial/Ethnic Group	Age	Attitude on Abortion Scale (High score = Strong opposition)
A	Single	White	18	10
B	Single	Hispanic	20	9
C	Widowed	White	21	8
D	Married	White	30	10
E	Married	Hispanic	25	7
F	Married	White	26	7
G	Divorced	Black	19	9
H	Widowed	White	29	6
I	Divorced	White	31	10
J	Married	Black	55	5
K	Widowed	Asian American	32	4
L	Married	American Indian	28	3
M	Divorced	White	23	2
N	Married	White	24	1
O	Divorced	Black	32	9

3.6 SOC The following are four variables for 30 cases from the General Social Survey. Age is reported in years. The variable "happiness" consists of answers to the question "Taken all together, would you say that you are (1) very happy, (2) pretty happy, or (3) not too happy?" Respondents were asked how many sex partners they had had over the past five years. Responses were measured on the following scale: 0–4 = actual numbers; 5 = 5–10 partners; 6 = 11–20 partners; 7 = 21–100 partners; 8 = more than 100.

For each variable, find the appropriate measure of central tendency and write a sentence reporting this statistical information as you would in a research report.

Respondent	Age	Happiness	No. of Partners	Religion
1	20	1	2	Protestant
2	32	1	1	Protestant
3	31	1	1	Catholic
4	34	2	5	Protestant
5	34	2	3	Protestant
6	31	3	0	Jewish
7	35	1	4	None
8	42	1	3	Protestant
9	48	1	1	Catholic
10	27	2	1	None
11	41	1	1	Protestant
12	42	2	0	Other
13	29	1	8	None
14	28	1	1	Jewish
15	47	2	1	Protestant
16	69	2	2	Catholic
17	44	1	4	Other
18	21	3	1	Protestant
19	33	2	1	None
20	56	1	2	Protestant
21	73	2	0	Catholic
22	31	1	1	Catholic
23	53	2	3	None
24	78	1	0	Protestant
25	47	2	3	Protestant
26	88	3	0	Catholic
27	43	1	2	Protestant
28	24	1	1	None
29	24	2	3	None
30	60	1	1	Protestant

3.7 SOC The following table lists the average weekly earnings for workers in 11 Canadian provinces in 1999 and 2002. Compute the mean and median for each year. Use these statistics to compare the two years and describe the changes you observe. Which measure of central tendency is greater for each year? Are the distributions skewed? In which direction?

Province	1999	2002
Newfoundland and Labrador	712	734
Prince Edward Island	667	709
Nova Scotia	570	697
New Brunswick	660	662
Quebec	648	745
Ontario	682	700
Manitoba	633	656
Saskatchewan	645	706
Alberta	617	680
British Columbia	712	741
Yukon	894	962

3.8 SOC The administration is considering a total ban on student automobiles. You have conducted a poll on this issue of 20 fellow students and 20 of the neighbors who live around the campus and have calculated scores for your respondents. On the scale you used, a high score indicates strong opposition to the proposed ban. The scores are presented here. Calculate an appropriate measure of central tendency and compare the two groups in a sentence or two.

Students		Neighbors	
10	11	0	7
10	9	1	6
10	8	0	0
10	11	1	3
9	8	7	4
10	11	11	0
9	7	0	0
5	1	1	10
5	2	10	9
0	10	10	0

3.9 SW As the head of a social services agency, you believe that your staff of 20 social workers is very much overworked compared to 10 years ago. The caseloads for each worker are reported below for each of the two years in question. Has the average caseload increased? What measure of central tendency is most appropriate to answer this question? Why?

1995		2005	
52	55	42	82
50	49	75	50
57	50	69	52
49	52	65	50
45	59	58	55
65	60	64	65
60	65	69	60
55	68	60	60
42	60	50	60
50	42	60	60

3.10 SOC The following table lists the approximate number of cars per 100 population for eight nations in 1999. Compute the mean and median for this data. Which measure of central tendency is greater in value? Is there a positive skew in this data? How do you know?

Nation	Number of Cars per 100 Population
United States	50
Canada	45
France	46
Germany	51
Japan	39
Mexico	10
Sweden	44
United Kingdom	37

3.11 SW For the test scores first presented in problem 2.6 (Chapter 2) and reproduced below, compute a median and mean for both the pretest and posttest. Interpret these statistics.

Case	Pretest	Posttest
A	8	12
B	7	13
C	10	12
D	15	19
E	10	8
F	10	17
G	3	12
H	10	11
I	5	7
J	15	12
K	13	20
L	4	5
M	10	15
N	8	11
O	12	20

3.12 SOC A sample of 25 freshmen at a major university completed a survey that measured their degree of racial prejudice (the higher the score, the greater the prejudice).

a. Compute the median and mean for these data.

10	43	30	30	45
40	12	40	42	35
45	25	10	33	50
42	32	38	11	47
22	26	37	38	10

b. These same 25 students completed the same survey during their senior year. Compute measures of central tendency for this second set of scores and compare them to the earlier set. What happened?

10	45	35	27	50
35	10	50	40	30
40	10	10	37	10
40	15	30	20	43
23	25	30	40	10

3.13 PA The data below represent the percentage of workers living in the central area of each city who used public transportation to commute to work in 2000.

City	Percent
Baltimore	20
Boston	32
Chicago	26
Dallas	6
Detroit	9
Houston	6
Los Angeles	10
Miami	11
Minneapolis	15
New Orleans	14
New York	53
Philadelphia	25
Phoenix	3
San Antonio	4
San Diego	4
San Francisco	31
Seattle	18
Washington, DC	33

a. Calculate the mean and median of this distribution.

b. Compare the mean and median. Which is the higher value? Why?

c. If you removed New York from this distribution and recalculated, what would happen to the mean? To the median? Why?

d. Report the statistics as you would in a formal research report.

3.14 Professional athletes are threatening to strike because they claim that they are underpaid. The team owners have released a statement that says, in part, "the average salary for players was $1.2 million last year." The players counter by issuing their own statement that says, in part, "the average player earned only $753,000 last year." Is either side necessarily lying? If you were a sports reporter and had just read Chapter 3 of this text, what questions would you ask about these statistics?

SPSS for Windows

Using *SPSS for Windows* for Measures of Central Tendency

Start *SPSS for Windows* by clicking the SPSS icon on your monitor screen. Load the 2004 GSS and when you see the message "SPSS Processor is Ready" on the bottom of the "closest" screen, you are ready to proceed.

SPSS DEMONSTRATION 3.1 Producing Measures of Central Tendency

The only procedure in SPSS that will produce all three commonly used measures of central tendency (mode, median, and mean) is **Frequencies.** We used this procedure to produce frequency distributions in Demonstrations 2.1 and 2.2 and in Appendix F. Here we will use **Frequencies** to calculate measures of central tendency for three variables: *relig* (religious denomination), *attend* (frequency of attendance at religious services), and *age.*

The three variables vary in level of measurement, and we could request only the appropriate measure of central tendency for each variable. However, it's actually more convenient to get all three measures for each variable and ignore the irrelevant output. Statistical packages typically generate more information than necessary, and it is common to disregard some of the output.

To produce modes, medians, and means, begin by clicking **Analyze** from the menu bar and then click **Descriptive Statistics** and **Frequencies.** In the **Frequencies** dialog box, find the variable names in the list on the left and click the arrow button in the middle of the screen to move the names (*relig, attend,* and *age*) to the **Variables** box on the right.

To request specific statistics, click the **Statistics** button at the bottom of the **Frequencies** dialog box, and the **Frequencies: Statistics** dialog box will open. Find the **Central Tendency** box on the right and click **Mean, Median,** and **Mode.** Click

Continue, and you will be returned to the **Frequencies** dialog box, where you might want to click the **Display Frequency Tables** box. When this box is *not* checked, SPSS will not produce frequency distribution tables, and only the statistics we request (mode, median, and mean) will appear in the **Output** window. Click **OK,** and SPSS will produce the following output:

Statistics[*]

	R'S RELIGIOUS PREFERENCES	HOW OFTEN R ATTENDS RELIGIOUS SERVICES	AGE OF RESPONDENT
N Valid	1334	1408	1411
Missing	81	7	4
Mean	1.81	3.77	46.07
Median	1.00	3.00	45.00
Mode	1	0	40

[*]This table has been slightly modified to improve clarity.

Looking only at the most appropriate measures for each variable, the mode for *relig* ("RS RELIGIOUS PREFERENCE") is "1" (see the bottom line of output). What does the value mean? You can find out by consulting either the codebook in Appendix G of the text or the online codebook. To use the latter, click **Utilities** and then click **Variables** and find *relig* in the variable list on the left. Either way, you will find that a score of 1 indicates Protestant. This was the most common religious affiliation in the sample and, thus, is the mode.

The median for *attend* ("HOW OFTEN R ATTENDS RELIGIOUS SERVICES"), an ordinal variable, is a value of "3.00." Either Appendix G or the online codebook will tell us that the category associated with this score is "several times a year." This means that the middle case in this distribution of 1,408 cases has a score of 3 (or that the middle case is in the interval "several times a year").

The output for *age* indicates that the respondents were, on the average, 46.07 years of age. Because age is an interval-ratio variable that has been measured in a defined unit (years), the value of the mean is numerically meaningful, and we do not need to consult the codebook to interpret its meaning.

Note that SPSS did not hesitate to compute means and medians for the two variables that were not interval-ratio. The program cannot distinguish between numerical codes (such as the scores for *relig*) and actual numbers—to SPSS, all numbers are the same. Also, SPSS cannot screen your commands to see if they are statistically appropriate. If you request nonsensical statistics (average sex or median religious denomination), SPSS will carry out your instructions without hesitation. The blind willingness of SPSS to simply obey commands makes it easy for you, the user, to request statistics that are completely meaningless. Computers don't care about meaning; they just crunch the numbers.

In this case, it was more convenient for us to produce statistics indiscriminately and then ignore the ones that are nonsensical. This will not always be the case and you should use SPSS wisely because, as the manager of your local computer center will be quick to remind you, computer resources (including paper) are not unlimited.

Before closing this demonstration, note that the mean for *age* is slightly greater than the median. Recalling Section 3.6, you understand that this indicates that the variable has a positive skew or a few extremely high (old) cases. You can verify this by doing a line chart for *age:* click **Graphs** and then click **Line** and then **Define.** The **Define Simple Line** dialog box appears with variable names listed on the left. Choose *age* from the variable list and click the arrow button in the middle of the screen to move the variable name to the **Category Axis** box. Click the **Options** button in the lower-left-hand corner and make sure that the button next to the **Display groups defined by missing values** option is *not* selected (or checked). Click **Continue** and then click **OK,** and the line chart for *age* will be produced. Although this figure is not as smooth as Figure 3.1, it does indicate that *age* is skewed in a positive direction (that is, it has a few very high scores).

SPSS DEMONSTRATION 3.2 Using the Descriptives Command to Produce Measures of Central Tendency and Dispersion

The **Descriptives** command in *SPSS for Windows* is designed to provide summary statistics for interval-ratio-level variables. By default (unless you tell it otherwise), **Descriptives** produces the mean, the minimum and maximum scores (that is, the lowest and highest scores, which can be used to compute the Range) and the standard deviation. Unlike **Frequencies,** this procedure will not produce frequency distributions.

To illustrate the use of **Descriptives,** let's run the procedure for *age, educ* (HIGHEST YEAR OF SCHOOL COMPLETED), and *tvhours* (HOURS PER DAY WATCHING TV). Click **Analyze, Descriptive Statistics,** and **Descriptives.** The **Descriptives** dialog box will open. This dialog box looks just like the **Frequencies** dialog box and works in the same way. Find the variable names in the list on the left, and, once they are highlighted, click the arrow button in the middle of the screen to transfer them to the **Variables** box on the right. Click **OK,** and the following output will be produced:

Descriptive Statistics[*]

	N	Minimum	Maximum	Mean	Std. Deviation
AGE OF RESPONDENT	1411	18	89	46.07	16.552
HIGHEST YEAR OF SCHOOL COMPLETED	1415	0	20	13.67	2.903
HOURS PER DAY WATCHING TV	422	0	20	2.97	2.960
Valid N (listwise)	420				

[*]This table has been slightly modified to improve clarity.

On the average, the sample is 46.07 years of age (this duplicates the **Frequencies** output in Demonstration 3.1), has almost 14 years of education, and watches a little less than three hours of TV daily.

Exercises

3.1 Use the **Frequencies** command to get all three measures of central tendency for *marital, income98, racecen1, region,* and five more variables of your own choosing. Select the most appropriate measure of central tendency for each variable. (*Hint: Use the level of measurement of the variable as your criterion.*) Write a sentence or two reporting each measure.

3.2 Use the **Descriptives** command to get means and standard deviations for *partnrs5* and *sexfreq.* Note that these variables are ordinal in level of measurement (see Appendix G) and, therefore, calculation of a mean is not fully justified. Such violations are common, however, in social science research. Write a sentence or two reporting and explaining the mean of each variable. Make sure you understand the coding scheme for each variable (see Appendix G) to help interpret the values.

4

Measures of Dispersion

LEARNING OBJECTIVES

By the end of this chapter, you will be able to:

1. Explain the purpose of measures of dispersion and the information they convey.
2. Compute and explain the range (R), the interquartile range (Q), the standard deviation (s), and the variance (s^2).
3. Select an appropriate measure of dispersion and correctly calculate and interpret the statistic.
4. Describe and explain the mathematical characteristics of the standard deviation.

4.1 INTRODUCTION

Chapters 2 and 3 presented a variety of ways of describing a variable, including frequency distributions, charts and graphs, and measures of central tendency. For a complete description of a distribution of scores, we must combine these with **measures of dispersion,** the subject of this chapter. While measures of central tendency describe the typical, average, or central score, measures of dispersion describe the variety, diversity, or heterogeneity of a set of scores.

The importance of the concept of **dispersion** might be easier to grasp if we consider a brief example. Suppose that the director of public safety wants to evaluate two ambulance services that have contracted with her city to provide emergency medical aid. As a part of the investigation, she has collected data on the response time of both services to calls for assistance. Data collected for the past year show that the average response time is 7.4 minutes for Service A and 7.6 minutes for Service B. These averages or means (calculated by adding up the response times to all calls and dividing by the number of calls) are so close to each other that they provide no basis for judging one service as more efficient than the other. Measures of dispersion, however, can reveal substantial differences even when the measures of central tendency are equivalent. For example, consider Figure 4.1, which displays the distribution of response times for the two services in the form of frequency polygons or line charts (see Chapter 2).

Compare the shapes of these two figures. Note that the line chart for Service B is much flatter than that for Service A. This is because the scores (or response times) for Service B are more spread out or more diverse than the scores for Service A. In other words, Service B was much more variable in response time and, compared to Service A, had more scores in the high and low ranges and fewer in the middle. Service A was more consistent in its response time and its scores are more clustered or grouped around the mean. Both distributions have essentially the same *average* response time, but there is considerably more variation or *dispersion* in the response times for Service B. If you were the director of public safety, would you be more likely to select an ambulance service that was always on the scene of an emergency in about the same

FIGURE 4.1   RESPONSE TIME FOR TWO AMBULANCE SERVICES

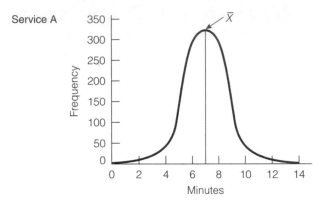

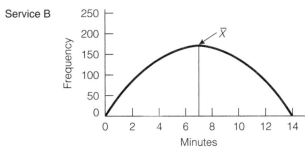

amount of time (Service A) or one that was sometimes very slow and sometimes very quick to respond (Service B)? Note that if you had not considered dispersion, an important difference in the performance of the two ambulance services might have gone unnoticed.

4.2 THE RANGE (R) AND INTERQUARTILE RANGE (Q)

We begin our treatment of measures of dispersion with the **range (R),** which is defined as the distance between the highest and lowest scores in a distribution:

FORMULA 4.1

$$R = \text{High score} - \text{Low score}$$

The range is easy to calculate and most useful as a quick and general indicator of dispersion. However, because it is based on only two scores, the range can tell us nothing about the scores between the two extremes. Furthermore, the range might be quite misleading as a measure of dispersion because almost any sizable distribution will contain some atypically high and low scores.

The **interquartile range (Q)** is a kind of range. It avoids some of the problems associated with R by considering only the middle 50% of the cases in a distribution. To find Q, arrange the scores from highest to lowest and then divide the distribution into quarters (as distinct from halves, as in locating the median). The first quartile (Q_1) is the point below which 25% of the cases fall and above which 75% of the cases fall. The second quartile (Q_2) divides the distribution into halves (thus, Q_2 is equal in value to the median). The third quartile (Q_3) is the point below which 75% of the cases fall and above which 25% of the cases fall. Thus, if line *LH* represents a distribution of scores, the quartiles are located as shown:

L 25%	25%	25%	25% *H*

Low	Q_1	Q_2	Q_3	High (Q_4)
Score		(Md)		Score

| $\leftarrow$ -------------- Q ---------------- $\rightarrow$ |

The interquartile range is the distance from the third to the first quartile, as stated in Formula 4.2.

FORMULA 4.2

$$Q = Q_3 - Q_1$$

The interquartile range essentially extracts the middle 50% of the distribution and, like *R*, is based on only two scores. While *Q* avoids the problem of being based on the most extreme scores, it has all the other disadvantages associated with *R*. Most importantly, *Q* also fails to yield any information about the variation of the scores other than the two upon which it is based.

4.3 COMPUTING THE RANGE AND INTERQUARTILE RANGE[1]

Table 4.1 presents per capita school expenditures for 20 states. What are the range and interquartile range of these data?

Note that the scores have already been ordered from high to low. This makes the range easy to calculate and is necessary for finding the interquartile

TABLE 4.1   PER CAPITA EXPENDITURES ON PUBLIC SCHOOLS, 2001

Rank	State	Expenditure
20 (highest)	Michigan	$1,873
19	California	$1,564
18	Wyoming	$1,563
17	North Carolina	$1,458
16	Alabama	$1,429
15	Oregon	$1,345
14	Idaho	$1,339
13	Virginia	$1,304
12	Mississippi	$1,281
11	Ohio	$1,278
10	Louisiana	$1,234
9	New Jersey	$1,221
8	Nebraska	$1,211
7	Texas	$1,161
6	Pennsylvania	$1,113
5	Maine	$1,085
4	Arizona	$1,077
3	New Hampshire	$1,049
2	Illinois	$1,033
1 (lowest)	Florida	$968

---

[1]This section is optional.

range. Of these 20 states, Michigan spent the most per capita on public education ($1,873) and Florida spent the least ($968). The range is therefore 1,873–968, or $905 ($R = \$905$).

To find Q, we must locate the first and third quartiles (Q_1 and Q_3). In a manner analogous to the technique for finding the median, we can define both of these points in terms of the scores associated with certain cases. Q_1 is determined by multiplying N by (.25). Because (20) × (.25) is 5, Q_1 is the score associated with the fifth case, counting up from the lowest score. The fifth case is Maine, with a score of $1,085. So, $Q_1 = 1,085$. The case that lies at the third quartile (Q_3) is given by multiplying N by (.75) and (20) × (.75) = 15th case. The 15th case, again counting up from the lowest score, is Oregon with a score of $1,345 ($Q_3 = 1,345$). Therefore:

$$Q = Q_3 - Q_1$$
$$Q = 1,345 - 1,085$$
$$Q = 260$$

In most situations, the locations of Q_1 and Q_3 will not be as obvious as they are when $N = 20$. For example, if N had been 157, then Q_1 would be (157)(.25), or the score associated with the 39.25th case, and Q_3 would be (157)(.75), or the score associated with the 117.75th case. Because fractions of cases are impossible, these numbers present some problems. The easy solution to this difficulty is to round off and take the score of the closest case to the numbers that mark the quartiles. Thus, Q_1 would be defined as the score of the 39th case and Q_3 as the score of the 118th case.

The more accurate solution would be to take the fractions of cases into account. For example, Q_1 could be defined as the score that is one-quarter of the distance between the scores of the 39th and 40th cases, and Q_3 could be defined as the score that is three-quarters of the distance between the scores of the 117th and 118th cases. (This procedure could be analogous to defining the median—which is also Q_2—as halfway between the two middle scores when N is even.) In most cases, the differences in the values of Q for these two methods would be quite small. *(For practice in finding and interpreting Q, see problem 4.8. You can try finding the range for any variable in the problems at the end of this chapter.)*

4.4 THE STANDARD DEVIATION AND VARIANCE

An important limitation of both Q and R is that they are based on only two scores. They do not use all the scores in the distribution and, thus, they do not capitalize on all the available information. Also, neither statistic provides any information on how far the scores are from each other or from some central point such as the mean. How can we design a measure of dispersion that would correct these faults? We can begin with some specifications; a good measure of dispersion should:

1. Use all the scores in the distribution. The statistic should use all the information available.

2. Describe the average or typical distance between the scores and some central point (such as the mean). The statistic should give us an idea about how far the scores are from each other or from the center of the distribution.

3. Increase in value as the distribution of scores becomes more diverse. This would be a very handy feature because it would indicate at a glance which distribution was more variable: the higher the numerical value of the statistic, the greater the dispersion.

One way to develop a statistic to meet these criteria would be to start with the distances between each score and the mean. The distances between the scores and the mean $(X_i - \overline{X})$ are called **deviations,** and these quantities will increase in value as the scores increase in their variety or heterogeneity. If the scores are more clustered around the mean (remember the graph for Service A in Figure 4.1), the deviations would be small. If the scores are more spread out or more varied (such as the scores for Service B in Figure 4.1), the deviations would be greater in value. How can we use the deviations of the scores around the mean to develop a useful statistic?

One course of action would be to use the sum of the deviations—$\Sigma(X_i - \overline{X})$—as the basis for a statistic, but, as we saw in Section 3.5, the sum of deviations will always be zero. To illustrate, consider a distribution of five scores: 10, 20, 30, 40, and 50 (see Table 4.2). If we sum the deviations of the scores from the mean, we would always wind up with a total of zero.

Still, the sum of the deviations is a logical basis for a statistic that measures the amount of variety in a set of scores, and statisticians have developed two ways around the fact that the positive deviations always equal the negative deviations. Both solutions eliminate the negative signs. The first does so by using the absolute values or by ignoring signs when summing the deviations. This is the basis for a statistic called the average deviation, a measure of dispersion that is rarely used and will not be mentioned further.

The second solution squares each of the deviations. This makes all values positive because a negative number multiplied by a negative number becomes positive. For example: $(-20) \times (-20) = 400$. In the example presented in Table 4.2, the sum of the squared deviations would be $(400 + 100 + 0 + 100 + 400)$ or 1,000. Thus, a statistic based on the sum of the squared deviations will have the properties we want in a good measure of dispersion.

Before we finish designing our measure of dispersion, we must deal with another problem. The sum of the squared deviations will increase with the size of the sample: the larger the *number* of scores, the greater the value of the measure. This would make it very difficult to compare the relative variability of distributions based on samples of different size. We can solve this problem by dividing the sum of the squared deviations by N (sample size) and thus standardizing for samples of different sizes.

TABLE 4.2   A DISTRIBUTION OF FIVE SCORES

Scores (X_i)	Deviations ($X_i - \overline{X}$)
10	$(10 - 30) = -20$
20	$(20 - 30) = -10$
30	$(30 - 30) = 0$
40	$(40 - 30) = 10$
50	$(50 - 30) = 20$
$\Sigma(X_i) = 150$	$\Sigma(X_i - \overline{X}) = 0$
$\overline{X} = 150/5 = 30$	

These procedures yield a statistic known as the **variance,** which is symbolized as s^2. The variance is used primarily in inferential statistics, although it is a central concept in the design of some measures of association. To describe the dispersion of a distribution, researchers typically use a closely related statistic called the **standard deviation** (symbolized as s). This statistic will be our focus for the remainder of the chapter.[2]

The formulas for the variance and standard deviation are:

FORMULA 4.3

$$s^2 = \frac{\Sigma(X_i - \overline{X})^2}{N}$$

FORMULA 4.4

$$s = \sqrt{\frac{\Sigma(X_i - \overline{X})^2}{N}}$$

Strictly speaking, Formulas 4.3 and 4.4 are for the variance and standard deviation of a population.[3] Slightly different formulas, with $N - 1$ instead of N in the denominator, should be used when we are working with random samples rather than entire populations. This is an important point because many of the electronic calculators and statistical software packages (including SPSS) that you might employ use "$N - 1$" in the denominator and, thus, produce results that differ at least slightly from Formulas 4.3 and 4.4. The size of the difference will decrease as sample size increases, but the problems and examples in this chapter use small samples and the differences between using N and $N - 1$ in the denominator can be considerable in such cases. Some calculators offer the choice of "$N - 1$" or "N" in the denominator. If you use the latter, the values calculated for the standard deviation should match the values in this text.

To compute the standard deviation, I suggest constructing a table such as Table 4.3 to organize computations. The five scores used in the previous example are listed in the left-hand column, the deviations are in the middle column, and the squared deviations are in the right-hand column.

The sum of the last column in Table 4.3 is the sum of the squared deviations and can be substituted into the numerator of Formula 4.4:

$$s = \sqrt{\frac{\Sigma(X_i - \overline{X})^2}{N}}$$

$$s = \sqrt{\frac{1,000}{5}}$$

$$s = \sqrt{200}$$

$$s = 14.14$$

To solve Formula 4.4, divide the sum of the squared deviations by N and take the square root of the result. To find the variance, square the standard deviation. For this problem, the variance is $s^2 = (14.14)^2 = 200$.

---

[2] The symbols for the variance and standard deviation of a sample are s^2 and s. The symbols for the variance and standard deviation of a population are σ^2 and σ (lowercase Greek letter sigma).
[3] See Chapter 1 for a discussion of populations and samples.

TABLE 4.3 COMPUTING THE STANDARD DEVIATION

Scores (X_i)	Deviations ($X_i - \bar{X}$)	Deviations Squared ($X_i - \bar{X}$)^2
10	$(10 - 30) = -20$	$(-20)^2 = 400$
20	$(20 - 30) = -10$	$(-10)^2 = 100$
30	$(30 - 30) = 0$	$(-0)^2 = 0$
40	$(40 - 30) = 10$	$(-10)^2 = 100$
50	$(50 - 30) = 20$	$(-20)^2 = 400$
$\Sigma(X_i) = 150$	$\Sigma(X_i - \bar{X}) = 0$	$\Sigma(X_i - \bar{X})^2 = 1{,}000$

STEP BY STEP	**Computing the Standard Deviation**

Step 1: Construct a computing table like Table 4.3 with columns for the scores (X_i), the deviations ($X_i - \bar{X}$), and the deviations squared ($X_i - X$)^2.

Step 2: Place the scores (X_i) in the left-hand column. Add up the scores and divide by N to find the mean.

Step 3: Find the deviations ($X_i - \bar{X}$) by subtracting the mean from each score, one at a time.

Step 4: Add up the deviations. The sum should equal zero (within rounding error). If the sum of the

deviations does not equal zero, you have made a computational error and need to repeat steps 2 and 3.

Step 5: Square each deviation one by one.

Step 6: Add up the squared deviations and transfer this sum to the numerator in Formula 4.4.

Step 7: Divide the sum of the squared deviations (see step 6) by N.

Step 8: Take the square root of the quantity you computed in step 7. This is the standard deviation.

Application 4.1

At a local preschool, 10 children were observed for an hour, and the number of aggressive acts committed by each was recorded in the following list. What is the standard deviation of this distribution? We will use Formula 4.4 to compute the standard deviation. If you use the preprogrammed function on a hand calculator to check these computations, remember to choose "divide by N" not "divide by $N-1$."

NUMBER OF AGGRESSIVE ACTS

(X_i)	($X_i - \bar{X}$)	($X_i - X$)^2
1	$1 - 4 = -3$	9
3	$3 - 4 = -1$	1
5	$5 - 4 = 1$	1
2	$2 - 4 = -2$	4
7	$7 - 4 = 3$	9
11	$11 - 4 = 7$	49
1	$1 - 4 = -3$	9
8	$8 - 4 = 4$	16
2	$2 - 4 = -2$	4
0	$0 - 4 = -4$	16
$\Sigma(X_i) = 40$	$\Sigma(X_i - \bar{X}) = 0$	$\Sigma(X_i - X)^2 = 118$

$$\bar{X} = \frac{\Sigma(X_i)}{N} = \frac{40}{10} = 4$$

Substituting into Formula 4.4, we have:

$$s = \sqrt{\frac{\Sigma(X_i - \bar{X})^2}{N}} = \sqrt{\frac{118}{10}} = \sqrt{11.8} = 3.44$$

The standard deviation for these data is 3.44.

Application 4.2

Five western and five eastern states were compared as part of a study of traffic safety. The states vary in size, and comparisons were made in terms of the rate of fatal accidents (number of fatal accidents per 100,000 licensed drivers). Columns for the computa- tion of the standard deviation have already been added to the following tables. Which group of states varies most in terms of this variable? Computations for both the mean and standard deviation appear below.

FATALITIES PER 100,000 LICENSED DRIVERS FOR 2001, EASTERN STATES

State	Fatalities (X_i)	Deviations $(X_i - \bar{X})$	Deviations Squared $(X_i - X)^2$
Pennsylvania	18.60	$18.60 - 13.57 = 5.03$	25.30
New York	14.05	$14.05 - 13.57 = .48$	.23
New Jersey	13.07	$13.07 - 13.57 = -.50$	.25
Connecticut	11.77	$11.77 - 13.57 = -1.80$	3.24
Massachusetts	10.35	$10.35 - 13.57 = -3.22$	10.37
	$\Sigma(X_i) = 67.84$	$\Sigma(X_i - \bar{X}) = -0.01$	$\Sigma(X_i - X)^2 = 39.39$

$$X = \frac{\Sigma(X_i)}{N} = \frac{67.84}{5} = 13.57$$

$$s = \sqrt{\frac{\Sigma(X_i - \bar{X})^2}{N}} = \sqrt{\frac{39.39}{5}} = \sqrt{7.88} = 2.81$$

FATALITIES PER 100,000 LICENSED DRIVERS FOR 2001, WESTERN STATES

State	Fatalities (X_i)	Deviations $(X_i - \bar{X})$	Deviations Squared $(X_i - X)^2$
Wyoming	50.17	$50.17 - 25.01 = 25.16$	633.03
Nevada	22.03	$22.03 - 25.01 = -2.98$	8.88
Oregon	19.25	$19.25 - 25.01 = -5.76$	33.18
California	18.29	$18.29 - 25.01 = -6.72$	45.16
Washington	15.31	$15.31 - 25.01 = -9.70$	94.09
	$\Sigma(X_i) = 125.05$	$\Sigma(X_i - \bar{X}) = 0.00$	$\Sigma(X_i - X)^2 = 814.34$

$$X = \frac{\Sigma(X_i)}{N} = \frac{125.05}{5} = 25.01$$

$$s = \sqrt{\frac{\Sigma(X_i - \bar{X})^2}{N}} = \sqrt{\frac{814.34}{5}} = \sqrt{162.87} = 12.76$$

With such small groups, you can tell by simply inspecting the scores that the western states have higher fatality rates. This impression is confirmed by both the median, which is 13.07 for the eastern states (New Jersey is the middle case) and 19.25 for the western states (Oregon is the middle case), and the mean (13.57 for the eastern states and 25.01 for the western states). For both groups, the mean is greater than the median, indicating a positive skew. Note also that the skew is much greater for the western states (that is, the mean is much higher than the me- dian for the western states than for the eastern states). This is caused by Wyoming, which has a much higher score than any other state. However, even with Wyoming removed, the western states still have higher rates. With Wyoming removed, the mean of the remaining four western states is 18.72.

The five western states are also much more vari- able and diverse than the eastern states. The range for the western states is 34.86 ($R = 50.17 - 15.31 = 34.86$), much higher than the range for the eastern states of 8.25 ($R = 18.60 - 10.35 = 8.25$). Similarly, the standard deviation for the western states (12.76) is nearly five times greater in value that the standard deviation for the eastern states (2.81). The greater dispersion in the western states is also largely a re- flection of Wyoming's extremely high score.

In summary, the five western states average higher fatality rates and are also more variable than the five eastern states.

4.5 COMPUTING THE STANDARD DEVIATION: AN ADDITIONAL EXAMPLE

An additional example will help to clarify the procedures for computing and interpreting the standard deviation. A researcher is comparing the student bodies of two campuses. One college is located in a small town and almost all students reside on campus. The other is located in a large city and the students are almost all part-time commuters. The researcher wants to compare the age structure of the two campuses and has compiled the information presented in Table 4.4. Which student body is older and more diverse in age? (These very small groups are much too small to be used for serious research and are used here only to simplify computations.)

We see from the means that the students from the residential campus are considerably younger than the students from the urban campus (19 versus 23 years of age). Which group is more diverse on this variable? Computing the standard deviation will answer this question.

To solve Formula 4.4, substitute the sum of the right-hand column ("Deviations Squared") in the numerator and N (5 in this case) in the denominator:

Residential Campus:

$$s = \sqrt{\frac{\Sigma(X_i - \overline{X})^2}{N}} = \sqrt{\frac{4}{5}} = \sqrt{.8} = .89$$

Urban Campus:

$$s = \sqrt{\frac{\Sigma(X_i - \overline{X})^2}{N}} = \sqrt{\frac{88}{5}} = \sqrt{17.6} = 4.20$$

TABLE 4.4   COMPUTING THE STANDARD DEVIATION FOR TWO CAMPUSES

Residential Campus		
Ages (X_i)	Deviations ($X_i - \overline{X}$)	Deviations Squared ($X_i - \overline{X})^2$
18	$(18 - 19) = -1$	$(-1)^2 = 1$
19	$(19 - 19) = 0$	$(-0)^2 = 0$
20	$(20 - 19) = 1$	$(1)^2 = 1$
18	$(18 - 19) = -1$	$(-1)^2 = 1$
20	$(20 - 19) = 1$	$(1)^2 = 1$
$\Sigma(X_i) = 95$	$\Sigma(X_i - \overline{X}) = 0$	$\Sigma(X_i - \overline{X})^2 = 4$

$$\overline{X} = \frac{\Sigma(X_i)}{N} = \frac{95}{5} = 19$$

Urban Campus		
Ages (X_i)	Deviations ($X_i - \overline{X}$)	Deviations Squared ($X_i - \overline{X})^2$
20	$(20 - 23) = -3$	$(-3)^2 = 9$
22	$(22 - 23) = -1$	$(-1)^2 = 1$
18	$(18 - 23) = -5$	$(-5)^2 = 25$
25	$(25 - 23) = 2$	$(2)^2 = 4$
30	$(30 - 23) = 7$	$(7)^2 = 49$
$\Sigma(X_i) = 115$	$\Sigma(X_i - \overline{X}) = 0$	$\Sigma(X_i - \overline{X})^2 = 88$

$$\overline{X} = \frac{\Sigma(X_i)}{N} = \frac{115}{5} = 23$$

STATISTICS IN EVERYDAY LIFE: Where's the Nicest Place to Live?

If you could live anywhere in the United States, where would it be? What criteria would you use to make your selection? Suppose you place a very high value on climate and consider mild temperatures (around 70° F.) to be especially desirable; you decide to use this criterion to select your next place of residence.

So far, so good but your decision about the "nicest place to live" could be sadly mistaken if you consider *only* the average temperature. What if the cities you examined had roughly the same average temperature of 70° but differed in the range or in the extremes of temperature they experienced? What if some cities had temperatures that ranged from blistering (a high of 103°) to freezing (a low of –10°) but others had nothing but mild, balmy temperatures year round. Clearly, you would want to consider dispersion as well as central tendency in choosing your residence.

What American city comes closest to the ideal of a constant 70° temperature year round? The table below presents data for five cities:

Compared to your ideal average temperature of 70°, Chicago and NYC are too cold and Dallas and Miami are too hot. This leaves San Diego, which matches your ideal average daily temperature almost exactly.

Considering dispersion, temperature fluctuations are greatest for Chicago (with a standard deviation of almost 20°) and also very considerable for NYC and Dallas. These cities experience all four seasons, and this is not what you want in your ideal residence. Miami and San Diego have much smaller temperature swings throughout the year, and the latter city, with its 71° average temperature and small standard deviation, comes closest to matching your ideal climate. Although there are many more places to investigate in the search for the perfect place to live (and, no doubt, other criteria to consider besides climate), San Diego is the clear winner of these five cities.

AVERAGE MONTHLY TEMPERATURE DATA FOR FIVE CITIES*

	Miami	Chicago	New York City	Dallas	San Diego
Mean	82.79	58.63	62.32	76.28	70.79
Std. Dev.	5.12	19.90	17.31	15.21	4.55
Range	13.80	54.70	47.60	42.50	11.90

*Statistics are computed from average monthly temperatures for the past 30 years. Data are available at http://www.met.utah.edu /jhorel/html/wx/climate/maxtemp.html.

The higher value of the standard deviation for the urban campus means that it is much more diverse. As you can see by scanning the scores, the students at the residential college are within a narrow age range ($R = 20 - 18 = 2$) while the students at the urban campus are more mixed and include students of age 25 and 30 ($R = 30 - 18 = 12$). *(For practice in computing and interpreting the standard deviation, see any of the problems at the end of this chapter. Problems with smaller data sets, such as 4.1 and 4.2, are recommended for practicing computations until you are comfortable with these procedures.)*

4.6 INTERPRETING THE STANDARD DEVIATION

It is very possible that the meaning of the standard deviation (that is, why we calculate it) is not completely obvious to you at this point. You might be asking: "Once I've gone to the trouble of calculating the standard deviation, what do I have?" The meaning of this measure of dispersion can be expressed in three ways. The first and most important involves the normal curve, and we will defer this interpretation until the next chapter.

A second way of thinking about the standard deviation is as an index of variability that increases in value as the distribution becomes more variable. In other words, the standard deviation is higher for more diverse distributions and lower for less diverse distributions. The lowest value the standard deviation can have is zero, which would occur for distributions with no dispersion (that is, if every single case in the sample had exactly the same score). Thus, zero is the lowest value possible for the standard deviation (although there is no upper limit).

A third way to get a feel for the meaning of the standard deviation is by comparing one distribution with another. We already did this when comparing the residential and urban campuses in the example we used in Section 4.5. You might also do this when comparing one group against another (e.g., men versus women, blacks versus whites) or the same variable at two different times. For example, suppose we found that the ages of the students on a particular campus had changed over time as indicated by the following summary statistics:

1970	2000
$\overline{X} = 21$	$\overline{X} = 25$
$s = 1$	$s = 3$

In 1970, students were, on the average, 21 years of age. By 2000, the average age had risen to 25. Clearly, the student body has grown older and, according to the standard deviation, it has also grown more diverse in terms of age. The lower standard deviation for 1970 indicates that the distribution of ages in that year would be more clustered around the mean (remember the distribution for Service A in Figure 4.1) while, in 2000, the distribution would be flatter and more spread out, like the distribution for Service B in Figure 4.1. In other words, compared to 2000, the student body in 1970 was more similar to each other and more clustered in a narrower age range. The standard deviation is extremely useful for making comparisons of this sort between distributions of scores.

SUMMARY

1. Measures of dispersion present information about the heterogeneity or variety in a distribution of scores. When combined with an appropriate measure of central tendency, these statistics convey a large volume of information in just a few numbers. While measures of central tendency locate the central points of the distribution, measures of dispersion indicate the amount of diversity in the distribution.

2. The range (R) is the distance from the highest to the lowest score in the distribution. The interquartile range (Q) is the distance from the third to the first quartile (the "range" of the middle 50% of the scores). We can use these two ranges with variables measured at either the ordinal or interval-ratio level.

3. The standard deviation (s) is the most important measure of dispersion because of its central role in many more advanced statistical applications. The standard deviation has a minimum value of zero (indicating no variation in the distribution) and increases in value as the variability of the distribution increases. It is used most appropriately with variables measured at the interval-ratio level.

SUMMARY OF FORMULAS

Range	4.1	$R = \text{High score} - \text{Low score}$
Interquartile range	4.2	$Q = Q_3 - Q_1$
Variance	4.3	$s^2 = \dfrac{\Sigma(X_i - \bar{X})^2}{N}$
Standard deviation	4.4	$s = \sqrt{\dfrac{\Sigma(X_i - \bar{X})^2}{N}}$

GLOSSARY

Deviation. The distance between a score and the mean.

Dispersion. The amount of variety or heterogeneity in a distribution of scores.

Interquartile range (Q). The distance from the third quartile to the first quartile.

Measure of dispersion. Statistic that indicates the amount of variety or heterogeneity in a distribution of scores.

Range (R). The highest score minus the lowest score.

Standard deviation. The square root of the squared deviations of the scores around the mean, divided

by N. The most important and useful descriptive measure of dispersion; s represents the standard deviation of a sample; σ, the standard deviation of a population.

Variance. The deviations of the scores around the mean divided by N. A measure of dispersion used primarily in inferential statistics and also in correlation and regression techniques; s^2 represents the variance of a sample; σ^2 the variance of a population.

MULTIMEDIA RESOURCES

The Wadsworth Sociology Resource Center: Virtual Society at
http://www.thomsonedu.com/sociology

Visit the companion website for the seventh edition of *Statistics: A Tool for Social Research* to access a wide range of student resources. Begin by clicking on the Student Resources section of the book's website to access the following study tools:

- Basic math review
- Flash cards

- Additional chapter problems
- Internet links
- Table of random numbers
- MicroCase and SPSS examples and exercises
- Hypothesis testing for variables measured at the ordinal level

PROBLEMS

4.1 Compute the range and standard deviation of 10 scores reported below. (*HINT: It will be helpful to organize your computations as in Table 4.3.*)

10, 12, 15, 20, 25, 30, 32, 35, 40, 50

4.2 Compute the range and standard deviation of the 10 test scores below.

77, 83, 69, 72, 85, 90, 95, 75, 55, 45

4.3. In problem 3.1 at the end of Chapter 3, you calculated measures of central tendency for six variables for freshmen and seniors. Three of those variables are reproduced here. Calculate mean, range, and standard deviation for each variable for each class. Write a paragraph summarizing the differences between freshmen and seniors

Out-of-Pocket Expenses		Number of Movies		Rating of Cafeteria Food	
Freshmen	Seniors	Freshmen	Seniors	Freshmen	Seniors
33	65	0	0	10	1
39	62	14	5	7	2
45	60	10	11	2	8
47	90	7	3	1	4
62	62	5	4	8	3
48	57	1	14	6	6
52	40	0	0	10	2
65	49	14	7	0	9
51	45	3	5	5	4
43	85	4	3	6	7
	78		5		4

4.4 In problem 3.5 at the end of Chapter 3, you calculated measures of central tendency for four variables for 15 respondents. Two of those variables are reproduced below. Calculate mean, range, and standard deviation for each variable. Write a paragraph summarizing this statistical information.

Respondent	Age	Attitude on Abortion Scale (High score = Strong opposition)
A	18	10
B	20	9
C	21	8
D	30	10
E	25	7
F	26	7
G	19	9
H	29	6
I	31	10
J	55	5
K	32	4
L	28	3
M	23	2
N	24	1
O	32	9

4.5 SOC In problem 3.7, you calculated means and medians for the average weekly earnings in 11

Canadian provinces in 1999 and 2002. If necessary, recalculate the mean earnings and then calculate the range and standard deviation for each year separately. What information do the measures of dispersion add to the measures of central tendency? Summarize this information in a paragraph.

Province	1999	2002
Newfoundland and Labrador	712	734
Prince Edward Island	667	709
Nova Scotia	570	697
New Brunswick	660	662
Quebec	648	745
Ontario	682	700
Manitoba	633	656
Saskatchewan	645	706
Alberta	617	680
British Columbia	712	741
Yukon	894	962

4.6 SOC Data on several variables measuring overall heath and well-being for 10 nations are reported below for 2000 with projections to 2010. Are these nations becoming more or less diverse on these variables? Calculate the mean, range, and standard deviation for each year for each variable. Summarize the results in a paragraph.

	Life Expectancy (years)		Infant Mortality Rate*		Fertility Rate#	
	2000	2010	2000	2010	2000	2010
Canada	80	81	5.0	4.5	1.6	1.6
U.S.	77	79	6.8	6.2	2.0	2.1
Mexico	72	74	25.4	18.5	2.6	2.3
Colombia	71	73	24.0	17.8	2.7	2.4
Japan	80	82	3.9	3.6	1.4	1.5
China	72	74	28.1	20.5	1.8	1.8
Sudan	57	61	68.7	55.2	5.4	4.2
Kenya	48	44	68.0	60.4	3.5	2.4
Italy	79	80	5.8	5.1	1.2	1.2
Germany	78	79	4.7	4.2	1.4	1.4

Source: U.S. Bureau of the Census. 2003. *Statistical Abstract of the United States,* 2002. Washington, DC: Government Printing Office, p. 829.

* Number of deaths of children under one year of age per 1,000 live births.

#Average number of children per female.

4.7 [SOC] Labor force participation rates (percent employed), percent high school graduates, and mean income for males and females in 10 states are reported below. Calculate a mean and a standard deviation for both groups for each variable and describe the differences. Are males and females unequal on any of these variables? How great is the gender inequality?

State	Percent Labor Force Participation		Percent High School Graduates		Mean Income (in dollars)	
	Male	Female	Male	Female	Male	Female
A	74	54	65	67	35,623	27,345
B	81	63	57	60	32,345	28,134
C	81	59	72	76	35,789	30,546
D	77	60	77	75	38,907	31,788
E	80	61	75	74	42,023	35,560
F	74	52	70	72	34,000	35,980
G	74	51	68	66	25,800	19,001
H	78	55	70	71	29,000	26,603
I	77	54	66	66	31,145	30,550
J	80	75	72	75	34,334	29,117

4.8 [CJ] Per capita expenditures for police protection for 20 cities are reported below for 1995 and 2000 in dollars. Compute a mean and standard deviation for each year, and describe the differences in expenditures for the five-year period. As an option, find the range and interquartile range for each year.

City	1995	2000
A	180	210
B	95	110
C	87	124
D	101	131
E	52	197
F	117	200
G	115	119
H	88	87
I	85	125
J	100	150
K	167	225
L	101	209
M	120	201
N	78	141
O	107	94
P	55	248
Q	78	140
R	92	131
S	99	152
T	103	178

4.9 [SOC] The following list shows rates of abortion per 100,000 women for 20 states in 1973 and 1975. Describe what happened to these distributions over the two-year period. Did the average rate increase or decrease? What happened to the dispersion of these distributions? What happened between 1973 and 1975 that might explain these changes in central tendency and dispersion? (*HINT: It was a Supreme Court decision.*)

State	1973	1975
Maine	3.5	9.5
Massachusetts	10.0	25.7
New York	53.5	40.7
Pennsylvania	12.1	18.5
Ohio	7.3	17.9
Michigan	18.7	20.3
Iowa	8.8	14.7
Nebraska	7.3	14.3
Virginia	7.8	18.0
South Carolina	3.8	10.3
Florida	15.8	30.5
Tennessee	4.2	19.2
Mississippi	0.2	0.6
Arkansas	2.9	6.3
Texas	6.8	19.1
Montana	3.1	9.9
Colorado	14.4	24.6
Arizona	6.9	15.8
California	30.8	33.6
Hawaii	26.3	31.6

Source: U.S. Bureau of the Census, 1978. *Statistical Abstracts of the United States: 1977.* Washington, DC: Government Printing Office.

4.10 [SW] One of your goals as the new chief administrator of a large social service bureau is to equalize workloads within the various divisions of the agency. You have gathered data on caseloads per worker within each division. Which division comes closest to the ideal of an equalized workload? Which is farthest away?

A	B	C	D
50	60	60	75
51	59	61	80
55	58	58	74
60	55	59	70
68	56	59	69
59	61	60	82
60	62	61	85
57	63	60	83
50	60	59	65
55	59	58	60

4.11 At St. Algebra College, the math department ran some special sections of the freshman math course using a variety of innovative teaching techniques. Students were randomly assigned to either the traditional sections or the experimental sections, and all students were given the same final exam. The results of the final are summarized below. What was the effect of the experimental course?

Traditional	Experimental
$\bar{X} = 77.8$	$\bar{X} = 76.8$
$s = 12.3$	$s = 6.2$
$N = 478$	$N = 465$

4.12 You're the governor of the state and must decide which of four metropolitan police departments will win the annual award for efficiency. The performance of each department is summarized in monthly arrest statistics as reported below. Which department will win the award? Why?

	Departments		
A	B	C	D
$\bar{X} = 601.30$	633.17	592.70	599.99
$s = 2.30$	27.32	40.17	60.23

SPSS for Windows

Using *SPSS for Windows* for Measures of Dispersion

DEMONSTRATION 4.1 Producing the Range and the Standard Deviation

In this demonstration, we will use **Descriptives** to find the range and standard deviation for *age* (YEARS OF AGE), *educ* (HIGHEST YEAR OF SCHOOL COMPLETED), and *tvhours* (HOURS PER DAY WATCHING TV).

From the main menu, click **Analyze, Descriptive Statistics,** and **Descriptives.** The **Descriptives** dialog box will open. Use the cursor to find the names of the three variables in the list on the left and click the right arrow button to transfer them to the **Variables** box. Click **OK,** and SPSS will produce the same output we analyzed in Demonstration 3.2. Now, however, we will consider dispersion rather than central tendency. The output looks like this:

Descriptive Statistics*

	N	Minimum	Maximum	Mean	Std. Deviation
AGE OF RESPONDENT	1411	18	89	46.07	16.552
HIGHEST YEAR OF SCHOOL COMPLETED	1415	0	20	13.67	2.903
HOURS PER DAY WATCHING TV	422	0	20	2.97	2.960
Valid N (listwise)	420				

*This table has been slightly modified to improve clarity.

The column labeled "Std. Deviation" reports the value of the standard deviation for each variable, and we can compute the range from the values given in the Minimum and Maximum columns. The standard deviation for *age* is 16.55 years, and the youngest and oldest respondents were 18 and 89, respectively. For *educ*, the standard deviation is 2.90, and scores ranged from 0 to 20. Respondents with scores of 20 have completed four years of formal education beyond the bachelor's level. The standard deviation is 2.96 for *tvhours*, and scores ranged from 0 hours of television watching to 20 (nearly the maximum possible in a single day).

At this point, the range is probably easier to understand and interpret than the standard deviation. As we saw in Section 4.6, the latter is more meaningful when we have a point of comparison. For example, suppose we were interested in the variable *tvhours* and how television-viewing habits have changed over the years. The **Descriptives** output for 2004 shows that people watched an average of 2.97 hours a day with a standard deviation of 2.96. Suppose that a sample from 1982 showed an average of 3.70 hours of television viewing a day with a standard deviation of 1.1. You could conclude that television watching had, on the average, decreased over the 20-year period but that Americans had also become much more diverse in their viewing habits.

DEMONSTRATION 4.2 Using the Compute Command to Create an "Attitude Toward Abortion" Scale

SPSS provides a variety of ways to transform and manipulate variables. In the SPSS exercises at the end of Chapter 2, the **Recode** command was introduced as a way of changing the values associated with a variable. In this demonstration, we will use the **Compute** command to create new variables and summary scales.

Let's begin by considering the two questions from the 2004 GSS that measure attitudes toward abortion, *abany* and *abhlth*. The items present two different situations under which an abortion might be desired and ask the respondent to react to each situation independently. *Abany* asks if a legal abortion should be possible for "any reason" at all, and *abhlth* asks if a legal abortion should be available for the more specific reason that the woman's health is in danger. Because these two situations are distinct, each item should be analyzed in its own right. Suppose, however, that you wanted to create a summary scale that indicated a person's *overall* feelings about abortion.

One way to do this would be to add the scores on the two variables together. This would create a new variable, which we will call *abscale,* with three possible scores. If a respondent was consistently "pro abortion" and answered "yes" (coded as "1") to both items, the respondent's score on the summary variable would be 2. A score of 3 would occur when a respondent answered "yes" to one item and "no" to the other. This might be labeled an "intermediate" or "moderate" position. The final possibility would be a score of 4, if the respondent answered "no" (coded as "2") to both items. This would be a consistent "antiabortion" position. The table below summarizes the scoring possibilities.

If response on *abany* is:	and	response on *abhlth* is:	score on *abscale* will be
1 (Yes)		1 (Yes)	2 (pro abortion)
1 (Yes)		2 (No)	3 (moderate)
2 (No)		1 (Yes)	3 (moderate)
2 (No)		2 (No)	4 (antiabortion)

The new variable, *abscale,* summarizes each respondent's overall position on the issue. Once you create this variable, you could analyze, transform, and manipulate *abscale* exactly like a variable actually recorded in the data file.

To use the **Compute** command, click **Transform** and then **Compute** from the main menu. The **Compute Variable** window will appear. Find the **Target Variable** box in the upper-left-hand corner of this window. The first thing we need to do is assign a name to the new variable we are about to compute (*abscale*) and type that name in this box. Next, we need to tell SPSS how to compute the new variable. In this case, *abscale* will be computed by adding the scores of *abany* and *abhlth*. Find *abany* in the variable list on the left and click the arrow button in the middle of the screen to transfer the variable name to the **Numeric Expression** box. Next, click the plus sign (+) on the calculator pad under the **Numeric Expression** box, and the sign will appear next to *abany.* Finally, highlight *abhlth* in the variable list and click the arrow button to transfer the variable name to the **Numeric Expression** box.

The expression in the **Numeric Expression** box should now read:

abany + abhlth

Click **OK,** and *abscale* will be created and added to the data set. If you want to keep this new variable permanently, click **Save** from the **File** menu, and the updated data set with *abscale* added will be saved to disk. If you are using the student version of SPSS, remember that your data set is limited to 50 variables.

We now have three variables that measure attitudes toward abortion — two items referring to specific situations and a more general, summary item. It is always a good idea to check the frequency distribution for computed and recoded variables to make sure that the computations were carried out as you intended. Use the **Frequencies** procedure (click **Analyze, Descriptive Statistics,** and **Frequencies**) to get tables for *abany, abhlth,* and *abscale.* Your output will look like this:

ABORTION IF WOMAN WANTS FOR ANY REASON*

		Frequency	Percent	Valid Percent	Cumulative Percent
Valid	YES	176	12.4	40.7	40.7
	NO	256	18.1	59.3	100.0
	Total	432	30.5	100.0	
Missing	System	983	69.5		
	Total	1415	100.0		

*This table has been slightly edited to improve readability.

ABORTION IF WOMAN'S HEALTH SERIOUSLY ENDANGERED*

		Frequency	Percent	Valid Percent	Cumulative Percent
Valid	YES	378	26.7	86.5	86.5
	NO	59	4.2	13.5	100.0
	Total	437	30.9	100.0	
Missing	System	978	69.1		
	Total	1415	100.0		

*This table has been slightly edited to improve readability.

ABSCALE*

		Frequency	Percent	Valid Percent	Cumulative Percent
Valid	2.00	173	12.2	41.1	41.1
	3.00	191	13.5	45.4	86.5
	4.00	57	4.0	13.5	100.00
	Total	421	29.8	100.00	
Missing	System	994	70.2		
	Total	1415	100.00		

*This table has been slightly edited to improve readability.

The level of approval for abortion differs markedly in the two specific situations. Less than half of the respondents (about 40%) approve "for any reason" but the huge majority approves when the health of the mother is a consideration. Looking at the combined scores on *abscale,* we see that about 41% of the sample approved of the legal right to an abortion in both cases (scored 2), and only about 14% disapproved in both cases (scored 4). About 45% approved in one situation but not in the other, and this reflects the pattern of approval we saw on the individual items.

Note that only 432 and 437 of the 1,415 total respondents to the 2004 GSS answered the two specific abortion items. Remember that no respondent is given the entire GSS, and the vast majority of the "missing cases" received a form of the GSS that did not include these two items. Now look at *abscale* and note that even fewer cases (421) are included in the summary scale than in either of the two original items. When SPSS executes a **Compute** statement, it automatically eliminates any cases that are missing scores on any of the constituent items. If these cases were not eliminated, a variety of errors and misclassifications could result. For example, if cases with missing scores were included, a person who scored a 2 ("antiabortion") on *abany* and then failed to respond to *abhlth* would have a total score of 2 on *abscale.* Thus, this case would be treated as "pro abortion" when the only information we have indicates that this respondent is "antiabortion." To eliminate this kind of error, cases with missing scores on any of the constituent variables are deleted from calculations.

Exercises

4.1 Use **Descriptives** to produce univariate descriptive statistics for *paeduc* (respondent's father's years of education) and *educ* (respondent's years of education). Compare the statistics for *paeduc* and *educ.* Describe the difference between the two generations. Are the respondents more or less educated than their fathers? Are the respondents more or less homogeneous on this variable than their fathers?

4.2 Use the **Compute** command to create a summary scale for *fepresch* and *fefam.* Get univariate descriptive statistics for the summary scale and for *fepresch* and *fefam.* Write a few sentences summarizing these tables using our description of the distributions for *abany, abhlth,* and *abscale* as a model.

5

The Normal Curve

LEARNING OBJECTIVES   By the end of this chapter, you will be able to:

1. Define and explain the concept of the normal curve.
2. Convert empirical scores to Z scores and use Z scores and the normal curve table (Appendix A) to find areas above, below, and between points on the curve.
3. Express areas under the curve in terms of probabilities.

5.1 INTRODUCTION

The **normal curve** is a concept of great importance in statistics. In combination with the mean and standard deviation, we can use the normal curve to construct precise descriptive statements about empirical distributions. In addition, as we shall see in Part II, it is also central to the theory that underlies inferential statistics. This chapter will conclude our treatment of descriptive statistics in Part I and lay important groundwork for Part II.

The normal curve is a theoretical model, a kind of frequency polygon or line chart that is unimodal (that is, it has a single mode or peak), perfectly smooth, and symmetrical (unskewed) so that its mean, median, and mode are all exactly the same value. It is bell shaped and its tails extend infinitely to the left and to the right. Of course, no empirical distribution has a shape that perfectly matches this ideal model, but many variables (e.g., test results from large classes, standardized test scores such as the GRE, people's height and weight) are close enough to permit the assumption of normality. In turn, this assumption makes possible one of the most important uses of the normal curve—the description of empirical distributions based on our knowledge of the theoretical normal curve.

The crucial point about the normal curve is that distances along the horizontal axis of the distribution, when measured in standard deviations from the mean, always encompass the same proportion of the total area under the curve. In other words, on any normal curve, the distance from any given point to the mean (when measured in standard deviations) will cut off exactly the same proportion of the total area.

To illustrate, Figures 5.1 and 5.2 present two hypothetical distributions of IQ scores, one for a group of males and one for a group of females, both normally distributed (or nearly so), such that:

Males	**Females**
$\bar{X} = 100$	$\bar{X} = 100$
$s = 20$	$s = 10$
$N = 1,000$	$N = 1,000$

FIGURE 5.1   IQ SCORES FOR A GROUP OF MALES

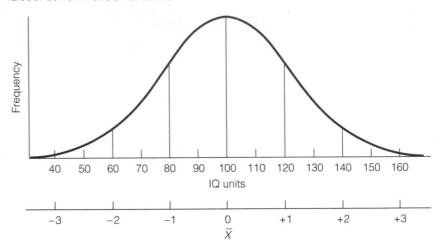

FIGURE 5.2   IQ SCORES FOR A GROUP OF FEMALES

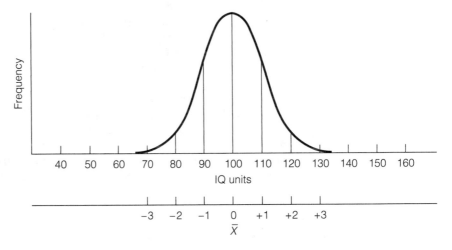

Figures 5.1 and 5.2 have two scales on the horizontal axis or abscissa of the graph. The upper scale is stated in "IQ units" and the lower scale in standard deviations from the mean. These scales are interchangeable and we can easily shift from one to the other. For example, for the males, an IQ score of 120 is one standard deviation (remember that, for the male group, $s = 20$) above (to the right of) the mean and an IQ of 140 is two standard deviations above (to the right of) the mean. Scores below or to the left of the mean are marked as negative values because they are less than the mean. For the males, an IQ of 80 is one standard deviation below the mean, an IQ score of 60 is two standard deviations less than the mean, and so forth. Figure 5.2 is marked in a similar way except that, because its standard deviation is a different value ($s = 10$), the markings occur at different points. For the female sample, one standard deviation above the mean is an IQ of 110, one standard deviation below the mean is an IQ of 90, and so forth.

Recall that the crucial point about the normal curve is that distances along the horizontal axis, when measured in standard deviations, always encompass exactly the same proportion of the total area under the curve. Specifically, the distance between one standard deviation above the mean and one standard deviation below the mean (or ±1 standard deviation) encompasses exactly 68.26% of the total area under the curve. This means that in Figure 5.1, 68.26% of the total area lies between the score of 80 (−1 standard deviation) and 120 (+1 standard deviation). The standard deviation for females is 10, so the same percentage of the area (68.26%) lies between the scores of 90 and 110. As long as an empirical distribution is normal, 68.26% of the total area will always be encompassed between ±1 standard deviation, regardless of the trait being measured and the number values of the mean and standard deviation.

It will be useful to familiarize yourself with the following relationships between distances from the mean and areas under the curve:

Between	Lies
±1 standard deviation	68.26% of the area
±2 standard deviations	95.44% of the area
±3 standard deviations	99.72% of the area

Figure 5.3 displays these relationships graphically.

The relationship between distance from the mean and area allows us to describe empirical distributions that are at least approximately normal. We can describe the position of individual scores with respect to the mean, the distribution as a whole, or any other score in the distribution.

The areas between scores can also be expressed, if desired, in numbers of cases rather than percentage of total area. For example, a normal distribution of 1,000 cases will contain about 683 cases (68.26% of 1,000 cases) between ±1 standard deviation of the mean, about 954 between ±2 standard deviations, and about 997 between ±3 standard deviations. Thus, for any normal distribution, only a few cases will be farther away from the mean than ±3 standard deviations.

FIGURE 5.3   AREAS UNDER THE THEORETICAL NORMAL CURVE

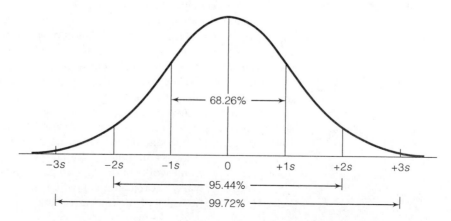

STATISTICS IN EVERYDAY LIFE: How Normal Is the Normal Curve?

The normal curve is a crucial concept is statistics (especially, as you will see in chapters to come, in inferential statistics) but how common is it in everyday life? What traits are normally distributed? What variables important to our everyday lives can be accurately described as unimodal, symmetrical, and bell shaped?

We have already seen, in the "Statistics in Everyday Life" feature in Chapter 3, that some variables—salaries, housing prices—are not normal. The shape of many other variables (e.g., support for legal abortion, approval of the president's performance) is unknown and may be highly volatile, changing in shape from month to month or even day to day as the intensity of public debate and concern waxes and wanes.

Variables are most likely to be normal when their scores are randomly distributed around their mean. For example, imagine a manufacturing process in which a product (e.g., a part for an automobile or piece of furniture) must be a certain length. The tolerance for errors might be very slight—hundreds or thousands of an inch—but there will be some deviation around the desired length: Sometimes the part will be slightly too long, sometimes a little too short. These deviations from the ideal length will be just as likely to be above the mean as below and will become increasingly uncommon as we get further from the desired size. Thus, the distribution of errors around the mean will approximate a normal curve.

Some measures of physical stature (e.g., the length of appendages among certain species) follow this pattern of random deviations from the average but, for most variables of interest to the social sciences, the shape of the distribution is either unknown or known to *not* be normal. Consider income to illustrate. Deviations from the mean for this variable are not random. Each individual has a greater or lesser chance to acquire high income or wealth depending on a number of socially powerful forces, many of which are related to the circumstances of birth: gender, race, and—most especially—the SES of parents.

On the other hand, many standardized tests of achievement and ability do have scores that are more or less normally distributed. This is because the tests are intentionally designed to generate this outcome. Hard and easy items are mixed in proportion so that lower scores will be roughly equal in number to higher scores and so that both high and low scores will decline in frequency as they move further from the mean. Note, however, that normality in the test scores does not mean that the underlying trait being measured is normal. For example, the scores on a particular IQ test may have a normal shape, but this does not necessarily mean that human intelligence itself is normally distributed.

All in all, then, the normal curve may not be especially common in nature. In statistics, the uncommon nature of normality is not a problem and, as you will see, we can still use the curve for a number of important purposes.

5.2 COMPUTING *Z* SCORES   To find the percentage of the total area (or number of cases) above, below, or between scores in an empirical distribution, the original scores must first be expressed in units of the standard deviation or converted into **Z scores.** The original scores could be in any unit of measurement (inches, IQ, dollars), but *Z* scores always have the same values for their mean (0) and standard deviation (1).

Think of converting the original scores into *Z* scores as a process of changing value scales, similar to changing from meters to yards, kilometers to miles, or quarts to liters. These units are different but equally valid ways of expressing distance, length, or volume. For example, a mile equals 1.61 kilometers so two towns that are 10 miles apart are also 16.1 kilometers apart and a "5k" race covers about 3.10 miles. Although you may be more familiar with miles than kilometers, either unit works perfectly well as a way of expressing distance.

In the same way, the original (or "raw") scores and *Z* scores are two equally valid but different ways of measuring distances under the normal curve. In

STEP BY STEP	Computing *Z* Scores

Step 1: Subtract the value of the mean ($\overline{X}$) from the value of the score (X_i).

Step 2: Divide the quantity found in step 1 by the value of the standard deviation (*s*).

Figure 5.1, for example, we could describe a particular score in terms of IQ units ("John's score was 120") or standard deviations ("John scored one standard deviation above the mean").

When we compute *Z* scores, we convert the original units of measurement (IQ scores, inches, dollars, and so on) to *Z* scores and, thus, "standardize" the normal curve to a distribution that has a mean of 0 and a standard deviation of 1. The mean of the empirical normal distribution will be converted to 0, its standard deviation to 1, and all values will be expressed in *Z*-score form. The formula for computing *Z* scores is:

FORMULA 5.1

$$Z = \frac{X_i - \overline{X}}{s}$$

This formula will convert any score (X_i) from an empirical normal distribution into the equivalent *Z* score. To illustrate with the men's IQ data (Figure 5.1), the *Z*-score equivalent of a raw score of 120 would be:

$$Z = \frac{120 - 100}{20} = +1.00$$

The *Z* score of positive 1.00 indicates that the original score lies one standard deviation unit above (to the right of) the mean. A negative score would fall below (to the left of) the mean. *(For practice in computing Z scores, see any of the problems at the end of this chapter.)*

5.3 THE NORMAL CURVE TABLE

Statisticians have thoroughly analyzed and described the theoretical normal curve. The areas related to any *Z* score have been precisely determined and organized into a table format. Appendix A presents this **normal curve table** or *Z*-score table, and Table 5.1 reproduces a small portion of it here for purposes of illustration.

The normal curve table consists of three columns, with *Z* scores in the left-hand column (a), areas between the *Z* score and the mean of the curve in the middle column (b), and areas beyond the *Z* score in the right-hand column (c). To find the area between any *Z* score and the mean, go down the column labeled "*Z*" until you find the *Z* score. For example, go down column a either in Appendix A or in Table 5.1 until you find a *Z* score of 1.00. The entry in column b ("Area Between Mean and *Z*") is 0.3413. The table presents all areas in proportions, but we can easily translate these into percentages by multiplying them by 100 (see Chapter 2). We could say either "a proportion of 0.3413 of the total area under the curve lies between a *Z* score of 1.00 and the mean," or "34.13% of the total area lies between a score of 1.00 and the mean."

To illustrate further, find the *Z* score of 1.50 either in column a of Appendix A or the abbreviated table in Table 5.1. This score is 1½ standard deviations

TABLE 5.1 AN ILLUSTRATION OF HOW TO FIND AREAS UNDER THE NORMAL CURVE USING APPENDIX A

(a) Z	(b) Area Between Mean and Z	(c) Area Beyond Z
0.00	0.0000	0.5000
0.01	0.0040	0.4960
0.02	0.0080	0.4920
0.03	0.0120	0.4880
.	.	.
.		.
1.00	0.3413	0.1587
1.01	0.3438	0.1562
1.02	0.3461	0.1539
1.03	0.3485	0.1515
.	.	.
.		.
1.50	0.4332	0.0668
1.51	0.4345	0.0655
1.52	0.4357	0.0643
1.53	0.4370	0.0630
.	.	.
.		.

to the right of the mean and corresponds to an IQ of 130 for the men's IQ data. The area in column b for this score is 0.4332. This means that a proportion of 0.4332, or a percentage of 43.32%, of all the area under the curve lies between this score and the mean.

The third column in the table presents "Areas Beyond Z." These are areas above (to the right of) positive scores or below (to the left of) negative scores. We will use this column when we want to find an area above or below certain Z scores, an application that I will explain in Section 5.4.

To conserve space, the normal curve table in Appendix A includes only positive Z scores. Because the normal curve is perfectly symmetrical, however, the area between a negative score and the mean (column b) will be exactly the same as those for a positive score of the same numerical value. For example, the area between a Z score of -1.00 and the mean will be 34.13%, exactly the same as the area we found previously for a score of $+1.00$. As will be repeatedly demonstrated later, however, the sign of the Z score is extremely important and should be carefully noted.

For practice in using Appendix A to describe areas under an empirical normal curve, verify that the Z scores and areas given below are correct for the men's IQ distribution. For each IQ score, compute the equivalent Z score using Formula 5.1, and then use Appendix A to find the area between the score and the mean. ($\overline{X} = 100$, $s = 20$ throughout.)

IQ Score	Z Score	Area Between Z and the Mean
110	+0.50	19.15%
125	+1.25	39.44%
133	+1.65	45.05%
138	+1.90	47.13%

The same procedures apply when the Z-score equivalent of an actual score happens to be a minus value (that is, when the raw score lies below the mean).

IQ Score	Z Score	Area Between Z and the Mean
93	−0.35	13.68%
85	−0.75	27.34%
67	−1.65	45.05%
62	−1.90	47.13%

Remember that the areas in Appendix A will be the same for Z scores of the same numerical value regardless of sign. The area between the score of 138 (+1.90) and the mean is the same as the area between 62 (−1.90) and the mean. *(For practice in using the normal curve table, see any of the problems at the end of this chapter.)*

5.4 FINDING TOTAL AREA ABOVE AND BELOW A SCORE

To this point, we have seen how to use the normal curve table (Appendix A) to find areas between a Z score and the mean. We can also use the table to find other kinds of areas in empirical distributions that are at least approximately normal in shape. For example, suppose you need to determine the total area below the scores of two male subjects in the distribution described in Figure 5.1. The first subject has a score of 117 ($X_1 = 117$), which is equivalent to a Z score of +0.85:

$$Z_1 = \frac{X_i - \bar{X}}{s} = \frac{117 - 100}{20} = \frac{17}{20} = +0.85$$

The plus sign of the Z score indicates that the score should be placed above (to the right of) the mean. To find the area below a positive Z score, the area between the score and the mean (given in column b) must be added to the area below the mean. As we noted earlier, the normal curve is symmetrical (unskewed) and its mean will be equal to its median. Therefore, the area below the mean (just like the median) will be 50%. Study Figure 5.4 carefully. We are interested in the shaded area.

By consulting the normal curve table, we find that the area between the score and the mean (see column b) is 30.23% of the total area. The area below

FIGURE 5.4   FINDING THE AREA BELOW A POSITIVE Z SCORE

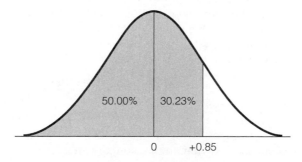

FIGURE 5.5   FINDING THE AREA BELOW A NEGATIVE *Z* SCORE

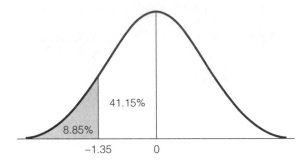

41.15%

8.85%

−1.35   0

FIGURE 5.6   FINDING THE AREA ABOVE A POSITIVE *Z* SCORE

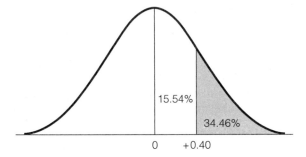

15.54%

34.46%

0   +0.40

a *Z* score of +0.85 is therefore 80.23% (50.00% + 30.23%). This subject scored higher than 80.23% of the persons tested.

The second subject has an IQ score of 73 ($X_2 = 73$), which is equivalent to a *Z* score of −1.35:

$$Z_2 = \frac{X_i - \overline{X}}{s} = \frac{73 - 100}{20} = -\frac{27}{20} = -1.35$$

To find the area below a negative score, we use the column labeled "Area Beyond *Z*." Figure 5.5 depicts the area of interest; we must determine the size of the shaded area. The area beyond a score of −1.35 is given as 0.0885, which we can express as 8.85%. The second subject ($X_2 = 73$) scored higher than 8.85% of the tested group.

In the preceding examples, we used the techniques for finding the area below a score. We use essentially the same techniques to find the area above a score. If we need to determine the area above an IQ score of 108, for example, we would first convert to a *Z* score:

$$Z = \frac{X_i - \overline{X}}{s} = \frac{108 - 100}{20} = \frac{8}{20} = +0.40$$

and then proceed to Appendix A. The shaded area in Figure 5.6 represents the area of interest to us. The area above a positive score appears in the "Area Beyond *Z*" column; in this case, the area is 0.3446, or 34.46%.

TABLE 5.2   FINDING AREAS ABOVE AND BELOW POSITIVE AND NEGATIVE *Z* SCORES

To Find Area	When the Score Is	
	Positive	Negative
Above *Z*	Look in column c.	Add column b area to .5000 or 50.00%.
Below *Z*	Add column b area to .5000 or 50.00%.	Look in column c.

Table 5.2 summarizes these procedures. To find the total area above a positive Z score or below a negative Z score, go down the "Z" column of Appendix A until you find the score. The area you are seeking will be in the "Area Beyond Z" column (column c). The value in column c can be left as a proportion or changed to a percentage (by multiplying be 100).

To find the total area below a positive Z score or above a negative score, locate the Z score in column a and then add the area in the "Area Between Mean and Z" (column b) to either .5000 (for proportions) or 50.00 (for percentages). Again, you can leave the value as a proportion or convert it to a percentage. These techniques might be confusing at first, and you will find it helpful to draw the curve and shade in the areas in which you are interested. *(For practice in finding areas above or below Z scores, see problems 5.1 to 5.7.)*

5.5 FINDING AREAS BETWEEN TWO SCORES

On occasion, you will need to determine the area between two scores rather than the total area above or below one score. When the scores are on opposite sides of the mean, you can find the area between the scores by adding the areas between each score and the mean. Using the men's IQ data as an example, if we wanted to know the area between the IQ scores of 93 and 112, we would convert both scores to Z scores, find the area between each score and the mean

---

STEP BY STEP   **Finding Areas Above and Below Positive and Negative *Z* Scores**

Step 1: Compute the Z score. Note whether the score is positive or negative.

Step 2: Find the Z score in column a of the Normal Curve Table (Appendix A) and do one of the following:

To find the total area below a positive *z* score

Step 3: Add the column b area for this score to .5000 or to 50.00%.

To find the total area above a positive *z* score

Step 4: Look in column c for this score for the area expressed as a proportion. Multiply the column c area by 100 to express the area as a percentage.

To find the total area below a negative *z* score

Step 5: Look in column c for this score for the area expressed as a proportion. Multiply the column c area by 100 to express the area as a percentage.

To find the total area above a negative *z* score

Step 6: Add the column b area for this score to .5000 or to 50.00%.

Application 5.1

You have just received your score on a test of intelligence. If your score was 78 and you know that the mean score on the test was 67 with a standard deviation of 5, how does your score compare with the distribution of all test scores?

If you can assume that the test scores are normally distributed, then you can compute a Z score and find the area below or above your score. The Z-score equivalent of your raw score would be:

$$Z = \frac{X_i - \overline{X}}{s} = \frac{78 - 67}{5} = \frac{11}{5} = +2.20$$

Turning to Appendix A, we find that the "Area Between Mean and Z" for a Z score of 2.20 is 0.4861, which could also be expressed as 48.61%. Because this is a positive Z score, we need to add this area to 50.00% to find the total area below (see Table 5.2). Your score is higher than (48.61 + 50.00), or 98.61%, of all the test scores. You did well!

---

from Appendix A, and add these two areas together. The first IQ score of 93 converts to a Z score of -0.35:

$$Z_1 = \frac{X_i - \overline{X}}{s} = \frac{93 - 100}{20} = -\frac{7}{20} = -0.35$$

The second IQ score (112) converts to $+0.60$:

$$Z_2 = \frac{X_i - \overline{X}}{s} = \frac{112 - 100}{20} = \frac{12}{20} = 0.60$$

Figure 5.7 includes both scores. We are interested in the total shaded area. The total area between these two scores is 13.68% + 22.57%, or 36.25%. Therefore, 36.25% of the total area (or about 363 of the 1,000 cases) lies between the IQ scores of 93 and 112.

When the scores of interest are on the same side of the mean, a different procedure must be followed to determine the area between them. For example, if we were interested in the area between the scores of 113 and 121, we would begin by converting these scores into Z scores:

$$Z_1 = \frac{X_i - \overline{X}}{s} = \frac{113 - 100}{20} = \frac{13}{20} = +0.65$$

$$Z_2 = \frac{X_i - \overline{X}}{s} = \frac{121 - 100}{20} = \frac{21}{20} = +1.05$$

FIGURE 5.7   FINDING THE AREA BETWEEN TWO SCORES

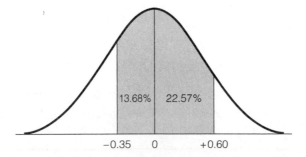

13.68%   22.57%

−0.35   0   +0.60

Figure 5.8 displays these scores; we are interested in the shadowed area. To find the area between two scores on the same side of the mean, find the area between each score and the mean (given in column b of Appendix A) and then subtract the smaller area from the larger. Between the Z score of $+0.65$ and the mean lies 24.22% of the total area. Between $+1.05$ and the mean lies 35.31% of the total area. Therefore, the area between these two scores is $35.31\% - 24.22\%$, or 11.09% of the total area. We would follow the same technique if both scores had been below the mean. Table 5.3 summarizes the procedures for finding areas between two scores. *(For practice in finding areas between two scores, see problems 5.3, 5.4, and 5.6 to 5.9.)*

FIGURE 5.8   FINDING THE AREA BETWEEN TWO SCORES

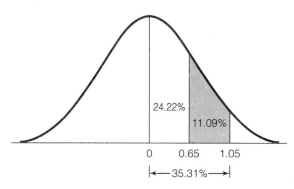

TABLE 5.3   FINDING AREAS BETWEEN SCORES

Situation	Procedure
Scores are on the SAME side of the mean	Find areas between each score and the mean in column b. Subtract the smaller area from the larger area.
Scores are on OPPOSITE sides of the mean	Find areas between each score and the mean in column b. Add the two areas together.

STEP BY STEP   **Finding Areas Between *Z* Scores**

Step 1: Compute the Z scores for both raw scores. Note whether the scores are positive or negative and do one of the following:

If the Scores are on the *Same* Side of the Mean

Step 2: Find areas between each score and the mean in column b. Subtract the smaller area from the larger area. Multiply this value by 100 to express it as a percentage.

If the Scores are on *Opposite* Sides of the Mean

Step 3: Find areas between each score and the mean in column b. Add the two areas together to get the total area between the scores. Multiply this value by 100 to express it as a percentage.

Application 5.2

All sections of Biology 101 at a large university were given the same final exam. Test scores were distributed normally with a mean of 72 and a standard deviation of 8. What percentage of student scored between 60 and 69 (a grade of D) and what percentage scored between 70 and 79 (a grade of C)? The first two scores are both below the mean. Using Table 5.3 as a guide, we must compute Z scores, find areas between each score and the mean, and then subtract the smaller area from the larger:

$$Z_1 = \frac{X_i - \overline{X}}{s} = \frac{60 - 72}{8} = -\frac{12}{8} = -1.50$$

$$Z_2 = \frac{X_i - \overline{X}}{s} = \frac{69 - 72}{8} = -\frac{3}{8} = -0.38$$

Using column b, we see that the area between $Z = -1.50$ and the mean is .4332 and the area between $Z = -0.38$ and the mean is .1480. Subtracting the smaller from the larger (.4332 − .1480) gives .2852. Changing to percentage format, we can say that 28.52% of the students received a D on the test.

To find the percentage of students who earned a C, we must add column b areas together since the scores (70 and 79) are on opposite sides of the mean (see Table 5.3):

$$Z_1 = \frac{X_i - \overline{X}}{s} = \frac{70 - 72}{8} = -\frac{2}{8} = -0.25$$

$$Z_2 = \frac{X_i - \overline{X}}{s} = \frac{79 - 72}{8} = \frac{7}{8} = 0.88$$

Using column b, we see that the area between $Z = -0.25$ and the mean is .0987, and the area between $Z = 0.88$ and the mean is .3106. Therefore, the total area between these two scores is (.0987 + .3106) or 0.4093. Translating to percentages again, we can say that 40.93% of the students earned a C on this test.

5.6 USING THE NORMAL CURVE TO ESTIMATE PROBABILITIES

To this point, we have thought of the theoretical normal curve as a way of describing the percentage of total area above, below, and between scores in an empirical distribution. We have also seen that these areas can be converted into the number of cases above, below, and between scores. In this section, we introduce the idea that the theoretical normal curve can also be thought of as a distribution of probabilities. Specifically, we can use the properties of the theoretical normal curve (Appendix A) to estimate the **probability** that a case randomly selected from an empirical normal distribution will have a score that falls in a certain range. In terms of techniques, we will find these probabilities in exactly the same way as we found areas. Before we consider these mechanics, however, let us examine what the concept of probability means.

Although we are rarely systematic or rigorous about it, we all attempt to deal with probabilities every day and, indeed, we base our behavior on our estimates of the likelihood that certain events will occur. We constantly ask (and answer) questions such as: What is the probability of rain? Of drawing to an inside straight in poker? Of the worn-out tires on my car going flat? Of passing a test if I don't study?

To estimate the probability of an event, we must first be able to define what would constitute a "success." The examples above contain several different definitions of a success (that is, rain, drawing a certain card, flat tires, and passing grades). To determine a probability, a fraction must be established, with the numerator equaling the number of events that would constitute a success and the denominator equaling the total number of possible events where a success could theoretically occur:

$$probability = \frac{\#successes}{\#events}$$

To illustrate, assume that we want to know the probability of selecting a specific card—say, the king of hearts—in one draw from a well-shuffled deck of cards. Our definition of a success is specific (drawing the king of hearts), and with the information given, we can establish a fraction. Only one card satisfies our definition of success, so the number of events that would constitute a success is 1; this value will be the numerator of the fraction. There are 52 possible events (that is, 52 cards in the deck), so the denominator will be 52. The fraction is thus 1/52, which represents the probability of selecting the king of hearts on one draw from a well-shuffled deck of cards. Our probability of success is 1 out of 52.

We can leave the fraction established above as it is, or we can express it in several other ways. For example, we can express it as an odds ratio by inverting the fraction, showing that the odds of selecting the king of hearts on a single draw are 52:1 (or 52 to 1). We can express the fraction as a proportion by dividing the numerator by the denominator. For our example above, the corresponding proportion is .0192, which is the proportion of all possible events that would satisfy our definition of a success. In the social sciences, probabilities are usually expressed as proportions, and we will follow this convention throughout the remainder of this section. Using p to represent "probability," the probability of drawing the king of hearts (or any specific card) can be expressed as:

$$p \text{ (king of hearts)} = \frac{\#successes}{\#events} = \frac{1}{52} = 0.192$$

As conceptualized here, probabilities have an exact meaning: Over the long run, the events that we define as successes will bear a certain proportional relationship to the total number of events. The probability of .0192 for selecting the king of hearts in a single draw really means that, over thousands of selections of one card at a time from a full deck of 52 cards, the proportion of successful draws would be .0192. Or, for every 10,000 draws, 192 would be the king of hearts, and the remaining 9,808 selections would be other cards. Thus, when we say that the probability of drawing the king of hearts in one draw is .0192, we are essentially applying our knowledge of what would happen over thousands of draws to a single draw.

Like proportions, probabilities range from 0.00 (meaning that the event has absolutely no chance of occurrence) to 1.00 (a certainty). As the value of the probability increases, the likelihood that the defined event will occur also increases. A probability of .0192 is close to zero, and this means that the event (drawing the king of hearts) is unlikely or improbable.

We can use these techniques to establish simple probabilities in any situation in which we can specify the number of successes and the total number of events. For example, a single die has six sides or faces, each with a different value ranging from 1 to 6. The probability of getting any specific number (say, a 4) in a single roll of a die is therefore:

$$p \text{ (rolling a four)} = \frac{1}{6} = .1667$$

Combining this way of thinking about probability with our knowledge of the theoretical normal curve allows us to estimate the likelihood of selecting a case that has a score within a certain range. For example, suppose we want to estimate the probability that a randomly chosen subject from the distribution of men's IQ scores would have an IQ score between 95 and the mean score of 100.

Our definition of a success here would be the selection of any subject with a score in the specified range. Normally, we would next establish a fraction with the numerator equal to the number of subjects with scores in the defined range and the denominator equal to the total number of subjects. However, if the empirical distribution is normal in form, we can skip this step because Appendix A already states the probabilities, in proportion form. That is, the areas in Appendix A can be interpreted as probabilities.

To determine the probability that a randomly selected case will have a score between 95 and the mean, we would convert the original score to a Z score:

$$Z = \frac{X_i - \overline{X}}{s} = \frac{95 - 100}{20} = \frac{-5}{20} = -.25$$

Using Appendix A, we see that the area between this score and the mean is 0.0987. This is the probability we are seeking. The probability that a randomly selected case will have a score between 95 and 100 is 0.0987 (or, rounded off, 0.1, or one out of 10). In the same fashion, we can estimate the probability of selecting a subject from any range of scores. Note that the techniques for estimating probabilities are exactly the same as those for finding areas. The only new information introduced in this section is the idea that the areas in the normal curve table can also be thought of as probabilities.

To consider an additional example, what is the probability that a randomly selected male will have an IQ less than 123? We will find probabilities in exactly the same way we found areas. The score (X_i) is above the mean and we will find the probability we are seeking by adding the area in column b to 0.5000. First, we find the Z score:

$$Z = \frac{X_i - \overline{X}}{s} = \frac{123 - 100}{20} = \frac{23}{20} = +1.15$$

Next, look in column b of Appendix A to find the area between this score and the mean. Then, add the area (0.3749) to 0.5000. The probability of selecting a male with an IQ of less than 123 is 0.3749 + 0.5000 or 0.8749. Rounding this value to .88, we can say that the odds are .88 (very high) that we will select a male with an IQ score in this range. Technically, remember that this probability expresses what would happen over the long run: For every 100 males selected from this group over an infinite number of trials, 88 would have IQ scores less than 123 and 12 would not.

Let me close by stressing a key point about probabilities and the normal curve. The probability is very high that any case randomly selected from a normal distribution will have a score close in value to that of the mean. The normal curve reaches its highest point at the mean (which is the same as the median and the mode), and, thus, there are more cases at this point than at any other. (Remember that the curve is a frequency polygon with scores arrayed along the horizontal axis and number of cases along the vertical axis.) For any normal curve, cases are clustered around the mean and decline in frequency as we move farther away—either to the right or to the left—from the mean value. In fact, given what we know about the normal curve, the probability that a randomly selected case will have a score within ±1 standard deviation of the mean is 0.6826. Rounding off, we can say that 68 out of 100 cases—or about two-thirds of all cases—selected over the long run will have a score between ±1 standard deviation or Z score of the mean. The probability is high that any randomly selected case will have a score close in value to the mean.

STEP BY STEP	**Finding Probabilities**

Step 1: Compute the Z score (or scores). Note whether the score is positive or negative.

Step 2: Find the Z score (or scores) in column a of the Normal Curve Table (Appendix A).

Step 3: Find the area above or below the score (or between the scores) as you would normally (see the two previous "Step by Step" boxes in this chapter) and express the result as a proportion. Typically, probabilities are expressed as a value between 0.00 and 1.00 rounded to two digits beyond the decimal point.

Application 5.3

The distribution of scores on a biology final exam used in Application 5.2 had a mean of 72 and a standard deviation of 8. What is the probability that a student selected at random will have a score less than 61? More than 80? Less than 98? To answer these questions, we must first calculate Z scores and then consult Appendix A. We are looking for probabilities, so we will leave the areas in proportion form. The Z score for a score of 61 is:

$$Z_1 = \frac{X_i - \bar{X}}{s} = \frac{61 - 72}{8} = -\frac{11}{8} = -1.38$$

This score is a negative value (below or to the left of the mean), and we are looking for the area below. Using Table 5.2 as a guide, we see that we must use column c to find area below a negative score. This area is .0838. Rounding off, we can say that the odds of selecting a student with a score less than 61 are only 8 out of 100. This low value tells us that this would be an unlikely event.

The Z score for the score of 80 is:

$$Z_2 = \frac{X_i - \bar{X}}{s} = \frac{80 - 72}{8} = \frac{8}{8} = 1.00$$

The Z score is positive; to find the area above (greater than) 80, we look in column c (see Table 5.2). This value is .1587. The odds of selecting a student with a score greater than 80 is roughly 16 out of 100, about twice as likely as selecting a student with a score of less than 61.

The Z score for the score of 98 is:

$$Z_3 = \frac{X_i - \bar{X}}{s} = \frac{98 - 72}{8} = \frac{26}{8} = 3.25$$

To find the area below a positive Z score, we add the area between the score and the mean (column b) to .5000 (see Table 5.2). This value is (.4994 + .5000) or .9994. It is extremely likely that a randomly selected student will have a score less than 98. Remember that scores more than ±3 standard deviations from the mean are very rare.

In contrast, the probability of the case having a score beyond 3 standard deviations from the mean is very small. Look in column c ("Area Beyond Z") for a Z score of 3.00 and you will find the value .0014. Adding the areas in the upper tail (beyond +3.00) to the area in the lower tail (beyond −3.00) gives us .0014 + .0014 for a total of .0028. The probability of selecting a case with a very high score or a very low score is .0028. If we randomly select cases from a normally distributed variable, we would select cases with Z scores beyond ±3.00 only 28 times out of every 10,000 trials.

The general point to remember is that cases with scores close to the mean are common and cases with scores far above or below the mean are rare. This relationship is central for an understanding of inferential statistics in Part II. *(For practice in using the normal curve table to find probabilities, see problems 5.8 to 5.10 and 5.13.)*

STATISTICS IN EVERYDAY LIFE: Applying the Laws of Probability

Many people enjoy playing games in which rational analysis is mixed with random chance. Games as disparate as Monopoly or Old Maid with the kids, bridge with the neighbors, or roulette in the glittering casinos of Las Vegas can be challenging, absorbing, and even addictive. The tension and drama generated by the element of chance—the possibility that a skilled expert could be ignominiously defeated or that some naïve greenhorn might stumble into victory—is what keeps many people coming back to the game board, the local bingo game, or the roulette wheel.

People who enjoy competing against the odds engage in an extraordinary array of behaviors to try to control their fate. Some rely on lucky charms or mantras, others pray to, bargain with, cajole, or even threaten the gods of bingo, poker, Monopoly, or state lotteries. While superstitious behaviors are very common, there are also more rational and scientific ways of increasing your chances of walking away a winner and a careful study of the role of probability in your preferred game can dramatically improve your chances for victory. (Of course, there are no guarantees—that's why they call it gambling.)

One example of the application of rigorous logic to understand and tame the odds is reported in the best seller *Bringing Down the House* (Mizrich, Ben. 2003. New York: Free Press). A group of MIT college students, along with some advisors and financial backers, applied the mathematics of probability to the game of blackjack (or twenty-one).[1] They studied the nature of the game, using computers and advanced math; based on their findings, they developed a way to beat the game over the long run. In other words, their system

virtually guaranteed the group a profit from gambling. They took their system to Las Vegas, Atlantic City, and other casinos and won—literally—millions of dollars.

Their application of probability theory goes far beyond the material presented in this chapter, but it is very much in the tradition of statistics. The "laws of probability" were first established and demonstrated several centuries ago by scientists and mathematicians trying to understand the outcomes of various gambling games and improve their chances of winning. The MIT students simply extended a long tradition of mathematical research.

What was their system? The book does not reveal all of their secrets but, basically, they counted cards (that is, they kept track of which cards had been dealt), played in carefully orchestrated teams, and practiced their strategies religiously. There was nothing illegal about anything they did, but the casinos they played in had the right to "refuse service" to anyone who, from their perspective, was abusing their hospitality. So, in addition to the math and strategizing, the MIT team had to develop elaborate disguises, secret code words, and other routines to obscure what they were doing from casino security. While they had a long run of fabulous success, security did eventually catch up with them and bring their career to a grinding (and sometimes bloody) stop.

Still, the story is fascinating and illustrates how a study of the odds can improve your chances of winning in every game from Old Maid to roulette. (The book also illustrates that, despite the most brilliant ploys, at the end of the day, the house will still find a way to win.)

[1]If you are not familiar with this game, here's a brief summary of the rules. Players are dealt two cards from an ordinary deck and may request additional cards. The winner of a hand is the player with the highest total less than or equal to 21. In calculating total value of a hand, cards two through 10 count as their face value, all picture cards (jack, queen, and king) count 10, and the ace counts as 1 or 11, whichever helps the most. If the total value of the hand exceeds 21, the player loses. Getting exactly 21 is called blackjack.

SUMMARY

1. We can use the normal curve, in combination with the mean and standard deviation, to construct precise descriptive statements about empirical distributions that are normally distributed. This chapter also lays some important groundwork for Part II.
2. To work with the theoretical normal curve, raw scores must be transformed into their equivalent Z scores. Z scores allow us to find areas under the theoretical normal curve (Appendix A).
3. We considered three uses of the theoretical normal curve: finding total areas above and below a score, finding areas between two scores, and expressing these areas as probabilities. This last use of the normal curve is especially germane because inferential statistics are centrally concerned with estimating the probabilities of defined events in a fashion very similar to the process introduced in Section 5.6.

SUMMARY OF FORMULAS

Z scores	5.1	$Z = \dfrac{X_i - \overline{X}}{s}$

GLOSSARY

Normal curve. A theoretical distribution of scores that is symmetrical, unimodal, and bell shaped. The standard normal curve always has a mean of 0 and a standard deviation of 1.
Normal curve table. Appendix A; a detailed description of the area between a Z score and the mean of any standardized normal distribution.

Probability. The likelihood that a defined event will occur.
Z scores. Standard scores; the way scores are expressed after they have been standardized to the theoretical normal curve.

MULTIMEDIA RESOURCES

The Wadsworth Sociology Resource Center: Virtual Society at
http://www.thomsonedu.com/sociology

Visit the companion website for the seventh edition of *Statistics: A Tool for Social Research* to access a wide range of student resources. Begin by clicking on the Student Resources section of the book's website to access the following study tools:

- Basic math review
- Flash cards

- Additional chapter problems
- Internet links
- Table of random numbers
- MicroCase and SPSS examples and exercises
- Hypothesis testing for variables measured at the ordinal level

PROBLEMS

5.1 Scores on a quiz were normally distributed and had a mean of 10 and a standard deviation of 3. For each score below, find the Z score and the percentage of area above and below the score.

X_i	Z Score	% Area Above	% Area Below
5			
6			
7			
8			
9			
11			
12			
14			
15			
16			
18			

5.2 Assume that the distribution of a college entrance exam is normal with a mean of 500 and a standard deviation of 100. For each score below, find the equivalent Z score, the percentage of the area above the score, and the percentage of the area below the score.

X_i	Z Score	% Area Above	% Area Below
650			
400			
375			
586			
437			
526			
621			
498			
517			
398			

5.3 The senior class has been given a comprehensive examination to assess its educational experience. The mean on the test was 74 and the standard deviation was 10. What percentage of the students had scores
a. between 75 and 85?
b. between 80 and 85?
c. above 80?
d. above 83?
e. between 80 and 70?
f. between 75 and 70?
g. below 75?
h. below 77?
i. below 80?
j. below 85?

5.4 For a normal distribution where the mean is 50 and the standard deviation is 10, what percentage of the area is
a. between the scores of 40 and 47?
b. above a score of 47?
c. below a score of 53?
d. between the scores of 35 and 65?
e. above a score of 72?
f. below a score of 31 and above a score of 69?
g. between the scores of 55 and 62?
h. between the scores of 32 and 47?

5.5 At St. Algebra College, the 200 freshmen enrolled in Introductory Biology took a final exam on which their mean score was 72 and their standard deviation was 6. The table below presents the grades of 10 students. Convert each into a Z score and determine the *number of people* who scored higher or lower than each of the 10 students. (*HINT: Multiply the appropriate proportion by N and round the result.*)

X_i	Z Score	Number of Students Above	Number of Students Below
60			
57			
55			
67			
70			
72			
78			
82			
90			
95			

5.6 If a distribution of test scores is normal with a mean of 78 and a standard deviation of 11, what percentage of the area lies
a. below 60?
b. below 70?
c. below 80?
d. below 90?
e. between 60 and 65?
f. between 65 and 79?
g. between 70 and 95?
h. between 80 and 90?
i. above 99?
j. above 89?
k. above 75?
l. above 65?

5.7 A scale measuring prejudice has been administered to a large sample of respondents. The distribution of scores is approximately normal with a mean of 31 and a standard deviation of 5. What percentage of the sample had scores
a. below 20?
b. below 40?
c. between 30 and 40?
d. between 35 and 45?
e. above 25?
f. above 35?

5.8 The average burglary rate for a jurisdiction has been 311 per year with a standard deviation of 50. What is the probability that next year the number of burglaries will be
a. less than 250?
b. less than 300?
c. more than 350?
d. more than 400?
e. between 250 and 350?
f. between 300 and 350?
g. between 350 and 375?

5.9 For a math test on which the mean was 59 and the standard deviation was 4, what is the probability that a student randomly selected from this class will have a score
a. between 55 and 65?
b. between 60 and 65?
c. above 65?
d. between 60 and 50?
e. between 55 and 50?
f. below 55?

5.10 SOC On the scale mentioned in problem 5.7, if a score of 40 or more is considered "highly prejudiced," what is the probability that a person selected at random will have a score in that range?

5.11 The local police force gives all applicants an entrance exam and accepts only those applicants who score in the top 15% on this test. If the mean score this year is 87 and the standard deviation is 8, would an individual with a score of 110 be accepted?

5.12 After taking the state merit examinations for the positions of social worker and employment counselor, you receive the following information on the tests and on your performance. On which of the tests did you do better?

Social Worker	Employment Counselor
$\bar{X} = 118$	$\bar{X} = 27$
$s = 17$	$s = 3$
Your score = 127	Your score = 29

5.13 In a distribution of scores with a mean of 35 and a standard deviation of 4, which event is more likely: that a randomly selected score will be between 29 and 31 or that a randomly selected score will be between 40 and 42?

5.14 To be accepted into an honor society, students must have GPAs in the top 10% of the school. If the mean GPA is 2.78 and the standard deviation is .33, which of the following GPAs would qualify? 3.20, 3.21, 3.25, 3.30, 3.35

Part II

Inferential Statistics

The chapters in Part II cover the techniques and concepts of inferential or inductive statistics. Generally speaking, these applications allow us to learn about large groups (populations) from small, carefully selected subgroups (samples). These statistical techniques are powerful and extremely useful. They are used to assess public opinion, research the potential market for new products, project the winners of elections, test the effects of new drugs, and in hundreds of other ways both inside and outside the social sciences.

Chapter 6 includes a brief description of sampling, but the most essential part of this chapter concerns the sampling distribution, the single most important concept in inferential statistics. The sampling distribution is normal in shape, and it is the key link between populations and samples. The chapter also covers estimation, the first of the two main applications of inferential statistics. In this section you will learn how to use statistical information from a sample (e.g., a mean or a proportion) to estimate the characteristics of a population. Public opinion polling and election projection are the most common uses of this technique.

Chapters 7 through 10 cover a second application of inferential statistics: hypothesis testing. Most of the relevant concepts for this material are introduced in Chapter 7, and each chapter covers a different situation for using hypothesis testing. For example, Chapter 8 presents the necessary techniques when we are comparing information from two different samples or groups (e.g., men versus women), whereas Chapter 9 covers applications involving more than two groups or samples (e.g., Republicans versus Democrats versus Independents).

Hypothesis testing is one of the more challenging aspects of statistics for beginning students, and this book includes an abundance of learning aids to ease the chore of assimilating this material. Hypothesis testing is also one of the most common and important statistical applications in social science research. Mastery of this material is essential for developing the ability to read the professional literature.

6

Introduction to Inferential Statistics, the Sampling Distribution, and Estimation

LEARNING OBJECTIVES

By the end of this chapter, you will be able to:

1. Explain the purpose of inferential statistics in terms of generalizing from a sample to a population.
2. Explain the principle of random sampling and these key terms: population, sample, parameter, statistic, representative, EPSEM.
3. Differentiate among the sampling distribution, the sample, and the population.
4. Explain the two theorems presented.
5. Explain the logic of estimation and the role of the sample, sampling distribution, and population.
6. Define and explain the concepts of bias and efficiency.
7. Construct and interpret confidence intervals for sample means and sample proportions.

6.1 INTRODUCTION

One of the goals of social science research is to test our theories and hypotheses using many different people, groups, societies, and historical eras. Obviously, we can have the greatest confidence in theories that have stood up to testing against the greatest variety of cases and social settings. A major problem we often face in social science research, however, is that the populations in which we are interested are too large to test. For example, a theory concerning political party preference among U.S. citizens would be most suitably tested using the entire electorate, but it is impossible to interview every member of this population (about 100 million people). Indeed, even for theories that could be reasonably tested with smaller populations—such as a local community or the student body at a university—the logistics of gathering data from every single case (entire populations) are staggering to contemplate.

If it is too difficult or expensive to do research with entire populations, how can we reasonably test our theories? To deal with this problem, social scientists select samples, or subsets of cases, from the populations of interest. Our goal in inferential statistics is to learn about the characteristics (or **parameters**) of a population based on what we can learn from our samples. This book discusses two applications of inferential statistics. In estimation procedures, covered in this chapter, researchers make a "guess" of the population parameter, based on what is known about the sample. In hypothesis testing, which Chapters 7 through 10 explore, the validity of a hypothesis about the population is tested against sample outcomes. Before we can address these applications, however, we need to consider sampling (the techniques for selecting cases for a sample) and a key concept in inferential statistics: the sampling distribution.

6.2 PROBABILITY SAMPLING

In this chapter, we will review the basic procedure for selecting probability samples, the only type of sample that fully supports the use of inferential statistical techniques to generalize to populations. These types of samples are often described as "random," and you may be more familiar with this terminology. Because of its greater familiarity, I will use the phrase "random sample" sometimes in the following chapters. The term "probability sample" is preferred, however, because, in everyday language, "random" frequently means "by coincidence" or suggests unpredictability. To the contrary, probability samples are selected by techniques that are careful and methodical and leave no room for haphazardness. Interviewing the people you happen to meet in a mall one afternoon may be "random" in some sense, but this technique will not result in a sample that could support inferential statistics.

Before considering probability sampling, let me point out that social scientists often use nonprobability samples. For example, researchers who are studying small group dynamics or the structure of attitudes or personal values might use the students enrolled in their classes as subjects. Such "convenience" samples are very useful for a number of purposes (e.g., exploring ideas or pretesting survey forms before embarking on a more ambitious project) and are typically inexpensive and easy to assemble. The major limitation of these samples is that results cannot be generalized beyond the group being tested. If a theory of prejudice, for example, has been tested only on the students who happen to have been enrolled in a particular section of Introductory Sociology at a particular university in a particular year, then the researcher cannot generalize the findings to other types of people in other social locations or at other times. Even when the evidence is very strong, we cannot place a lot of confidence in the generalizability of theories tested on nonprobability samples only.

The goal of probability sampling is to select cases so that the final sample is **representative** of the population from which it was drawn. A sample is representative if it reproduces the important characteristics of the population. For example, if the population consists of 60% females and 40% males, the sample should contain essentially the same proportions. In other words, a representative sample is very much like the population, only smaller. It is crucial for inferential statistics that samples be representative: If they are not, generalizing to the population becomes, at best, extremely hazardous.

How can we ensure that our samples are representative? Unfortunately, it is not possible to guarantee that samples are representative. However, we can maximize the chances of drawing a representative sample by following the principle of **EPSEM** (the "**E**qual **P**robability of **SE**lection **M**ethod"). To use this method, we select the sample so that every case in the population has an equal probability of being selected for the sample. Our goal is to select a representative sample, and the technique we use to maximize the chance of achieving that goal is to follow the rule of EPSEM.

The most basic EPSEM sampling technique produces a **simple random sample.** There are numerous variations and refinements on this technique but, in this text, we will consider only the most straightforward application. To draw a simple random sample, we need a list of all cases in the population and a system for selecting cases that ensures that every case has an equal chance of being selected for the sample. The selection process could be based on a number of different kinds of operations (for example, drawing cards from a well-shuffled deck, flipping coins, throwing dice, drawing numbers from a hat, and so on). Cases are often selected by using a list of random numbers that has been generated by a computer program or that is presented in a table of random numbers. Either way, the numbers are random in the sense that they have no order or pattern: Each number in the list is just as likely as any other number. An example of a table of random numbers is available at the website for this text.

To use random numbers to select a simple random sample, first assign a unique identification number to each case on the population list. Then, select cases for the sample when their identification number corresponds to the number chosen from the table. This procedure will produce an EPSEM sample because the numbers in the table are in random order and any number is just as likely as any other number. Stop selecting cases when you have reached your desired sample size and, if an identification number is selected more than once, ignore the repeats.[1]

Remember that the EPSEM selection technique and the representativeness of the final sample are two different things. In other words, the fact that a sample is selected according to EPSEM does not guarantee that it will be an exact representation or microcosm of the population. The probability is very high that an EPSEM sample will be representative but, just as a perfectly honest coin will sometimes show 10 heads in a row when flipped, an EPSEM sample will occasionally present an inaccurate picture of the population. One of the great strengths of inferential statistics is that they allow the researcher to estimate the probability of this type of error and interpret results accordingly.

To summarize this section: The purpose of inferential statistics is to acquire knowledge about populations based on the information derived from samples of that population. Each of the applications of inferential statistics that I will present in this text requires that samples be selected according to EPSEM. While even the most painstaking and sophisticated sampling techniques will not guarantee representativeness, the probability is high that EPSEM samples will be representative of the populations from which they are selected.

6.3 THE SAMPLING DISTRIBUTION

Once we have selected a probability sample, what do we know? Remember that our goal is to learn more about the population from which the sample was selected. We can gather information from the cases in the sample and then use that information to help us learn more about the population: Sample information is important primarily insofar as it allows us to generalize to the population.

When we use inferential statistics, we generally measure some variable (e.g., age, political party preference, or opinions about abortion) in the sample and then use the information from the sample to learn more about that variable in the population. In Part I of this text, you learned that three types of information are generally necessary to adequately characterize a variable: (1) the shape of its distribution, (2) some measure of central tendency, and (3) some measure

---

[1] Ignoring identification numbers when they are repeated is called "sampling without replacement." Technically, this practice compromises the randomness of the selection process. However, if the sample is a small fraction of the total population, we will be unlikely to select the same case twice, and ignoring repeats will not bias our conclusions.

of dispersion. Clearly, we can gather all three kinds of information about the cases in the sample. Just as clearly, none of the information is available for the population. The means and standard deviations of variables in the population, as well as their shapes, are unknown: If we had this information for the population, inferential statistics would be unnecessary.

In statistics, we link information from the sample to the population with a device known as the **sampling distribution,** which is the theoretical, probabilistic distribution of a statistic for all possible samples of a certain sample size (N). That is, the sampling distribution includes statistics that represent every conceivable combination of cases from the population. A crucial point about the sampling distribution is that its characteristics are based on the laws of probability, not on empirical information, and are very well known. In fact, the sampling distribution is the central concept in inferential statistics, so let's thoroughly examine its characteristics.

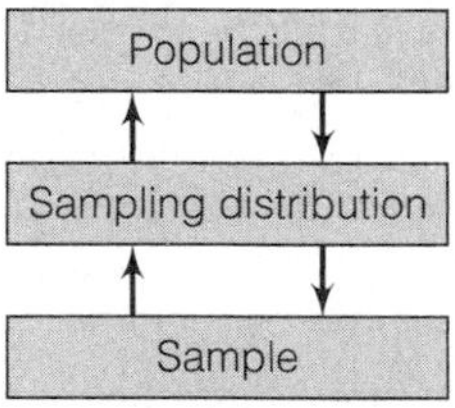

As Figure 6.1 illustrates, we move between the sample and population by means of the sampling distribution. Thus, every application of inferential statistics involves three separate and distinct distributions:

FIGURE 6.1   THE RELATIONSHIPS AMONG THE SAMPLE, SAMPLING DISTRIBUTION, AND POPULATION

1. The sample distribution, which is empirical (that is, it exists in reality) and known in the sense that the shape, central tendency, and dispersion of any variable can be ascertained for the sample. Remember that the information from the sample is important primarily insofar as it allows the researcher to learn about the population.

2. The population distribution, which, while empirical, is unknown. Amassing information about or making inferences to the population is the sole purpose of inferential statistics.

3. The sampling distribution, which is theoretical or non-empirical. Because of the laws of probability, a great deal is known about this distribution. Specifically, the shape, central tendency, and dispersion of the distribution can be deduced and, therefore, the distribution can be adequately characterized.

The utility of the sampling distribution is implied by its definition. Because it encompasses all possible sample outcomes, the sampling distribution enables us to estimate the probability of any particular sample outcome, a process that will occupy our attention for the remainder of this chapter and the next four chapters to come.

The sampling distribution is theoretical, which means that it is never obtained in reality by the researcher. However, to understand better the structure and function of the distribution, let's consider an example of how one might be constructed. Suppose that we wanted to gather some information about the age of a particular community of 10,000 individuals. We draw an EPSEM sample of 100 residents, ask all 100 respondents their age, and use those individual scores to compute a mean age of 27. The graph in Figure 6.2 indicates this score. Note that this sample is one of countless possible combinations of 100 people taken from this population of 10,000, and the mean of 27 is one of millions of possible sample outcomes.

Now, replace the 100 respondents in the first sample and draw another sample of the same size ($N = 100$) and again compute the average age.

FIGURE 6.2   CONSTRUCTING A SAMPLE DISTRIBUTION

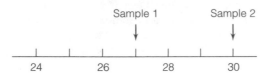

Assume that the mean for the second sample is 30 and note this sample outcome in Figure 6.2. This second sample is another of the countless possible combinations of 100 people taken from this population of 10,000, and the sample mean of 30 is another of the millions of possible sample outcomes. Replace the respondents from the second sample and draw still another sample, calculate and note the mean, replace this third sample, and draw a fourth sample, continuing these operations an infinite number of times, calculating and noting the mean of each sample. Now, try to imagine what Figure 6.2 would look like after tens of thousands of individual samples had been collected and the mean had been computed for each sample. What shape, mean, and standard deviation would this distribution of sample means have after we had collected all possible combinations of 100 respondents from the population of 10,000?

For one thing, we know that each sample will be at least slightly different from every other sample because it is very unlikely that we will sample exactly the same 100 people twice. Because each sample will almost certainly be a unique combination of individuals, each sample mean will be at least slightly different in value. We also know that even though the samples are chosen according to EPSEM, they will not be representative of the population in every single case. For example, if we continue taking samples of 100 people long enough, we will eventually choose a sample that includes only the very youngest residents. Such a sample would have a mean much lower than the true population mean. Likewise, by random chance alone, some of our samples will include only senior citizens and will have means that are much higher than the population mean. Common sense suggests, however, that such unrepresentative samples will be rare and that most sample means will cluster around the true population value.

To illustrate further, assume that we somehow learn that the true mean age of the population is 30. As we have seen, if the population mean is 30, most of the sample means will also be approximately 30 and the sampling distribution of these sample means should peak at 30. Some of the sample means will be much too low or much too high, but the frequency of such misses should decline as we get farther away from 30. That is, the distribution should slope to the base as we move away from the population value: Sample means of 29 or 31 should be common; means of 20 or 40 should be rare. Because the samples are random, the means should miss an equal number of times on either side of the population value, and the distribution itself should therefore be roughly symmetrical. In other words, the sampling distribution of all possible sample means should be approximately normal and will resemble the distribution that Figure 6.3 shows. Recall from Chapter 5 that, on any normal curve, cases close to the mean (say, within ±1 standard deviation) are common and cases far away from the mean (say, beyond ±3 standard deviations) are rare.

FIGURE 6.3   A SAMPLING DISTRIBUTION OF SAMPLE MEANS

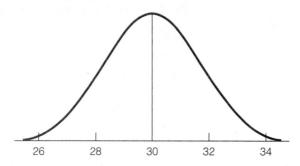

Two theorems state these commonsense notions about the shape of the sampling distribution and other very important information about central tendency and dispersion. The first says that:

> If repeated random samples of size N are drawn from a normal population with mean μ and standard deviation σ, then the sampling distribution of sample means will be normal with a mean μ and a standard deviation of $\sigma/\sqrt{N}$.

To translate: If we begin with a trait that is normally distributed across a population (IQ, height, or weight, for example) and take an infinite number of equally sized random samples from that population, then the sampling distribution of sample means will be normal. If we know that the variable is distributed normally in the population, then we can assume that the sampling distribution will be normal.

The theorem tells us more than the shape of the sampling distribution of all possible sample means, however. It also defines its mean and standard deviation of the sampling distribution. In fact, it says that the mean of the sampling distribution will be exactly the same value as the mean of the population. That is, if we know that the mean IQ of the entire population is 100, then we know that the mean of any sampling distribution of sample mean IQs will also be 100. Exactly why this is so is not a matter that can be fully explained at this level. Recall, however, that most sample means will cluster around the population value over the long run. Thus, the fact that the mean of the population and the mean of the sampling distribution of all possible sample means are equal should have intuitive appeal. As for dispersion, the theorem says that the standard deviation of the sampling distribution, also called the **standard error of the mean,** will equal the standard deviation of the population divided by the square root of N (symbolically: $\sigma/\sqrt{N}$).

If the mean and standard deviation of a normally distributed population are known, the theorem allows us to compute the mean and standard deviation of the sampling distribution.[2] Thus, we will know exactly as much about the sampling distribution (shape, central tendency, and dispersion) as we ever knew about any empirical distribution.

---

[2] In the typical research situation, the values of the population mean and standard deviation are, of course, unknown. However, these values can be estimated from sample statistics, as we shall see in the chapters that follow.

The first theorem requires a normal population distribution. What happens when the distribution of the variable in question is unknown or is known to not be normal in shape (such as income, which always has a positive skew)? These eventualities (very common, in fact) are covered by a second theorem, called the **Central Limit Theorem:**

> If repeated random samples of size N are drawn from any population, with mean μ and standard deviation σ, then, as N becomes large, the sampling distribution of sample means will approach normality, with mean μ and standard deviation $\sigma/\sqrt{N}$.

To translate: For *any* trait or variable, even those that are not normally distributed in the population, as sample size grows larger, the sampling distribution of sample means will become normal in shape. When N is large, the mean of the sampling distribution will equal the population mean and its standard deviation (or the standard error of the mean) will be equal to $\sigma/\sqrt{N}$.

The importance of the Central Limit Theorem is that it removes the constraint of normality in the population. Whenever sample size is large, we can assume that the sampling distribution is normal, with a mean equal to the population mean and a standard deviation equal to $\sigma/\sqrt{N}$ regardless of the shape of the variable in the population. Thus, even if we are working with a variable that we know has a skewed distribution (such as income), we can still assume a normal sampling distribution.

The issue remaining, of course, is to define large sample. A good rule of thumb is that if sample size (N) is 100 or more, the Central Limit Theorem applies, and you can assume that the sampling distribution is normal in shape. When N is less than 100, you must have good evidence of a normal population distribution before you can assume that the sampling distribution is normal. Thus, you can ensure a normal sampling distribution by using fairly large samples.

6.4 THE SAMPLING DISTRIBUTION: AN ADDITIONAL EXAMPLE

Developing an understanding of the sampling distribution—what it is and why it's important—is often one of the more challenging tasks for beginning students of statistics. It may be helpful to briefly list the most important points about the sampling distribution:

1. Its definition: *The sampling distribution is the distribution of a statistic (such as a mean or a proportion) for all possible sample outcomes of a certain size.*
2. Its shape: *normal* (see Appendix A).
3. Its central tendency and dispersion: *The mean of the sampling distribution is the same value as the mean of the population. The standard deviation of the sampling distribution—or the standard error—is equal to the population standard deviation divided by the square root of* N. (See the theorems.)
4. The role of the sampling distribution in inferential statistics: *It links the sample with the population.*

To reinforce these points, let's consider an additional example of how the sampling distribution works together with the sample and the population. Consider the General Social Survey (GSS), the database used for computer exercises in this text. The GSS has been administered to randomly selected samples of adult Americans since 1972, and it explores a broad range of characteristics and issues, including confidence in the Supreme Court, attitudes about assisted suicide, number of siblings, and level of education. The GSS has its limits, of course, but it has proven to be a very valuable resource for testing theories and learning more about American society. Focusing on this survey, let's review the roles played by the population, sample, and sampling distribution when we use this database.

We'll start with the population or the group we are actually interested in and want to learn more about. In the case of the GSS, the population consists of all adult (older than 18) Americans, which includes about 225 million people. Clearly, we can never interview all of these people and learn what they are like or what they are thinking about abortion, capital punishment, gun control, affirmative action, sex education in the public schools, or any number of other issues. We should also note that this information is worth having. It could help inform public debates, provide some basis in fact for the discussion of many controversial issues (e.g., the polls show consistently that the majority of Americans favor some form of gun control), and assist people in clarifying their personal beliefs. If the information is valuable, what can be done to learn more about this huge population?

This brings us to the sample, a carefully chosen subset of the population. The GSS is administered to about 3,000 people, each of whom is chosen by a sophisticated technology based on the principle of EPSEM. A key point to remember is that samples chosen by this method are very likely to be representative of the populations from which they were selected. In other words, whatever is true of the sample will also be true of the population (with some limits and qualifications, of course).

The 3,000 respondents are contacted at home and asked for background information (religion, gender, years of education, and so on) as well as their opinions and attitudes. When all of this information is collated, the GSS database includes information (shape, central tendency, dispersion) on hundreds of variables (age, level of prejudice, marital status) for the people in the sample. So, we have a lot of information about the variables for the sample (the 3,000 or so people who actually responded to the survey), but no information about these variables for the population (the 225 million adult Americans). How do we get from the known characteristics of the sample to the unknown population? This is the central question of inferential statistics and the answer, as you should realize by now, is "by using the sampling distribution."

Remember that, unlike the sample and the population, the sampling distribution is theoretical. We can work with the sampling distribution because the theorems presented earlier in this chapter define its shape, central tendency, and dispersion. For any variable from the GSS, we know that the sampling distribution will be normal in shape because the sample is "large" (N is much greater than 100). Second, the theorems tell us that the mean of the sampling distribution will be the same value as the mean of the population. If *all* adult Americans have completed an average of 13.5 years of schooling ($\mu = 13.5$), the mean of the sampling distribution will also be 13.5. Third, the theorems tell us that the standard deviation (or standard error) of the sampling distribution is equal to the population standard deviation (σ) divided by the square root of N. Thus, the theorems tell us the statistical characteristics of the sampling distribution (shape, central tendency, and dispersion), and this information allows us to link the sample to the population.

How does the sampling distribution link the sample to the population? The fact that the sampling distribution will be normal when N is large is crucial. This means that more than two-thirds (about 68%) of all samples will be within ± 1 Z of the mean (which is the same value as the population mean), about 95% are within ± 2 Z scores, and so forth. We do not (and cannot) know the actual value of the mean of the sampling distribution, but we do know that the probabilities are very high that our sample is approximately equal to this parameter. Similarly, the theorems give us crucial information about the mean and the standard error of the sampling distribution that we can use, as you will see, to link information from the sample to the population.

To summarize, our goal is to infer information about the population (in the case of the GSS: all adult Americans). When populations are too large to test (and contacting 225 million adult Americans is far beyond the capacity of even the most energetic pollster), we use information from randomly selected samples, carefully drawn from the population of interest, to estimate that population's characteristics. In the case of the GSS, the full sample consists of about 3,000 adult Americans who have responded to the questions on the survey. The sampling distribution, the theoretical distribution whose characteristics are defined by the theorems, links the known sample to the unknown population.

6.5 SYMBOLS AND TERMINOLOGY

In inferential statistics, we work with three entirely different distributions (sample, population, and sampling distribution). Furthermore, we will be concerned with several different kinds of sampling distributions—including the sampling distribution of sample means and the sampling distribution of sample proportions.

To distinguish clearly among these various distributions, we will often use symbols. Chapters 3 and 4 introduced the symbols used for the means and standard deviations of samples and populations. Table 6.1 includes some of the symbols used for the sampling distribution in summary form for quick reference.

Note that the mean and standard deviation of a sample are denoted with English letters ($\overline{X}$ and s) while the mean and standard deviation of a population are denoted with the Greek letter equivalents (μ and σ). Proportions calculated on samples are symbolized as P-sub-s (s for sample) while population proportions are denoted as P-sub-u (u for "universe" or population). The symbols for the sampling distribution are Greek letters with English letter subscripts. The mean and standard deviation of a sampling distribution of sample means are $\mu_{\overline{X}}$ "mew-sub-ex-bar" and $\sigma_{\overline{X}}$ "sigma-sub-ex-bar." The mean and standard

TABLE 6.1   SYMBOLS FOR MEANS AND STANDARD DEVIATIONS OF THREE DISTRIBUTIONS

	Mean	Standard Deviation	Proportion
1. Samples	$\overline{X}$	s	P_s
2. Populations	μ	σ	P_u
3. Sampling distributions			
Of means	$\mu_{\overline{X}}$	$\sigma_{\overline{X}}$	
Of proportions	μ_p	σ_p	

deviation of a sampling distribution of sample proportions are μ_p "mew-sub-p" and σ_p "sigma-sub-p."

6.6 INTRODUCTION TO ESTIMATION

The object of this branch of inferential statistics is to estimate population values or parameters from statistics computed from samples. Although these techniques may be new to you, you are certainly familiar with their most common applications: public-opinion polls and election projections. Polls and surveys on every conceivable issue—from the sublime to the trivial—have become a staple of the mass media and popular culture. The techniques you will learn in this chapter are essentially the same as those the most reputable, sophisticated, and scientific pollsters use.

The standard procedure for estimating population values is to construct a **confidence interval:** a mathematical statement that says that the parameter lies within a certain range of values, or interval. A confidence interval estimate might be phrased as "between 71% and 77% of Americans approve of capital punishment." The interval places the population value (the percentage of ALL Americans who support capital punishment) between 71% and 77%, but does not specify an exact value.

6.7 BIAS AND EFFICIENCY

Estimation procedures are based on sample statistics. Which of the many available sample statistics should be used? Researchers follow two criteria in selecting estimators: **bias** and **efficiency.** Estimates should be based on sample statistics that are unbiased and relatively efficient. We cover each of these criteria separately.

Bias. An estimator is unbiased if the mean of its sampling distribution is equal to the population value of interest. We know from the theorems presented earlier in this chapter that sample means conform to this criterion. The mean of the sampling distribution of sample means (which we will note symbolically as $\mu_{\bar{X}}$) is the same as the population mean (μ).

Sample proportions (P_s) are also unbiased. That is, if we calculate sample proportions from repeated random samples of size N and then array them in a line chart or frequency polygon, the sampling distribution of sample proportions will have a mean (μ_p) equal to the population proportion (P_u). Thus, if we are concerned with coin flips and sample honest coins 10 at a time ($N = 10$), the sampling distribution will have a mean equal to 0.5, which is the probability that an honest coin will be heads (or tails) when flipped. All statistics other than sample means and sample proportions are biased (that is, have sampling distributions with means not equal to the population value).[3]

Knowing that sample means and proportions are unbiased is crucial because it allows us to determine the probability that they fall within a given

---

[3] In particular, the sample standard deviation (s) is a biased estimator of the population standard deviation (σ). As you might expect, there is less dispersion in a sample than in a population and, as a consequence, s will underestimate σ. As we shall see, however, the sample standard deviation can be corrected for this bias and still serve as an estimate of the population standard deviation for large samples.

distance of the population values we are trying to estimate. To illustrate, consider a specific problem. Assume that we want to estimate the average income of a community. We take a random sample of 500 households ($N = 500$) and compute a sample mean of $35,000. In this example, the population mean is the average income of *all* households in the community, and the sample mean is the average income for the 500 households selected for our sample. Note that we do not know the value of the population mean (μ)—if we did, we wouldn't need the sample—but μ is what interests us. The sample mean of $35,000 is important and interesting primarily insofar as it can give us information about the population mean.

The two theorems presented earlier in this chapter give us a great deal of information about the sampling distribution of all possible sample means. Because N is large, we know that the sampling distribution is normal and that its mean is equal to the population mean. We also know that all normal curves contain about 68% of the cases (the cases here are sample means) within ± 1 Z, 95% of the cases within ± 2 Zs, and more than 99% of the cases within ± 3 Zs of the mean. Remember that we are discussing the sampling distribution here—the distribution of all possible sample outcomes or, in this instance, sample means. Thus, the probabilities are very good (approximately 68 out of 100 chances) that our sample mean of $35,000 is within ± 1 Z, excellent (95 out of 100) that it is within ± 2 Zs, and overwhelming (99 out of 100) that it is within ± 3 Zs of the mean of the sampling distribution (which is the same value as the population mean). Figure 6.4 graphically depicts these relationships.

If an estimator is unbiased, it is almost certainly an accurate estimate of the population parameter (μ in this case). However, in less than 1% of the cases, a sample mean will be more than ± 3 Zs away from the mean of the sampling distribution (very inaccurate) by random chance alone. We literally have no idea if our particular sample mean of $35,000 is in this small minority. We do know, however, that the odds are high that our sample mean is considerably closer than ± 3 Zs to the mean of the sampling distribution and, thus, to the population mean.

Efficiency. The second desirable characteristic of an estimator is efficiency, which is the extent to which the sampling distribution clusters around its mean.

FIGURE 6.4 AREAS UNDER THE SAMPLING DISTRIBUTION OF SAMPLE MEANS

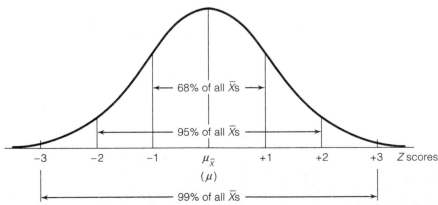

Efficiency or clustering is essentially a matter of dispersion, as we saw in Chapter 4 (see Figure 4.1). The smaller the standard deviation of a sampling distribution, the greater the clustering and the higher the efficiency. Remember that the standard deviation of the sampling distribution of sample means, or the standard error of the mean, is equal to the population standard deviation divided by the square root of N. Therefore, the standard deviation of the sampling distribution is an inverse function of N ($\sigma_{\bar{X}} = \sigma/\sqrt{N}$). As sample size increases, $\sigma_{\bar{X}}$ will decrease. We can improve the efficiency (or decrease the standard deviation of the sampling distribution) for any estimator by increasing sample size.

An example should make this clearer. Consider two samples of different sizes:

Sample 1	Sample 2
$\bar{X} = \$35,000$	$\bar{X} = \$35,000$
$N_1 = 100$	$N_2 = 1,000$

Both sample means are unbiased, but which is the more efficient estimator? Consider sample 1 and assume, for the sake of illustration, that the population standard deviation (σ) is \$500.[4] In this case, the standard deviation of the sampling distribution of all possible sample means with an N of 100 would be $\sigma/\sqrt{N}$ or $500/\sqrt{100}$ or \$50.00. For sample 2, the standard deviation of all possible sample means with an N of 1,000 would be much smaller. Specifically, it would be equal to $500/\sqrt{1,000}$ or \$15.81.

Sampling distribution 2 is much more clustered than sampling distribution 1. In fact, distribution 2 contains 68% of all possible sample means within ±15.81 of μ while distribution 1 requires a much broader interval of ±50.00 to do the same. The estimate based on a sample with 1,000 cases is much more likely to be close in value to the population parameter than is an estimate based on a sample of 100 cases. Figures 6.5 and 6.6 illustrate these relationships graphically.

FIGURE 6.5   A SAMPLING DISTRIBUTION WITH $N = 100$ AND $\sigma_{\bar{X}} = \$50.00$

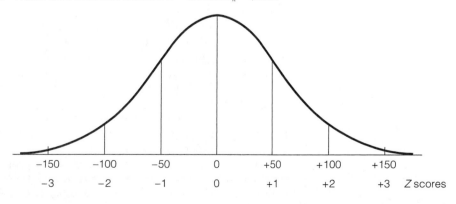

−150	−100	−50	0	+50	+100	+150
−3	−2	−1	0	+1	+2	+3 Z scores

---

[4] In reality, of course, the value of σ would be unknown.

FIGURE 6.6   A SAMPLING DISTRIBUTION WITH $N = 1,000$ AND $\sigma_{\bar{x}} = \$15.81$

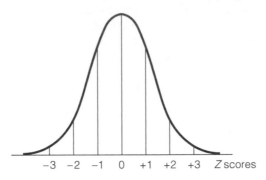

To summarize: The standard deviation of all sampling distributions is an inverse function of N. Therefore, the larger the sample, the greater the clustering and the higher the efficiency. In part, these relationships between sample size and the standard deviation of the sampling distribution do nothing more than underscore our commonsense notion that we can place much more confidence in large samples than in small (as long as both have been randomly selected).

6.8 ESTIMATION PROCEDURES: INTRODUCTION

The first step in constructing an interval estimate is to decide on the risk that you are willing to take of being wrong. An interval estimate is wrong if it does not include the population parameter. This probability of error is called **alpha** (symbolized α). The exact value of alpha will depend on the nature of the research situation, but a 0.05 probability is commonly used. Setting alpha equal to 0.05, also called using the 95% **confidence level,** means that over the long run the researcher is willing to be wrong only 5% of the time. Or, to put it another way, if an infinite number of intervals were constructed at this alpha level (and with all other things being equal), 95% of them would contain the population value and 5% would not. In reality, of course, only one interval is constructed and, by setting the probability of error very low, we are setting the odds in our favor that the interval will include the population value.

The second step is to picture the sampling distribution, divide the probability of error equally into the upper and lower tails of the distribution, and then find the corresponding Z score. For example, if we decided to set alpha equal to 0.05, we would place half (0.025) of this probability in the lower tail and half in the upper tail of the distribution. Figure 6.7 illustrates this division of the sampling distribution.

We need to find the Z score that marks the beginnings of the shaded areas in Figure 6.7. In Chapter 5, we learned how to first calculate a Z score and then find an area under the normal curve. Here, we will reverse that process. We need to find the Z score beyond which lies a proportion of .0250 of the total area. To do this, go down column c of Appendix A until you find this proportional value (.0250). The associated Z score is 1.96. Because the curve is symmetrical and we are interested in both the upper and lower tails, we designate the Z score that corresponds to an alpha of .05 as ± 1.96 (see Figure 6.8).

FIGURE 6.7   THE SAMPLING DISTRIBUTION WITH ALPHA (α) EQUAL TO 0.05

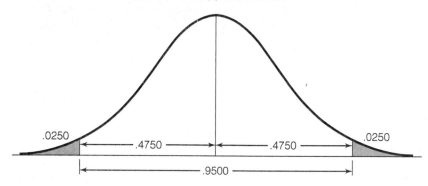

FIGURE 6.8   FINDING THE Z SCORE THAT CORRESPONDS TO AN ALPHA (α) OF 0.05

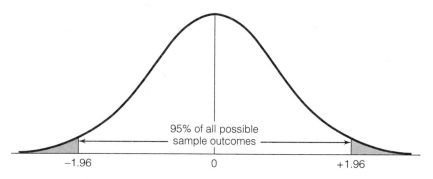

We now know that 95% of all possible sample outcomes fall within ±1.96 Z-score units of the population value. In reality, of course, there is only one sample outcome but, if we construct an interval estimate based on ±1.96 Zs, the probabilities are that 95% of all such intervals will trap the population value. Thus, we can be 95% confident that our interval contains the population value.

Besides the 95% level, there are three other commonly used confidence levels: the 90% level ($\alpha = .10$), the 99% level ($\alpha = 0.01$), and the 99.9% level ($\alpha = .001$). To find the corresponding Z scores for these levels, follow the procedures outlined above for an alpha of 0.05. Table 6.2 summarizes all the information you will need.

Turn to Appendix A and confirm for yourself that the Z scores in Table 6.2 do indeed correspond to these alpha levels. As you do, note that, in the cases

TABLE 6.2 Z SCORES FOR VARIOUS LEVELS OF ALPHA (α)

Confidence Level	Alpha	$\alpha/2$	Z score
90%	.10	.0500	±1.65
95%	.05	.0250	±1.96
99%	.01	.0050	±2.58
99.9%	.001	.0005	±3.29

where alpha is set at 0.10 and 0.01, the precise areas we seek do not appear in the table. For example, with an alpha of 0.10, we would look in column c ("Area Beyond") for the area .0500. Instead we find an area of .0505 ($Z = \pm 1.64$) and an area of .0495 ($Z = \pm 1.65$). The Z score we are seeking is somewhere between these two other scores. When this condition occurs, take the larger of the two scores as Z. This will make the interval as wide as possible under the circumstances and is thus the most conservative course of action. In the case of an alpha of 0.01, we encounter the same problem (the exact area .0050 is not in the table), resolve it the same way, and take the larger score as Z. Finally, note that in the case where alpha is set at .001, we can choose from several Z scores. Although our table is not detailed enough to show it, the closest Z score to the exact area we want is ± 3.291, which we can round off to ± 3.29. *(For practice in finding Z scores for various levels of confidence, see problem 6.3.)*

The third step is to actually construct the confidence interval. In the sections that follow, we see how to construct an interval estimate first with sample means and then with sample proportions.

6.9 INTERVAL ESTIMATION PROCEDURES FOR SAMPLE MEANS (LARGE SAMPLES)

FORMULA 6.1

Formula 6.1 summarizes how to construct a confidence interval based on sample means:

$$c.i. = \overline{X} \pm Z \left(\frac{\sigma}{\sqrt{N}} \right)$$

Where $c.i.$ = confidence interval

$\overline{X}$ = the sample mean

Z = the Z score as determined by the alpha level

$\dfrac{\sigma}{\sqrt{N}}$ = the standard deviation of the sampling distribution or the standard error of the mean

As an example, suppose you wanted to estimate the average IQ of a community and had randomly selected a sample of 200 residents, with a sample mean IQ of 105. Assume that the population standard deviation for IQ scores is about 15, so we can set σ equal to 15. If we are willing to run a 5% chance of being wrong and set alpha at 0.05, the corresponding Z score will be 1.96. These values can be directly substituted into Formula 6.1, and an interval can be constructed:

$$c.i. = \overline{X} \pm Z \left(\frac{\sigma}{\sqrt{N}} \right)$$

$$c.i. = 105 \pm 1.96 \left(\frac{15}{\sqrt{200}} \right)$$

$$c.i. = 105 \pm 1.96 \left(\frac{15}{14.14} \right)$$

$$c.i. = 105 \pm (1.96)(1.06)$$

$$c.i. = 105 \pm 2.08$$

That is, our estimate is that the average IQ for the population in question is somewhere between 102.92 ($105 - 2.08$) and 107.08 ($105 + 2.08$). Because 95%

of all possible sample means are within ± 1.96 Zs (or 2.08 IQ units in this case) of the mean of the sampling distribution, the odds are very high that our interval will contain the population mean. In fact, even if the sample mean is as far off as ± 1.96 Zs (which is unlikely), our interval will still contain $\mu_{\bar{X}}$ and, thus, μ. Only if our sample mean is one of the few that is more than ± 1.96 Zs from the mean of the sampling distribution will we have failed to include the population mean.

Note that in the example above, the value of the population standard deviation was supplied. Needless to say, it is unusual to have such information about a population. In the great majority of cases, we will have no knowledge of σ. In such cases, however, we can estimate σ with s, the sample standard deviation. Unfortunately, s is a biased estimator of σ, and the formula must be changed slightly to correct for the bias. For larger samples, the bias of s will not affect the interval very much. The revised formula for cases in which σ is unknown is:

FORMULA 6.2
$$c.i. = \bar{X} \pm Z\left(\frac{s}{\sqrt{N-1}}\right)$$

In comparing this formula with 6.1, note that there are two changes. First, σ is replaced by s and, second, the denominator of the last term is the square root of $N - 1$ rather than the square root of N. The latter change corrects the bias of s.

Let me stress here that the substitution of s for σ is permitted only for large samples (that is, samples with 100 or more cases). For smaller samples, when the value of the population standard deviation is unknown, the standardized normal distribution summarized in Appendix A cannot be used in the estimation process. To construct confidence intervals from sample means with samples smaller the 100, we must use a different theoretical distribution, called the Student's t distribution, to find areas under the sampling distribution. We will defer the presentation of the t distribution until Chapter 7 and confine our attention here to estimation procedures for large samples only.

Let's close this section by working through a sample problem with Formula 6.2. Average income for a random sample of a particular community is $35,000, with a standard deviation of $200. What is the 95% interval estimate of the population mean, μ? Given that:

$$\bar{X} = \$35,000$$
$$s = \$200$$
$$N = 500$$

and using an alpha of 0.05, the interval can be constructed:

$$c.i. = \bar{X} \pm Z\left(\frac{s}{\sqrt{N-1}}\right)$$

$$c.i. = 35,000 \pm 1.96\left(\frac{200}{\sqrt{499}}\right)$$

$$c.i. = 35,000 \pm 17.55$$

The average income for the community as a whole is between $34,982.45 (35,000 − 17.55) and $35,017.55 (35,000 + 17.55). Remember that this interval

STEP BY STEP	Constructing Confidence Intervals for Sample Means

Step 1: Select an alpha level. Commonly used alpha levels are appear in Table 6.2; the 0.05 level is particularly common.

Step 2: Divide the value of alpha in half. For example, if alpha is 0.05, half of alpha would be 0.025. Find this value in column c of Appendix A to find the Z score that corresponds to your selected alpha level. If alpha = 0.05, $Z = \pm 1.96$. See Table 6.2 for the Z scores associated with other commonly used alpha levels.

Step 3: Substitute the sample values into the proper formula. If the value of the population standard deviation (σ) is *known*, use Formula 6.1: $c.i. = \bar{X} \pm Z(\sigma/\sqrt{N})$. To solve this formula:
  a. Find the square root of N.
  b. Divide the value you found in step a into sigma (σ).

  c. Multiply this value by Z.

If the value of σ is *unknown* (this will be the usual case), use Formula 6.2: $c.i. = \bar{X} \pm Z(s/\sqrt{N - 1})$. To solve this formula:
  a. Find the square root of $N - 1$.
  b. Divide the value you found in step a into sigma (σ).
  c. Multiply this value by Z.

Step 4: State the confidence interval in a sentence or two that identifies:
  a. the sample mean
  b. the upper and lower limits of the interval
  c. the alpha level (α) or the confidence level
  d. the sample size (N)

---

has only a 5% chance of being wrong (that is, of not containing the population mean).

In social science research, we need to complete one more step after constructing the confidence interval. The fourth and final step in the process is to express our results in a way that can be easily understood and that identifies each of the following elements: the sample mean, the upper and lower limits of the interval, the alpha level, and the sample size (N). The results for the example used in this section might be expressed as: "Average income for this community is $35,000 \pm $17.55. This estimate is based on a sample of 500 respondents and we can be 95% confident that the interval estimate is correct." *(For practice in constructing and expressing confidence intervals for sample means, see problems 6.1, 6.4–6.7, and 6.18 a–c.)*

6.10 INTERVAL ESTIMATION PROCEDURES FOR SAMPLE PROPORTIONS (LARGE SAMPLES)

Estimation procedures for sample proportions are essentially the same as those for sample means. The major difference is that, because proportions are different statistics, we must use a different sampling distribution. In fact, again based on the Central Limit Theorem, we know that sample proportions have sampling distributions that are normal in shape with means (μ_p) equal to the population value (P_u) and standard deviations (σ_p) equal to $\sqrt{P_u(1 - P_u)/N}$. The formula for constructing confidence intervals based on sample proportions is:

FORMULA 6.3

$$c.i. = P_s \pm Z\sqrt{\frac{P_u(1 - P_u)}{N}}$$

Application 6.1

A study of the leisure activities of Americans was conducted on a sample of 1,000 households. The respondents identified television viewing as a major form of recreation. If the sample reported an average of 6.2 hours of television viewing a day, what is the estimate of the population mean? The information from the sample is:

$$\overline{X} = 6.2$$
$$s = 0.7$$
$$N = 1,000$$

If we set alpha at 0.05, the corresponding Z score will be ±1.96, and the 95% confidence interval will be:

$$c.i. = \overline{X} \pm Z\left(\frac{s}{\sqrt{N-1}}\right)$$
$$c.i. = 6.2 \pm 1.96\left(\frac{0.7}{\sqrt{1,000-1}}\right)$$
$$c.i. = 6.2 \pm 1.96\left(\frac{0.7}{\sqrt{31.61}}\right)$$
$$c.i. = 6.2 \pm 1.96(0.02)$$
$$c.i. = 6.2 \pm 0.04$$

Based on this result, we would estimate that the population spends an average of 6.2 ± .04 hours per day viewing television. The lower limit of our interval estimate (6.2 − .04) is 6.16, and the upper limit (6.2 + .04) is 6.24. Thus, another way to state the interval would be:

$$6.16 \leq \mu \leq 6.24$$

The population mean is greater than or equal to 6.16 and less than or equal to 6.24. Because alpha was set at the .05 level, this estimate has a 5% chance of being wrong (that is, of not containing the population mean).

The values for P_s and N come directly from the sample, and the value of Z is determined by the confidence level, as was the case with sample means. This leaves one unknown in the formula, P_u—the same value we are trying to estimate. This dilemma can be resolved by setting the value of P_u at 0.5. Because the second term in the numerator under the radical $(1 - P_u)$ is the reciprocal of P_u, the entire expression will always have a value of 0.5 × 0.5, or 0.25, which is the maximum value this expression can attain. That is, if we set P_u at any value other than 0.5, the expression $P_u(1 - P_u)$ will decrease in value. If we set P_u at 0.4, for example, the second term $(1 - P_u)$ would be 0.6, and the value of the entire expression would decrease to 0.24. Setting P_u at 0.5 ensures that the expression $P_u(1 - P_u)$ will be at its maximum possible value and, consequently, the interval will be at maximum width. This is the most conservative solution possible to the dilemma posed by having to assign a value to P_u in the estimation equation.

To illustrate these procedures, assume that you want to estimate the proportion of students at your university who missed at least one day of classes because of illness last semester. Out of a random sample of 200 students, 60 reported that they had been sick enough to miss classes at least once during the previous semester. The sample proportion upon which we will base our estimate is thus 60/200, or 0.30. At the 95% level, the interval estimate will be:

$$c.i. = P_s \pm Z\sqrt{\frac{P_u(1 - P_u)}{N}}$$

STEP BY STEP	**Constructing Confidence Intervals for Sample Proportions**

Step 1: Select an alpha level. Commonly used alpha levels are 0.10, 0.05, 0.01, 0.001; the 0.05 level is particularly common.

Step 2: Divide the value of alpha in half. For example, if alpha is 0.05, half of alpha would be 0.025. Find this value in column c of Appendix A to find the Z score that corresponds to your selected alpha level. If alpha = 0.05, $Z = \pm 1.96$.

(NOTE: See Table 6.2 for the Z scores for the common alpha levels.)

Step 3: Substitute the sample values into Formula 6.3: $c.i. = P_s \pm Z\sqrt{P_u(1 - P_u)/N}$. To solve this formula:

   a. Substitute 0.5 for P_u. This will make the second term in the numerator $(1 - P_u)$

also 0.5. Thus, the numerator will have a value of 0.25.

   b. Divide N into 0.25.

   c. Find the square root of the quantity you found in step b.

   d. Multiply the quantity you found in step c by the value of Z.

Step 4: State the confidence interval in a sentence or two that identifies:

   a. the sample proportion

   b. the upper and lower limits of the interval

   c. the alpha level (α) or the confidence level

   d. the sample size (N)

$$c.i. = 0.30 \pm 1.96\sqrt{\frac{(0.5)(0.5)}{200}}$$

$$c.i. = 0.30 \pm 1.96\sqrt{\frac{0.25}{200}}$$

$$c.i. = 0.30 \pm 1.96\sqrt{0.00125}$$

$$c.i. = 0.30 \pm (1.96)(0.035)$$

$$c.i. = 0.30 \pm 0.07$$

Based on this sample proportion of 0.30, you would estimate that the proportion of students who missed at least one day of classes because of illness was between 0.23 and 0.37. The estimate could, of course, also be phrased in percentages by reporting that between 23% and 37% of the student body was affected by illness at least once during the past semester.

As was the case with sample means, the final step in the process is to express the confidence interval in a way that is easy to understand and that includes the value of the sample proportion, the upper and lower limits of the interval, the alpha level, and the sample size (N). The results for the example used in this section might be expressed as: "The percentage of students sick enough to miss at least one day of class is 30% ± 7%. This estimate is based on a sample of 200 respondents, and we can be 95% confident that the interval estimate is correct." *(For practice with confidence intervals for sample proportions, see problems 6.2, 6.8–6.12, 6.16–6.17, and 6.18d–g.)*

Application 6.2

A total of 1,609 adult Canadians were randomly selected to participate in a study of attitudes toward homosexuality and same-sex marriages. Some results are reported below. What is the level of support in the population? The sample information expressed in terms of the proportion agreeing is:

"Gays and lesbians should have the same rights as heterosexuals."

$$P_s = .72$$
$$N = 1,609$$

"Marriage should be expanded to include same-sex unions."

$$P_s = .60$$
$$N = 1,609$$

For the first item, the confidence interval estimate to the population at the 95% confidence level is:

$$c.i. = P_s \pm Z \sqrt{\frac{P_u(1 - P_u)}{N}}$$

$$c.i. = .72 \pm 1.96 \sqrt{\frac{(0.5)(0.5)}{1609}}$$

$$c.i. = .72 \pm 1.96 \sqrt{.00016}$$

$$c.i. = .72 \pm (1.96)(.013)$$
$$c.i. = .72 \pm .03$$

Expressing these results in terms of percentages, we can conclude that, at the 95% confidence level, between 69% and 75% of adult Canadians support equal rights for gays and lesbians.

For the second survey item, the confidence interval estimate to the population at the 95% confidence level is:

$$c.i. = P_s \pm Z \sqrt{\frac{P_u(1 - P_u)}{N}}$$

$$c.i. = .60 \pm 1.96 \sqrt{\frac{(0.5)(0.5)}{1,609}}$$

$$c.i. = .60 \pm 1.96 \sqrt{.00016}$$

$$c.i. = .60 \pm (1.96)(.013)$$
$$c.i. = .60 \pm .03$$

Again expressing results in terms of percentages, we can conclude that, at the 95% confidence level, between 57% and 63% of adult Canadians support same-sex marriages. (Note that the width of the second confidence interval is exactly the same as the first. This is because we are using the same values for Z score and sample size in both estimates.)

6.11 A SUMMARY OF THE COMPUTATION OF CONFIDENCE INTERVALS

To this point, we have covered the construction of confidence intervals for sample means and sample proportions. In both cases, the procedures assume large samples (N greater than 100). The procedures for constructing confidence intervals for small samples are not covered in this text. Table 6.3 presents the three formulas for confidence intervals organized by the situations in which they are used. For sample means, when the population standard deviation is known, use Formula 6.1. When the population standard deviation is unknown (which is the usual case), use Formula 6.2. For sample proportions, always use Formula 6.3.

TABLE 6.3   CHOOSING FORMULAS FOR CONFIDENCE INTERVALS

If the sample statistic is a	and		Use formula
mean	the population standard deviation is known	6.1	$c.i. = \bar{X} \pm Z\left(\dfrac{\sigma}{\sqrt{N}}\right)$
mean	the population standard deviation is unknown	6.2	$c.i. = \bar{X} \pm Z\left(\dfrac{s}{\sqrt{N - 1}}\right)$
proportion		6.3	$c.i. = P_s \pm Z\sqrt{\dfrac{P_u(1 - P_u)}{N}}$

STATISTICS IN EVERYDAY LIFE: Election Projections, Polls, and Surveys

The statistical techniques covered in Sections 6.8 through 6.10 have become a part of everyday life in the United States and in many other societies. In politics, for example, estimation techniques are used to track public sentiment, measure how citizens perceive the performance of our leaders, and project the likely winners of upcoming elections. In this installment of "Statistics in Everyday Life," we'll examine the accuracy of election projections for the 2004 presidential election and the polls that have measured the approval ratings of incumbent U.S. presidents since the middle of the 20th century. Both kinds of polls use the same formulas introduced in this chapter to construct confidence intervals (although the samples on which the polls are based were assembled according to a complex and sophisticated technology that is beyond the scope of this text).

Forecasting Presidential Elections

Although pollsters have become very accurate in predicting the outcomes of presidential elections, 95% confidence intervals are accurate only to ±3 percentage points so the polls cannot distinguish the likely victor in very close races. The margin of victory was quite small in the 2004 election, and the polls indicated a statistical dead heat right up to the end of the campaigns. The table below shows CNN's "poll of polls"—or averages of polls from various sources—for the final weeks of the 2004 campaign and the final breakdown of votes for the two major candidates. The polls that CNN included were based on sample sizes of about 1,000.

Polling Results and Actual Vote, 2004 Presidential Election

	Actual Vote	
	BUSH	KERRY
	51%	48%

	Election Projections:	
	Percentage of Sample Estimated to Vote for:	
Date of Poll	BUSH	KERRY
November 1	48%	46%
October 25	49%	46%
October 18	50%	45%

The final polls in October and November show that the difference between the candidates was so small that a winner could not be projected. For example, in the November 1 poll, Bush's support could have been as low as 45% (48% − 3%) and Kerry's could have been as high as 49% (46% + 3%). When the confidence intervals overlap, the race is said to be "too close to call" and "a statistical dead heat." Although the race was too close for the pollsters to identify a likely winner, note that the polls were within the ± 3% margin of error that is associated with interval estimates based on the 95% confidence level and sample sizes of about 1,000.

The Ups and Downs of Presidential Popularity

Once you get to be president, you are not free of polls and confidence intervals. Since the middle of the 20th century, pollsters have tracked the

(continued)

6.12 CONTROLLING THE WIDTH OF INTERVAL ESTIMATES

We can partly control the width of a confidence interval for either sample means or sample proportions by manipulating two terms in the equation. First, we can raise or lower the confidence level and, second, we can widen or narrow the interval by gathering samples of different size. The researcher alone determines the risk he or she is willing to take of being wrong (that is, of not including the population value in the interval estimate). The exact confidence level (or alpha level) will depend, in part, on the purpose of the research. For example, if potentially harmful drugs were being tested, the

STATISTICS IN EVERYDAY LIFE *(continued)*

president's popularity by asking randomly selected samples of adult Americans if they approve or disapprove of the way the president is handling his job. President G. W. Bush's approval ratings (the percent of the sample that approves) are presented here. For purposes of clarity, point estimates are used for this graph, but remember that the confidence interval estimate would range about ± 3% (at the 95% confidence level) around these points.

The single most dramatic feature of this graph is the huge increase in approval that followed the September 11 terrorist attacks on the World Trade Center and the Pentagon in 2001. This burst of support reflects the increased solidarity, strong emotions, and high levels of patriotism with which

Americans responded to the attacks. The president's approval rating reached an astounding 90% shortly after the attacks but then, inevitably, began to trend down as the society (and politics) gradually returned to the routines of everyday business. By the spring of 2003, the president's approval rating had fallen back into the high 50s, much lower than the peaks of Fall 2001 but still higher than the historical average. President Bush's approval received another boost with the invasion of Iraq in March 2003, an echo of the "rally" effect that followed 9/11, but then began to sink to prewar levels and below. By the end of 2005, the approval rating had fallen to the high 30s and low 40s, a reflection of the growing concern about the war, the poor federal response to Hurricane Katrina, and other issues.

APPROVAL RATINGS OF PRESIDENT BUSH, JANUARY 2001 TO DECEMBER 2005

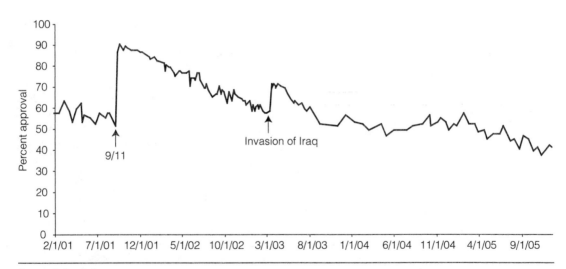

Source: Gallup Polls.

researcher would naturally demand very high levels of confidence (99.99% or even 99.999%). On the other hand, if intervals are being constructed only for loose "guesstimates," then much lower confidence levels can be tolerated (such as 90%).

The relationship between interval size and confidence level is that intervals widen as confidence levels increase. This relationship should make intuitive sense. Wider intervals are more likely to trap the population value; hence, we can place more confidence in them.

To illustrate this relationship, let's return to the example where we estimated the average income for a community. In this problem, we were working with a sample of 500 residents, and the average income for this sample was $35,000, with a standard deviation of $200. We constructed the 95% confidence interval and found that it extended 17.55 around the sample mean (that is, the interval was $35,000 ± 17.55).

If we had constructed the 90% confidence interval for these sample data (a lower confidence level), the Z score in the formula would have decreased to ±1.65, and the interval would have been narrower:

$$c.i. = \overline{X} \pm Z \left(\frac{s}{\sqrt{N-1}} \right)$$

$$c.i. = 35,000 \pm 1.65 \left(\frac{200}{\sqrt{499}} \right)$$

$$c.i. = 35,000 \pm 1.65(8.95)$$

$$c.i. = 35,000 \pm 14.77$$

On the other hand, if we had constructed the 99% confidence interval, the Z score would have increased to ±2.58, and the interval would have been wider:

$$c.i. = \overline{X} \pm Z \left(\frac{s}{\sqrt{N-1}} \right)$$

$$c.i. = 35,000 \pm 2.58 \left(\frac{200}{\sqrt{499}} \right)$$

$$c.i. = 35,000 \pm 2.58(8.95)$$

$$c.i. = 35,000 \pm 23.09$$

At the 99.9% confidence level, the Z score would be ±3.29, and the interval would be wider still:

$$c.i. = \overline{X} \pm Z \left(\frac{s}{\sqrt{N-1}} \right)$$

$$c.i. = 35,000 \pm 3.29 \left(\frac{200}{\sqrt{499}} \right)$$

$$c.i. = 35,000 \pm 3.29(8.95)$$

$$c.i. = 35,000 \pm 29.45$$

We can readily see the increase in interval size in Table 6.4, which groups these four intervals. Although I have used sample means to illustrate the relationship between interval width and confidence level, exactly the same relationships apply to sample proportions. (*To further explore the relationship between alpha and interval width, see problem 6.13.*)

Sample size bears the opposite relationship to interval width. As sample size increases, interval width decreases. Larger samples give more precise (narrower) estimates. Again, an example should make this clearer. In Table 6.5,

TABLE 6.4   INTERVAL ESTIMATES FOR FOUR CONFIDENCE LEVELS ($\bar{X}$ = $35,000, s = $200, N = 500 THROUGHOUT)

Alpha	Confidence Level	Interval	Interval Width
.10	90%	$35,000 ± 14.77	$29.54
.05	95%	$35,000 ± 17.55	$35.10
.01	99%	$35,000 ± 23.09	$46.18
.001	99.9%	$35,000 ± 29.45	$58.90

TABLE 6.5   INTERVAL ESTIMATES FOR FOUR DIFFERENT SAMPLES ($\bar{X}$ = $35,000, s = $200, ALPHA = 0.05 THROUGHOUT)

Sample 1 (N = 100)	Sample 2 (N = 500)
$c.i. = \$35,000 \pm 1.96 \left(\dfrac{200}{\sqrt{99}} \right)$	$c.i. = \$35,000 \pm 1.96 \left(\dfrac{200}{\sqrt{499}} \right)$
$c.i.$ = $35,000 ± 39.40	$c.i.$ = $35,000 ± 17.55

Sample 3 (N = 1,000)	Sample 4 (N = 10,000)
$c.i. = \$35,000 \pm 1.96 \left(\dfrac{200}{\sqrt{999}} \right)$	$c.i. = \$35,000 \pm 1.96 \left(\dfrac{200}{\sqrt{9,999}} \right)$
$c.i.$ = $35,000 ± 12.40	$c.i.$ = $35,000 ± 3.92

Sample	N	Interval Width
1	100	$78.80
2	500	$35.10
3	1,000	$24.80
4	10,000	$ 7.84

confidence intervals for four samples of various sizes are constructed and then grouped together for comparison. The sample data are the same as in Table 6.4, and the confidence level is 95% throughout. The relationships that Table 6.5 illustrates also hold true, of course, for sample proportions. *(To further explore the relationship between sample size and interval width, see problem 6.14.)*

Notice that the decrease in interval width (or, increase in precision) does not bear a constant or linear relationship with sample size. For example, sample 2 is five times larger than sample 1 but the interval constructed with the larger sample size is not five times as narrow. This is an important relationship because it means that N might have to be increased many times over to appreciably improve the accuracy of an estimate. Because the cost of a research project is directly related to sample size, this relationship implies a point of diminishing returns in estimation procedures. A sample of 10,000 will cost about twice as much as a sample of 5,000, but estimates based on the larger sample will not be twice as precise.

SUMMARY

1. Because populations are almost always too large to test, a fundamental strategy of social science research is to select a sample from the defined population and then use information from the sample to generalize to the population. This involves either estimation or hypothesis testing.
2. Researchers create simple random samples by selecting cases from a list of the population following the rule of EPSEM (each case has an equal probability of being selected). Samples selected by the rule of EPSEM have a very high probability of being representative.
3. The sampling distribution, the central concept in inferential statistics, is a theoretical distribution of all possible sample outcomes. Because its overall shape, mean, and standard deviation are known (under the conditions specified in the two theorems), the sampling distribution can be adequately characterized and used by researchers.
4. The two theorems introduced in this chapter state that when the variable of interest is normally distributed in the population or when sample size is large, the sampling distribution will be normal in shape, its mean will be equal to the population

mean, and its standard deviation (or standard error) will be equal to the population standard deviation divided by the square root of N.
5. Population values can be estimated with sample values. With confidence intervals, which can be based on either proportions or means, we estimate that the population value falls within a certain range of values. The width of the interval is a function of the risk we are willing to take of being wrong (the alpha level) and the sample size. The interval widens as our probability of being wrong decreases and as sample size decreases.
6. Estimates based on sample statistics must be unbiased and relatively efficient. Of all the sample statistics, only means and proportions are unbiased. The means of the sampling distributions of these statistics are equal to the respective population values. Efficiency is largely a matter of sample size. The greater the sample size, the lower the value of the standard deviation of the sampling distribution, the more tightly clustered the sample outcomes will be around the mean of the sampling distribution, and the more efficient the estimate.

SUMMARY OF FORMULAS

Confidence interval for a sample mean, large samples, population standard deviation known	6.1	$c.i. = \bar{X} \pm Z\left(\dfrac{\sigma}{\sqrt{N}}\right)$
Confidence interval for a sample mean, large samples, population standard deviation unknown	6.2	$c.i. = \bar{X} \pm Z\left(\dfrac{s}{\sqrt{N-1}}\right)$
Confidence interval for a sample proportion, large samples	6.3	$c.i. = P_s \pm Z\sqrt{\dfrac{P_u(1-P_u)}{N}}$

GLOSSARY

Alpha (α). The probability of error or the probability that a confidence interval does not contain the population value. Alpha levels are usually set at 0.10, 0.05, 0.01, or 0.001.

Bias. A criterion used to select sample statistics as estimators. A statistic is unbiased if the mean of its sampling distribution is equal to the population value of interest.

Central Limit Theorem. A theorem that specifies the mean, standard deviation, and shape of the

sampling distribution, given that the sample is large.

Confidence interval. An estimate of a population value in which a range of values is specified.

Confidence level. A frequently used alternate way of expressing alpha, the probability that an interval estimate will not contain the population value. Confidence levels of 90%, 95%, 99%, and 99.9% correspond to alphas of 0.10, 0.05, 0.01, and 0.001, respectively.

Efficiency. The extent to which the sample outcomes are clustered around the mean of the sampling distribution.

EPSEM. The **E**qual **P**robability of **SE**lection **M**ethod for selecting samples. Every element or case in the population must have an equal probability of selection for the sample.

μ. ("mew"). The mean of a population.

$\mu_{\bar{x}}$. ("mew-sub-ex-bar"). The mean of a sampling distribution of sample means.

μ_p. ("mew-sub-p"). The mean of a sampling distribution of sample proportions.

Parameter. A characteristic of a population.

P_s. ("P-sub-s"). Any sample proportion.

P_u. ("P-sub-u"). Any population proportion.

Representative. The quality a sample is said to have if it reproduces the major characteristics of the population from which it was drawn.

Sampling distribution. The distribution of a statistic for all possible sample outcomes of a certain size. Under conditions specified in two theorems, the sampling distribution will be normal in shape with a mean equal to the population value and a standard deviation equal to the population standard deviation divided by the square root of N.

Simple random sample. A method for choosing cases from a population by which every case and every combination of cases have an equal chance of being included.

Standard error of the mean. The standard deviation of a sampling distribution of sample means.

MULTIMEDIA RESOURCES

The Wadsworth Sociology Resource Center: Virtual Society at
http://www.thomsonedu.com/sociology

Visit the companion website for the seventh edition of *Statistics: A Tool for Social Research* to access a wide range of student resources. Begin by clicking on the Student Resources section of the book's website to access the following study tools:

■ Basic math review
■ Flash cards

■ Additional chapter problems
■ Internet links
■ Table of random numbers
■ MicroCase and SPSS examples and exercises
■ Hypothesis testing for variables measured at the ordinal level

PROBLEMS

6.1 For each set of sample outcomes below, construct the 95% confidence interval for estimating μ, the population mean.

a. $\bar{X} = 5.2$
   $s = .7$
   $N = 157$

b. $\bar{X} = 100$
   $s = 9$
   $N = 620$

c. $\bar{X} = 20$
   $s = 3$
   $N = 220$

d. $\bar{X} = 1{,}020$
   $s = 50$
   $N = 329$

e. $\bar{X} = 7.3$
   $s = 1.2$
   $N = 105$

f. $\bar{X} = 33$
   $s = 6$
   $N = 220$

6.2 For each set of sample outcomes below, construct the 99% confidence interval for estimating P_u.

a. $P_s = .14$
   $N = 100$

b. $P_s = .37$
   $N = 522$

c. $P_s = .79$
   $N = 121$

d. $P_s = .43$
   $N = 1{,}049$

e. $P_s = .40$
   $N = 548$

f. $P_s = .63$
   $N = 300$

6.3 For each confidence level below, determine the corresponding Z score.

Confidence Level	Alpha	Area Beyond Z	Z Score
95%	.05	.0250	± 1.96
94%			
92%			
97%			
98%			
99.9%			

6.4 $\boxed{\text{SW}}$ You have developed a series of questions to measure "burnout" in social workers. A random sample of 100 social workers in greater metropolitan Shinbone, Kansas, has an average score of 10.3, with a standard deviation of 2.7. What is your estimate of the average burnout score for the population as a whole? Use the 95% confidence level.

6.5 SOC A researcher has gathered information from a random sample of 178 households. For each variable below, construct confidence intervals to estimate the population mean. Use the 90% level.
 a. The average number of residents in each household was 2.3. Standard deviation is .35.
 b. Each household contained an average of 2.1 television sets ($s = .10$) and .78 telephones ($s = .55$).
 c. The households averaged 6.0 hours of television viewing per day ($s = 3.0$).

6.6 SOC A random sample of 100 television programs contained an average of 2.37 acts of physical violence per program. At the 99% level, what is your estimate of the population value?

$$\overline{X} = 2.37$$
$$s = 0.30$$
$$N = 100$$

6.7 SOC A random sample of 429 college students was interviewed about a number of matters.
 a. They reported that they had spent an average of $178.23 on textbooks during the previous semester. If the sample standard deviation for these data is $15.78, construct an estimate of the population mean at the 99% level.
 b. They also reported that they had visited the health clinic an average of 1.5 times a semester. If the sample standard deviation is 0.3, construct an estimate of the population mean at the 99% level.
 c. On the average, the sample had missed 2.8 days of classes per semester because of illness. If the sample standard deviation is 1.0, construct an estimate of the population mean at the 99% level.
 d. On the average, the sample had missed 3.5 days of classes per semester for reasons other than illness. If the sample standard deviation is 1.5, construct an estimate of the population mean at the 99% level.

6.8 CJ A random sample of 500 residents of Shinbone, Kansas, shows that exactly 50 of the respondents had been the victims of violent crime over the past year. Estimate the proportion of victims for the population as a whole, using the 90% confidence level. (*HINT: Calculate the sample proportion P_s before using Formula 6.3. Remember that proportions are equal to frequency divided by N.*)

6.9 SOC The survey mentioned in problem 6.5 found that 25 of the 178 households consisted of unmarried couples who were living together. What is your estimate of the population proportion? Use the 95% level.

6.10 PA A random sample of 324 residents of a community revealed that 30% were very satisfied with the quality of trash collection. At the 99% level, what is your estimate of the population value?

6.11 SOC A random sample of 1,496 respondents of a major metropolitan area was questioned about a number of issues. Construct estimates for the population at the 90% level for each of the results reported below. Express the final confidence interval in percentages (e.g., "between 40% and 45% agreed that premarital sex was always wrong").
 a. When asked to agree or disagree with the statement "Explicit sexual books and magazines lead to rape and other sex crimes," 823 agreed.
 b. When asked to agree or disagree with the statement "Handguns should be outlawed," 650 agreed.
 c. 375 of the sample agreed that marijuana should be legalized.
 d. 1,023 of the sample said that they had attended church or synagogue at least once within the past month.
 e. 800 agreed that public elementary schools should have sex education programs starting in the fifth grade.

6.12 SW A random sample of 100 patients treated in a program for alcoholism and drug dependency over the past 10 years was selected. Researchers determined that 53 of the patients had been readmitted to the program at least once. At the 95% level, construct an estimate for the population proportion.

6.13 For the sample data below, construct four different interval estimates of the population mean, one each for the 90%, 95%, 99%, and 99.9% level. What happens to the interval width as confidence level increases? Why?

$$\overline{X} = 100$$
$$s = 10$$
$$N = 500$$

6.14 For each of the three sample sizes below, construct the 95% confidence interval. Use a sample proportion of 0.40 throughout. What happens to interval width as sample size increases? Why?

$$P_s = 0.40$$
Sample A: $N = 100$
Sample B: $N = 1,000$
Sample C: $N = 10,000$

6.15 | PS | Two individuals are running for mayor of Shinbone. You conduct a survey a week before the election and find that 51% of the respondents prefer candidate A. Can you predict a winner? Use the 99% level. *(HINT: In a two-candidate race, what percentage of the vote would the winner need? Does the confidence interval indicate that candidate A has a sure margin of victory? Remember that while the population parameter is probably ($\alpha = .01$) in the confidence interval, it may be anywhere in the interval.)*

$$P_s = 0.51$$
$$N = 578$$

6.16 | SOC | The World Values Survey is administered periodically to random samples from societies around the globe. The number of respondents in each nation who said that they are "very happy" is listed here. Compute sample proportions and construct confidence interval estimates for each nation at the 95% level.

Nation	Year	Number "Very Happy"	Sample Size	Confidence Interval
Great Britain	1998	496	1,495	
Japan	1995	505	1,476	
Brazil	1996	329	1,492	
Nigeria	1995	695	1,471	
China	1995	338	1,493	

6.17 | SOC | The fraternities and sororities at St. Algebra College have been plagued by declining membership over the past several years and want to know if the incoming freshman class will be a fertile recruiting ground. Not having enough money to survey all 1,600 freshmen, they commission you to survey the interests of a random sample. You find that 35 of your 150 respondents are "extremely" interested in social clubs. At the 95% level, what is your estimate of the number of freshmen who would be extremely interested? *(HINT: The high and low values of your final confidence interval are proportions. How can proportions also be expressed as numbers?)*

6.18 | SOC | The results listed below are from a survey given to a random sample of the American public. For each sample statistic, construct a confidence interval estimate of the population parameter at the 95% confidence level. Sample size (N) is 2,987 throughout.

a. The average occupational prestige score was 43.87, with a standard deviation of 13.52.
b. The respondents reported watching an average of 2.86 hours of TV per day with a standard deviation of 2.20.
c. The average number of children was 1.81, with a standard deviation of 1.67.
d. Of the 2,987 respondents, 876 identified themselves as Catholic.
e. The number of respondents who said they had never married was 535.
f. The proportion of respondents who said they voted for Bush in the 2000 presidential election was 0.52.
g. When asked about capital punishment, 2,425 of the respondents said that they favored the death penalty for murder.

7

Hypothesis Testing I
The One-Sample Case

By the end of this chapter, you will be able to:

1. Explain the logic of hypothesis testing.
2. Define and explain the conceptual elements involved in hypothesis testing, especially the null hypothesis, the sampling distribution, the alpha level, and the test statistic.
3. Explain what it means to "reject the null hypothesis" or "fail to reject the null hypothesis."
4. Identify and cite examples of situations in which one-sample tests of hypotheses are appropriate.
5. Test the significance of single-sample means and proportions using the five- step model and correctly interpret the results.
6. Explain the difference between one- and two-tailed tests and specify when each is appropriate.
7. Define and explain Type I and Type II errors and relate each to the selection of an alpha level.

7.1 INTRODUCTION

Chapter 6 introduced the techniques for estimating population parameters from sample statistics. In Chapters 7 through 10, we will investigate a second application of inferential statistics called **hypothesis testing** or **significance testing.** This chapter introduces the techniques for hypothesis testing in the one-sample case. Researchers can use these procedures in situations such as the following:

1. A researcher has selected a sample of 789 older citizens who live in a particular state and also has information on the percentage of the entire population of the state that was victimized by crime during the past year. Are older citizens, as represented by this sample, more or less likely to be victimized than the population in general?
2. Are the grade-point averages (GPAs) of college athletes different from the GPAs of the student body as a whole? To investigate, the academic records of a random sample of 235 student athletes from a large state university are compared with the overall GPA of all students.
3. A sociologist has been hired to assess the effectiveness of a rehabilitation program for alcoholics in her city. The program serves a large area, and she does not have the resources to test every single client. Instead, she draws a random sample of 127 people from the list of all clients and questions them on a variety of issues. She notices that, on the average, the people in her sample miss fewer days of work each year than do workers in the city as a whole. Are alcoholics treated by the program more reliable than workers in general?

In each of these situations, we have randomly selected samples (of senior citizens, athletes, or treated alcoholics) that we want to compare to a population (the entire state, student body, or city). Note that we are not interested in the sample per se but in the larger group from which it was selected (*all* senior citizens in the state, *all* athletes on this campus, or *all* people who have completed the treatment program). Specifically, we want to know if the groups represented by the samples are different from the populations on a specific trait or variable (victimization rates, GPAs, or absenteeism).

Of course, it would be better if we could include all senior citizens, athletes, or treated alcoholics rather than these smaller samples. However, as we have seen, researchers usually do not have the resources necessary to test everyone in a large group and must use random samples instead. In these situations, conclusions are based on a comparison of a single sample (which represents a larger group) and the population. For example, if we found that the rate of victimization for the sample of senior citizens was higher than the state rate, we might conclude that "senior citizens are significantly more likely to be crime victims." The word "significantly" is a key word: It means that the difference between the sample's victimization rate and the population's rate is very unlikely to be the result of random chance alone. In other words, it is very likely that *all* senior citizens (not just the 789 people in the sample) have a higher victimization rate than do residents of the state as a whole. On the other hand, if we found no significant difference between the GPAs of the sample of athletes and the student body as a whole, we might conclude that athletes (*all* athletes on this campus) are essentially the same as other students in terms of academic achievement.

Thus, we can use samples to represent larger groups (senior citizens, athletes, or treated alcoholics) and compare and contrast the characteristics of the sample with the population and be extremely confident in our conclusions. Remember, however, that the EPSEM procedure does not guarantee that samples will be representative and, thus, there will always be a small amount of uncertainty in our conclusions. One of the great advantages of inferential statistics, however, is that we will be able to estimate the probability of error and evaluate our decisions accordingly.

7.2 AN OVERVIEW OF HYPOTHESIS TESTING

We'll begin with a general overview of hypothesis testing, using our third research situation above as an example, and introduce the more technical considerations and proper terminology throughout the remainder of the chapter. Let's examine this situation in some detail. First of all, the main question here is "Are people treated in this program more reliable workers than people in the community in general?" In other words, what the researcher would really like to do is compare information gathered from ALL clients (the *population* of alcoholics treated in the program) with information about the entire metropolitan area. If she had information for both of these groups (all clients and the entire metro population), she could answer the question easily, completely, and finally.

The problem is that the researcher does not have the time and or money to gather information on the thousands of people who have been treated by the program. Instead, she has drawn a random sample, following the rule of EPSEM, of 127 clients from agency records. The absentee rates for the sample and the community are:

Community	Sample of Treated Alcoholics
μ = 7.2 days per year	$\overline{X}$ = 6.8 days per year
σ = 1.43	N = 127

We can see that there is a difference in rates of absenteeism and that the average rate of absenteeism for the sample is lower than the rate for the community. Although it's tempting, we can't make any conclusions yet because we are working with a random sample of the population we are interested in, not the population itself (all people treated in the program).

Figure 7.1 should clarify these relationships. The largest circle symbolizes the community because it is the largest group. The population of all treated alcoholics is also symbolized by a large circle because it is a sizeable group, although only a small fraction of the community as a whole. The random sample of 127, the smallest of the three groups, is indicated by the smallest circle.

The labels on the arrows connecting the circles summarize the major questions and connections in this research situation. As we noted earlier, the main question the researcher wants to answer is "are all treated alcoholics the same as or different from the community" in terms of absenteeism? The population of treated alcoholics, too large to test, is represented by the randomly selected sample of 127. The main question related to the sample concerns the cause of the observed difference between its mean of 6.8 and the community mean of 7.2. There are two possible explanations for this difference, and we will consider them one at a time.

The first explanation, which we will call explanation A, is that the difference between the community mean of 7.2 days and the sample mean of 6.8 days reflects a real difference in absentee rates between the population of all treated alcoholics and the community. The difference is "statistically significant" in the sense that it is very unlikely to have occurred by random chance alone.

FIGURE 7.1   A TEST OF HYPOTHESIS FOR SINGLE SAMPLE MEANS

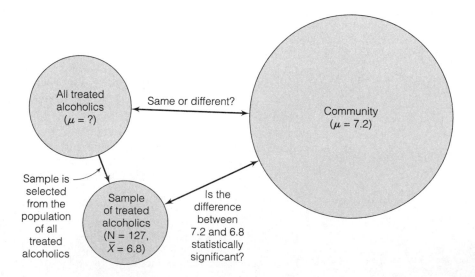

If explanation A is true, the population of all treated alcoholics is different from the community and the sample did *not* come from a population with a mean absentee rate of 7.2 days.

The second explanation, or explanation B, is that mere random chance caused the observed difference between sample and community means. In other words, there is no important difference between treated alcoholics and the community as a whole, and the difference between the sample mean of 6.8 days and the mean of 7.2 days of absenteeism for the community is trivial and due to random chance. If explanation B is true, the population of treated alcoholics is just like everyone else and has a mean absentee rate of 7.2 days.

Which explanation is correct? As long as we are working with a sample rather than the entire group of treated alcoholics, we cannot be absolutely (100%) sure about the answer to this question. However, we can set up a decision-making procedure so conservative that one of the two explanations can be chosen, with the knowledge that the probability of choosing the incorrect explanation is very low.

This decision-making process begins with the assumption that explanation B is correct. Symbolically, the assumption that the mean absentee rate for all treated alcoholics is the same as the rate for the community as a whole can be stated as:

$$\mu = 7.2 \text{ days per year}$$

Remember that this μ refers to the mean for *all* treated alcoholics, not just the 127 in the sample. This assumption, $\mu = 7.2$, can be tested statistically.

If explanation B ("*the population of treated alcoholics is not different from the community as a whole and has a μ of 7.2*") is true, then we can find the probability of getting the observed sample outcome ($\overline{X} = 6.8$). Let's add an objective decision rule in advance. If the odds of getting the observed difference are less than .05 (5 out of 100, or 1 in 20), we will reject explanation B. If this explanation were true, a difference of this size (7.2 days versus 6.8 days) would be a very rare event, and in hypothesis testing, we always bet *against* rare events.

How can we estimate the probability of the observed sample outcome ($\overline{X} = 6.8$) if explanation B is correct? We can determine this value by using our knowledge of the sampling distribution of all possible sample outcomes. Looking back at the information we have and applying the Central Limit Theorem (see Chapter 6), we can assume that the sampling distribution is normal in shape and has a mean of 7.2 (because $\mu_{\overline{X}} = \mu$) and a standard deviation of $1.43/\sqrt{127}$ because $\sigma_{\overline{X}} = \sigma/\sqrt{N}$. We also know that the standard normal distribution can be interpreted as a distribution of probabilities (see Chapter 5) and that the particular sample outcome noted above ($\overline{X} = 6.8$) is one of thousands of possible sample outcomes. Figure 7.2 depicts the sampling distribution and notes the sample outcome.

Using our knowledge of the standardized normal distribution, we can add further useful information to this sampling distribution of sample means. Specifically, with Z scores, we can depict the decision rule stated previously: Any sample outcome with probability less than 0.05 (assuming that explanation B is true) will cause us to reject explanation B. The probability of 0.05 can be translated into an area and divided equally into the upper and lower tails of the sampling distribution. Using Appendix A, we find that the Z-score equivalent of this area is ±1.96. Figure 7.3 shows the areas and Z scores.

Now we can rephrase the decision rule. Any sample outcome falling in the shaded areas depicted in Figure 7.3 by definition has a probability of occurrence

FIGURE 7.2 THE SAMPLING DISTRIBUTION OF ALL POSSIBLE SAMPLE MEANS

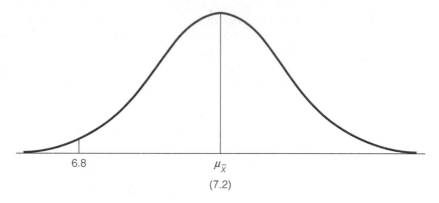

$$6.8 \qquad \mu_{\bar{X}}$$

(7.2)

of less than 0.05. Such an outcome would be a rare event and would cause us to reject explanation B.

All that remains is to translate our sample outcome into a Z score so we can see where it falls on the curve. To do this, we use the standard formula for locating any particular raw score under a normal distribution. When we use known or empirical distributions, this formula is expressed as:

$$Z = \frac{X_i - \bar{X}}{s}$$

Or, to find the equivalent Z score for any raw score, subtract the mean of the distribution from the raw score and divide by the standard deviation of the distribution. Because we are now concerned with the sampling distribution of all sample means rather than an empirical distribution, the symbols in the formula will change, but the form remains exactly the same:

FORMULA 7.1
$$Z = \frac{\bar{X} - \mu}{\sigma / \sqrt{N}}$$

Or, to find the equivalent Z score for any sample mean, subtract the mean of the sampling distribution, which is equal to the population mean or μ, from

FIGURE 7.3 THE SAMPLING DISTRIBUTION OF ALL POSSIBLE SAMPLE MEANS

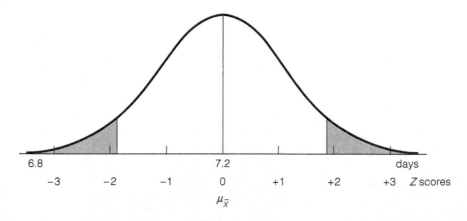

FIGURE 7.4 THE SAMPLING DISTRIBUTION OF SAMPLE MEANS WITH THE SAMPLE OUTCOME ($\bar{x} = 6.8$) NOTED IN Z SCORES

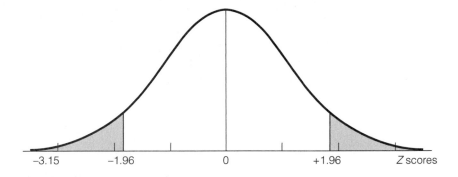

the sample mean and divide by the standard deviation of the sampling distribution.

Recalling the data given on this problem, we can now find the Z-score equivalent of the sample mean:

$$Z = \frac{6.8 - 7.2}{1.43/\sqrt{127}}$$

$$Z = \frac{-0.40}{0.127}$$

$$Z = -3.15$$

In Figure 7.4, this Z score of -3.15 is noted on the distribution of all possible sample means, and we see that the sample outcome does fall in the shaded area. If explanation B is true, this particular sample outcome has a probability of occurrence of less than 0.05 (how much less than 0.05 is irrelevant because the decision rule was established in advance). The sample outcome ($\bar{X} = 6.8$ or $Z = -3.15$) would therefore be rare if explanation B were true, and the researcher may reject explanation B. If explanation B were true, this sample outcome would be extremely unlikely. The sample of 127 treated alcoholics comes from a population that is significantly different from the community on the trait of absenteeism. Or, to put it another way, the sample does not come from a population that has a mean of 7.2 days of absences.

Keep in mind that our decisions in significance testing are based on information gathered from random samples. On rare occasions, a sample may not be representative of the population from which it was selected. The decision-making process outlined above has a very high probability of resulting in correct decisions, but, as long as we must work with samples rather than populations, we face an element of risk. That is, the decision to reject explanation B might be incorrect if this sample happens to be one of the few that is unrepresentative of the population of alcoholics treated in this program. One important strength of hypothesis testing is that we can estimate the probability of making an incorrect decision. In the example at hand, explanation B was rejected and the probability of this decision being incorrect is 0.05—the decision rule established at the beginning of the process. To say that the probability of rejecting explanation B incorrectly is 0.05 means that, if we repeated this same test an infinite number of times, we would incorrectly reject explanation B only 5 times out of every 100.

7.3 THE FIVE-STEP MODEL FOR HYPOTHESIS TESTING

All the formal elements and concepts used in hypothesis testing were surreptitiously sneaked into the preceding discussion. This section presents their proper names and introduces a **five-step model** for organizing all hypothesis testing:

Step 1. Making assumptions and meeting test requirements

Step 2. Stating the null hypothesis

Step 3. Selecting the sampling distribution and establishing the critical region

Step 4. Computing the test statistic

Step 5. Making a decision and interpreting the results of the test

We will look at each step individually, using the problem from Section 7.2 as an example throughout.

Step 1. Making Assumptions and Meeting Test Requirements. Any application of statistics requires that we make certain assumptions. In other words, for the test to be valid, the elements of the test and the variables involved must have certain characteristics. Specifically, three assumptions have to be satisfied when conducting a test of hypothesis with a single sample mean. First, we must be sure that we are working with a random sample, one that has been selected according to the rules of EPSEM (see Chapter 6). Second, to justify computation of a mean, we must assume that the variable being tested is interval-ratio in level of measurement. Finally, we must assume that the sampling distribution of all possible sample means is normal in shape so that we may use the standardized normal distribution to find areas under the sampling distribution. We can be sure that this assumption is satisfied by using large samples (see the Central Limit Theorem in Chapter 6).

Usually, we will state these assumptions in abbreviated form as a mathematical model for the test. For example:

Model: Random sampling
Level of measurement is interval-ratio
Sampling distribution is normal

Step 2. Stating the Null Hypothesis (H_0). The **null hypothesis** is the formal name for explanation B and is always a statement of "no difference." The exact form of the null hypothesis will vary depending on the test being conducted. In the single-sample case, the null hypothesis states that the sample comes from a population with a certain characteristic. In our example, the null hypothesis is that the population of treated alcoholics is "no different" from the community as a whole, that their average days of absenteeism is also 7.2, and that random chance caused the difference between 7.2 and the sample mean of 6.8. Symbolically, the null would be stated as

$$H_0: \mu = 7.2$$

where μ refers to the mean of the population of treated alcoholics. The null hypothesis is the central element in any test of hypothesis because the entire process is aimed at rejecting or failing to reject the H_0.

Usually, consistent with "explanation A," the researcher believes that there is a significant difference and wants to reject the null hypothesis. At this point in the five-step model, the researcher's belief is stated in a **research hypothesis (H_1),** a statement that directly contradicts the null hypothesis. Thus, the researcher's goal in hypothesis testing is often to gather evidence for the research hypothesis by rejecting the null hypothesis.

The research hypothesis can be stated in several ways. One form would simply assert that the population from which the sample was selected lacked a certain characteristic or, in terms of our example, had a mean that was not equal to a specific value:

$$(H_1: \mu \neq 7.2)$$
$$\text{where } \neq \text{ means "not equal to"}$$

Symbolically, this statement asserts that the sample does not come from a population with a mean of 7.2, or that the population of all treated alcoholics is different from the community as a whole. The research hypothesis is enclosed in parentheses to emphasize that it has no formal standing or role in testing the hypothesis (except, as we shall see in the next section, in choosing between one-tailed and two-tailed tests). It serves as a reminder of what the researcher believes to be the truth.

Step 3. Selecting the Sampling Distribution and Establishing the Critical Region. The sampling distribution is, as always, the probabilistic yardstick against which a particular sample outcome is measured. By assuming that the null hypothesis is true (and *only* by this assumption), we can attach values to the mean and standard deviation of the sampling distribution and thus measure the probability of any specific sample outcome. There are several different sampling distributions, but for now we will confine our attention to the sampling distribution described by the standard normal curve, which Appendix A summarizes.

The **critical region** consists of the areas under the sampling distribution that include unlikely sample outcomes. Before we test the hypothesis, we must define what we mean by *unlikely*. That is, we must specify in advance those sample outcomes so unlikely that they will lead us to reject the H_0. This decision rule will establish the critical region or **region of rejection.** The word *region* is used because, essentially, we are describing those areas under the sampling distribution that contain unlikely sample outcomes. In the example above, this area corresponded to a Z score of ± 1.96, called **Z(critical),** displayed in Figure 7.3. The shaded area is the critical region. Any sample outcome for which the Z-score equivalent fell in this area (that is, below -1.96 or above $+1.96$) would have caused us to reject the null hypothesis.

By convention, the size of the critical region is reported as alpha (α), the proportion of all of the area included in the critical region. In the example above, our **alpha level** was 0.05. Other commonly used alphas are 0.10, 0.01, and 0.001.

In abbreviated form, all the decisions that this step involves are noted below. The Z scores mark the beginnings of the critical region:

$$\text{Sampling distribution} = Z \text{ distribution}$$
$$\alpha = 0.05$$
$$Z(\text{critical}) = \pm 1.96$$

(For practice in finding Z(critical) scores, see problem 7.1a.)

Step 4. Computing the Test Statistic. To evaluate the probability of any given sample outcome, we must convert the sample value into a Z score. Solving the equation for Z-score equivalents is called computing the **test statistic,** and the resultant value is referred to as **Z(obtained)** to differentiate the test statistic from the critical region. In our example above, we found a Z(obtained) of -3.15. Note that, in many research situations, the value of the population standard deviation (σ) will not be known. In this case, substitute the value of the sample standard deviation (s) for σ if sample size is larger than 100. For small samples, see section 7.6. *(For practice in computing obtained Z scores for means, see problems 7.1c, 7.2 to 7.7, and 7.15e and f.)*

Step 5. Making a Decision and Interpreting the Results of the Test. As the last step in the hypothesis-testing process, the test statistic is compared with the critical region. If the test statistic falls into the critical region, our decision will be to reject the null hypothesis. If the test statistic does not fall into the critical region, we fail to reject the null hypothesis. In our example, the two values were:

$$Z(\text{critical}) = \pm 1.96$$
$$Z(\text{obtained}) = -3.15$$

and we saw that the Z(obtained) fell in the critical region (see Figure 7.4). Our decision was to reject the null hypothesis, which stated that treated alcoholics have a mean absentee rate of 7.2 days or, in other words, that there is no difference between treated alcoholics and other workers in the community. When we reject this statement, we are saying that treated alcoholics do *not* have a mean absentee rate of 7.2 days and that there *is* a difference between them and the community as a whole. We can also conclude that the difference between the sample mean of 6.8 and the community mean of 7.2 is statistically significant or unlikely to be caused by random chance alone. On the trait of absenteeism, treated alcoholics are different from the community as a whole.

Note that, to complete step 5, you have to do two things. First, you make a decision about the null hypothesis: If the test statistic falls in the critical region, you reject H_0. If the test statistic does not fall in the critical region, you fail to reject the H_0.

Second, you need to say what that decision means. In this case, the null hypothesis was rejected, which means there is a significant difference between the mean of the sample and the mean of the community and, therefore, treated alcoholics are different from the community as a whole.

This five-step model will serve us as a framework for decision making throughout the hypothesis-testing chapters. The exact nature and method of expression for our decisions will vary for different situations. However, familiarity with the five-step model will assist you in mastering this material by providing a common frame of reference for all significance testing.

STEP BY STEP	**Testing the Significance of the Difference Between a Sample Mean and a Population Mean (Large Samples): Computing Z(obtained) and Interpreting Results**

In step 4 of the five-step model, we compute the test statistic and use Formula 7.1 to convert the sample mean into a Z score called Z(obtained). Then, in step 5, we compare Z(obtained) with Z(critical). If the test statistic is in the critical region, we reject the null hypothesis. Otherwise, we fail to reject the null hypothesis.

To solve Formula 7.1: $Z = \dfrac{\bar{X} - \mu}{\sigma/\sqrt{N}}$

Step 1: Find the square root of N.

Step 2: Divide the quantity you found in step 1 into the population standard deviation (σ). *NOTE: If the population standard deviation (σ) is unknown and sample size is larger than 100, substitute the sample standard deviation (s) for σ. If sample size is less than 100, see Section 7.6.*

Step 3: Subtract the population mean from the sample mean.

Step 4: Divide the quantity you found in step 3 by the quantity you found in step 2. This is Z(obtained).

To decide whether or not to reject the null hypothesis:

Step 5: Compare Z(obtained) to Z(critical). If Z(obtained) is *in* the critical region, reject the null hypothesis. If Z(obtained) is *not* in the critical region, fail to reject the null hypothesis.

Step 6: Interpret the decision to reject or fail to reject the null hypothesis in terms of the original question. For example: Our conclusion for the example problem used in Section 7.3 was "Treated alcoholics miss significantly fewer days of work than the community as a whole."

7.4 ONE-TAILED AND TWO-TAILED TESTS OF HYPOTHESIS

Introduction.   The five-step model for hypothesis testing is fairly rigid, and the researcher has little room for making choices. Follow the steps one by one as specified above, and you cannot make a mistake. Nonetheless, the researcher must still make two crucial choices. First, he or she must decide between a one-tailed and a two-tailed test. Second, the researcher must select an alpha level. In this section, we will discuss the former decision and, in Section 7.5, the latter.

The choice between a one- and two-tailed test is based on the researcher's expectations about the population from which the sample was selected. These expectations are reflected in the research hypothesis (H_1), which is contradictory to the null hypothesis and usually states what the researcher believes to be "the truth." In most situations, the researcher will want to support the research hypothesis by rejecting the null hypothesis.

The format for the research hypothesis may take either of two forms, depending on the relationship between what the null hypothesis states and what the researcher believes to be the truth. The null hypothesis states that the population has a specific characteristic. In the example that has served us throughout this chapter, the null stated, in symbols, "all treated alcoholics have the *same* absentee rate (7.2 days) as the community as a whole." The researcher might believe that the population of treated alcoholics actually has *less* absenteeism (their population mean is *lower than* the value stated in the null hypothesis), *more* absenteeism (their population mean is *greater than* the

Application 7.1

For a random sample of 152 felony cases tried in a local court, the average prison sentence was 27.3 months. Is this significantly different from the average prison term for felons nationally (28.7 months)? We will use the five-step model to organize the decision-making process.

Step 1. Making Assumptions and Meeting Test Requirements.

> Model: Random sampling
> > Level of measurement is interval-ratio
> > Sampling distribution is normal

From the information given (this is a large sample with $N > 100$ and length of sentence is an interval-ratio variable), we can conclude that the model assumptions are satisfied.

Step 2. Stating the Null Hypothesis (H_0).
The null hypothesis would say that the average sentence locally (for *all* felony cases) is equal to the national average. In symbols:

$$H_0: \mu = 28.7$$

The question does not specify a direction; it only asks if the local sentences are "different from" (not higher or lower than) national averages. This seems to suggest a two-tailed test (see Section 7.4):

$$(H_1: \mu \neq 28.7)$$

Step 3. Selecting the Sampling Distribution and Establishing the Critical Region.

> Sampling distribution = Z distribution
> > $\alpha = 0.05$
> > $Z(\text{critical}) = \pm 1.96$

Step 4. Computing the Test Statistic.
The necessary information for conducting a test of the null hypothesis is

$$\overline{X} = 27.3 \qquad \mu = 28.7$$
$$s = 3.7$$
$$N = 152$$

The test statistic, Z(obtained), would be:

$$Z(\text{obtained}) = \frac{\overline{X} - \mu}{s/\sqrt{N - 1}}$$

$$Z(\text{obtained}) = \frac{27.3 - 28.7}{3.7/\sqrt{152 - 1}}$$

$$Z(\text{obtained}) = \frac{-1.40}{3.7/\sqrt{151}}$$

$$Z(\text{obtained}) = \frac{-1.40}{0.30}$$

$$Z(\text{obtained}) = -4.67$$

Step 5. Making a Decision and Interpreting Test Results.
With alpha set at 0.05, the critical region would begin at Z(critical) $= \pm 1.96$. With an obtained Z score of -4.67, the null would be rejected. This means that the difference between the prison sentences of felons convicted in the local court and felons convicted nationally is statistically significant. The difference is so large that we may conclude that it did not occur by random chance. The decision to reject the null hypothesis has a 0.05 probability of being wrong.

value stated in the null hypothesis), or he or she might be unsure about the direction of the difference.

If the researcher is unsure about the direction, the research hypothesis would state only that the population mean is "not equal" to the value stated in the null hypothesis. The research hypothesis stated in Section 7.3 ($\mu \neq 7.2$) was in this format. This is called a **two-tailed test** of significance because it means that the researcher will be equally concerned with the possibilities that the true population value is greater than the value specified in the null hypothesis and that the true population value is less than the value specified in the null hypothesis.

In other situations, the researcher might be concerned only with differences in a specific direction. If the direction of the difference can be predicted, or if the researcher is concerned only with differences in one direction, he or she can use a **one-tailed test.** A one-tailed test may take one of two forms, depending on the researcher's expectations about the direction of the difference. If the researcher believes that the true population value is greater than the value specified in the null, the research hypothesis would reflect that belief. In our example, if we had predicted that treated alcoholics had *higher* absentee rates than the community (or, averaged *more* than 7.2 days of absenteeism), our research hypothesis would have been:

$$(H_1: \mu > 7.2)$$
where $>$ signifies "greater than"

If we predicted that treated alcoholics had lower absentee rates than the community (or, averaged *fewer* than 7.2 days of absenteeism), our research hypothesis would have been:

$$(H_1: \mu < 7.2)$$
where $<$ signifies "less than"

One-tailed tests are often appropriate when programs designed to solve a problem or improve a situation are being evaluated. If the program for treating alcoholics made them *less* reliable workers with *higher* absentee rates, for example, the program would be considered a failure, at least on that one criterion. In a situation like this, researchers may focus only on outcomes that indicate that the program is a success (that is, when treated alcoholics have lower rates) and conduct a one-tailed test with a research hypothesis in the form: $H_1: \mu < 7.2$. Or consider the evaluation of a program designed to reduce unemployment. The evaluators would be concerned only with outcomes that show a decrease in the unemployment rate. If the rate shows no change, or if unemployment increases, the program is a failure, and the researchers might consider both of these outcomes equally negative. Thus, the researchers could legitimately use a one-tailed test that stated that unemployment rates for graduates of the program would be less than ($<$) rates in the community.

In terms of the five-step model, the choice of a one-tailed or two-tailed test determines what we do with the critical region under the sampling distribution in step 3. As you recall, in a two-tailed test, we split the critical region equally into the upper and lower tails of the sampling distribution. In a one-tailed test, we place the entire critical area in one tail of the sampling distribution. If we believe that the population characteristic is greater than the value stated in the null hypothesis (if the H_1 includes the $>$ symbol), we place the entire critical region in the upper tail. If we believe that the characteristic is less than the value stated in the null hypothesis (if the H_1 includes the $<$ symbol), the entire critical region goes in the lower tail.

For example, in a two-tailed test with alpha equal to 0.05, the critical region begins at Z(critical) = ± 1.96. In a one-tailed test at the same alpha level, the Z(critical) is $+1.65$ if the upper tail is specified and -1.65 if the lower tail

TABLE 7.1   ONE- VERSUS TWO-TAILED TESTS, α = .05

If the Research Hypothesis Uses	The Test Is	And Concern Is with	Z(critical) =
≠	Two-tailed	Both tails	±1.96
>	One-tailed	Upper tail	+1.65
<	One-tailed	Lower tail	−1.65

FIGURE 7.5   ESTABLISHING THE CRITICAL REGION, ONE-TAILED TESTS VERSUS TWO-TAILED TESTS (alpha = 0.05)

A. The two-tailed test, Z (critical) = ±1.96

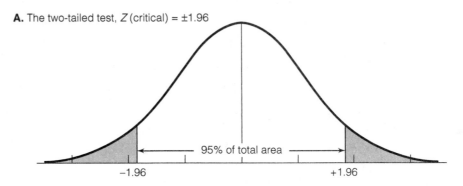

B. The one-tailed test for upper tail, Z (critical) = +1.65

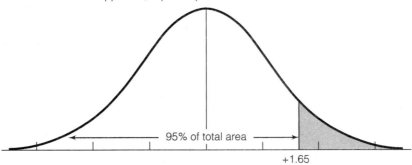

C. The one-tailed test for lower tail, Z (critical) = −1.65

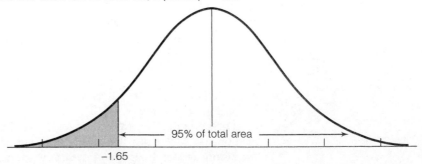

TABLE 7.2   FINDING CRITICAL Z SCORES FOR ONE-TAILED TESTS (single-sample means)

Alpha	Two-Tailed Value	One-Tailed Values Upper Tail	Lower Tail
.10	±1.65	+1.29	−1.29
.05	±1.96	+1.65	−1.65
.01	±2.58	+2.33	−2.33
.001	±3.29	+3.10	−3.10

is specified. Table 7.1 summarizes the procedures to follow in terms of the nature of the research hypothesis. Figure 7.5 graphically summarizes the difference in placing the critical region, and Table 7.2 gives the critical Z scores for the most common alpha levels for both one- and two-tailed tests.

Note that the critical Z values for one-tailed tests for all values of alpha are smaller in value and closer to the mean of the sampling distribution. Thus, a one-tailed test is more likely to reject the H_0 without changing the alpha level (assuming that we have specified the correct tail). One-tailed tests are a way of statistically both having and eating your cake and should be used whenever (1) the direction of the difference can be confidently predicted or (2) the researcher is concerned only with differences in one tail of the sampling distribution.

A Test of Hypothesis Using a One-Tailed Test.   After many years of work, a sociologist has noted that sociology majors seem more sophisticated, charming, and cosmopolitan than the rest of the student body. A "Sophistication Scale" test has been administered to the entire student body and to a random sample of 100 sociology majors. Here are the results:

Student Body	Sociology Majors
$\mu = 17.3$	$\overline{X} = 19.2$
$\sigma = 7.4$	$N = 100$

We will use the five-step model to test the H_0 of no difference between sociology majors and the general student body.

Step 1. Making Assumptions and Meeting Test Requirements. Because we are using a mean to summarize the sample outcome, we must assume that the Sophistication Scale generates interval-ratio-level data. With a sample size of 100, the Central Limit Theorem applies, and we can assume that the sampling distribution is normal in shape.

> Model: Random sampling
> Level of measurement is interval-ratio
> Sampling distribution is normal

Step 2. Stating the Null Hypothesis (H_0). The null hypothesis states that there is no difference between sociology majors and the general student body. The research hypothesis (H_1) will also be stated at this point. The researcher has predicted a direction for the difference ("Sociology majors are *more* sophisticated"), so a one-tailed test is justified. The two hypotheses may be stated as:

$$H_0: \mu = 17.3$$
$$(H_1: \mu > 17.3)$$

Step 3. Selecting the Sampling Distribution and Establishing the Critical Region. We will use the standardized normal distribution (Appendix A) to find areas under the sampling distribution. If alpha is set at 0.05, the critical region will begin at the Z score $+1.65$. That is, the researcher has predicted that sociology majors are *more* sophisticated and that this sample comes from a population that has a mean *greater than* 17.3, so he will be concerned only with sample outcomes in the upper tail of the sampling distribution. If sociology majors are *the same as* other students in terms of sophistication (if the H_0 is true), or if they are *less* sophisticated (and come from a population with a mean less than 17.3), the theory is disproved. These decisions may be summarized as:

$$\text{Sampling distribution} = Z \text{ distribution}$$
$$\alpha = 0.05$$
$$Z(\text{critical}) = +1.65$$

Step 4. Computing the Test Statistic.

$$Z(\text{obtained}) = \frac{\overline{X} - \mu}{\sigma/\sqrt{N}}$$

$$Z(\text{obtained}) = \frac{19.2 - 17.3}{7.4/\sqrt{100}}$$

$$Z(\text{obtained}) = +2.57$$

Step 5. Making a Decision and Interpreting Test Results. Comparing the $Z(\text{obtained})$ with the $Z(\text{critical})$,

$$Z(\text{critical}) = +1.65$$
$$Z(\text{obtained}) = +2.57$$

we see that the test statistic falls into the critical region, as Figure 7.6 shows. We will reject the null hypothesis because, if the H_0 were true, a difference of this

FIGURE 7.6   *Z* (OBTAINED) VERSUS *Z* (CRITICAL) (alpha = 0.05, one-tailed test)

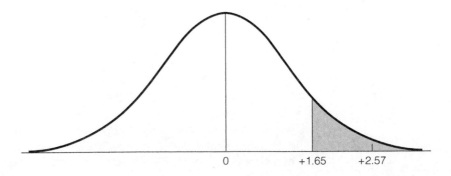

size would be very unlikely. There is a significant difference between sociology majors and the general student body in terms of sophistication. Because we have rejected the null hypothesis, the research hypothesis (sociology majors are more sophisticated) is supported. *(For practice in dealing with tests of significance for means that may call for one-tailed tests, see problems 7.2, 7.3, 7.6, 7.8, and 7.17.)*

7.5 SELECTING AN ALPHA LEVEL

In addition to deciding between one-tailed and two-tailed tests, the researcher must also select an alpha level. We have seen that the alpha level plays a crucial role in hypothesis testing. When we assign a value to alpha, we define what we mean by an "unlikely" sample outcome. If the probability of the observed sample outcome is lower than the alpha level (if the test statistic falls into the critical region), then we reject the null hypothesis as untrue. Thus, the alpha level will have important consequences for our decision in step 5.

How can we make reasonable decisions with respect to the value of alpha? Recall that, in addition to defining what *unlikely* will mean, the alpha level is also the probability that the decision to reject the null hypothesis, if the test statistic falls into the critical region, will be incorrect. In hypothesis testing, the error of incorrectly rejecting the null hypothesis or rejecting a null hypothesis that is actually true is called **Type I error, or alpha error.** To minimize this type of error, use very small values for alpha.

To elaborate: When an alpha level is specified, the sampling distribution is divided into two sets of possible sample outcomes. The critical region includes all unlikely or rare sample outcomes. Outcomes in this region will cause us to reject the null hypothesis. The remainder of the area consists of all sample outcomes that are not rare. The lower the level of alpha, the smaller the critical region and the greater the distance between the mean of the sampling distribution and the beginnings of the critical region. Compare, for the sake of illustration, the following alpha levels and values for Z(critical) for two-tailed tests. As you may recall, Table 6.2 also presented this information.

If Alpha Equals	The Two-Tailed Critical Region Will Begin at Z(critical) Equal to
0.10	±1.65
0.05	±1.96
0.01	±2.58
0.001	±3.29

As alpha goes down, the critical region becomes smaller and moves farther away from the mean of the sampling distribution. The lower the alpha level, the harder it will be to reject the null hypothesis and, because a Type I error can occur only if our decision in step 5 is to reject, the lower the probability of Type I error. To minimize the probability of rejecting a null hypothesis that is in fact true, use very low alpha levels.

However, there is a complication. As the critical region decreases in size (as alpha levels decrease), the noncritical region—the area between the two Z(critical) scores in a two-tailed test—must become larger. All other things being equal, the lower the alpha level, the less likely that the sample outcome will fall into

TABLE 7.3 DECISION MAKING AND THE NULL HYPOTHESIS

	Decision	
The H_0 Is Actually:	Reject	Fail to Reject
True	Type I or α error	OK
False	OK	Type II or β error

the critical region. This raises the possibility of a second type of incorrect decision, called **Type II error,** or **beta error:** failing to reject a null that is, in fact, false. The probability of Type I error decreases as alpha level decreases, but the probability of Type II error increases. Thus, the two types of error are inversely related, and it is not possible to minimize both in the same test. As the probability of one type of error decreases, the other increases, and vice versa.

It may be helpful to clarify the relationships between decision making and errors in table format. Table 7.3 lists the two decisions we can make in step 5 of the five-step model: We either reject or fail to reject the null hypothesis. The other dimension of Table 7.3 lists the two possible conditions of the null hypothesis: It is either actually true or actually false. The table combines these possibilities into a total of four possible combinations, two of which are desirable ("OK") and two of which indicate that an error has been made.

The two desirable outcomes are rejecting null hypotheses that are actually false and failing to reject null hypotheses that are actually true. The goal of any scientific investigation is to verify true statements and reject false statements. The remaining two combinations are errors or situations that, naturally, we want to avoid. If we reject a null hypothesis that is in fact true, we are saying that a true statement is false. Similarly, if we fail to reject a null hypothesis that is in fact false, we are saying that a false statement is true. Obviously, we would prefer to always wind up in one of the boxes labeled "OK" in Table 7.3—to always reject false statements and accept the truth when we find it. Remember, however, that hypothesis testing always carries an element of risk and that we cannot minimize the chances of both Type I and Type II error simultaneously.

What all of this means, finally, is that you must think of selecting an alpha level as an attempt to balance the two types of error. Higher alpha levels will minimize the probability of Type II error (saying that false statements are true), and lower alpha levels will minimize the probability of Type I error (saying that true statements are false). Normally, in social science research, we will want to minimize Type I error and will use lower alpha levels (.05, .01, .001 or lower). The 0.05 level in particular has emerged as a generally recognized indicator of a significant result. However, the widespread use of the 0.05 level is simply a convention, and there is no reason that alpha cannot be set at virtually any sensible level (such as 0.04, 0.027, 0.083). The researcher has the responsibility of selecting the alpha level that seems most reasonable in terms of the goals of the research project.

7.6 THE STUDENT'S t DISTRIBUTION

Introduction. To this point, we have considered only one type of hypothesis test. Specifically, we have focused on situations involving single-sample means where the value of the population standard deviation (σ) was known. In most research situations, the value of σ will not be known. However, a value for σ is necessary to compute the standard error of the mean $(\sigma/\sqrt{N})$, convert our sample outcome into a Z score, and place the Z(obtained) on the sampling distribution (step 4). How can we reasonably obtain a value for the population standard deviation?

It might seem sensible to estimate σ with s, the sample standard deviation. As we noted in Chapter 6, s is a biased estimator of σ, but the degree of bias decreases as sample size increases. For large samples (that is, samples with 100 or more cases), the sample standard deviation yields an adequate estimate of σ. Thus, for large samples, we simply substitute s for σ in the formula for Z(obtained) in step 4 and continue to use the standard normal curve to find areas under the sampling distribution.[1]

For smaller samples, however, when σ is unknown, we must use an alternative distribution called the **Student's t distribution** to find areas under the sampling distribution and establish the critical region. The shape of the t distribution varies as a function of sample size. Figure 7.7 depicts the relative shapes of the t and Z distributions. For small samples, the t distribution is much flatter than the Z distribution, but, as sample size increases, the t distribution comes to resemble the Z distribution more and more until the two are essentially identical when sample size is greater than 120. As N increases, the sample standard deviation (s) becomes a more and more adequate estimator of the population standard deviation (σ), and the t distribution becomes more and more like the Z distribution.

The Distribution of Student's t: Using Appendix B.   The t distribution is summarized in Appendix B. The t table differs from the Z table in several ways. First, note the column at the left of the table labeled df for "degrees of freedom."[2] As mentioned above, the exact shape of the t distribution and thus

FIGURE 7.7   THE t DISTRIBUTION AND THE Z DISTRIBUTION

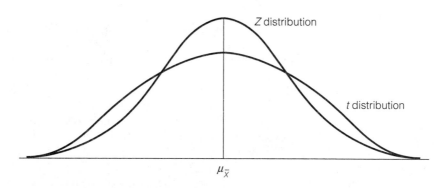

[1] Even though its effect will be minor and will decrease with sample size, we will always correct for the bias in s by using the term $N - 1$ rather than N in the computation for the standard deviation of the sampling distribution when σ is unknown.

[2] Degrees of freedom refer to the number of values in a distribution that are free to vary. For a sample mean, a distribution has $N - 1$ degrees of freedom. This means that, for a specific value of the mean, $N - 1$ scores are free to vary. For example, if the mean is 3 and $N = 5$, the distribution of five scores would have $N - 1$, or four, degrees of freedom. When the value of four of the scores is known, the value of the fifth is fixed. If four scores are 1, 2, 3, and 4, the fifth must be 5 and no other value.

the exact location of the critical region for any alpha level varies as a function of sample size. We must compute degrees of freedom, which are equal to $N - 1$ in the case of a single-sample mean, before we can locate the critical region for any alpha. Second, alpha levels are arrayed across the top of Appendix B in two rows, one row for the one-tailed tests and one for two-tailed tests. To use the table, begin by locating the selected alpha level in the appropriate row.

The third difference is that the entries in the table are the actual scores, called **t(critical),** that mark the beginnings of the critical regions and not areas under the sampling distribution. To illustrate the use of this table with single-sample means, find the critical region for alpha equal to 0.05, two-tailed test, for $N = 30$. The degrees of freedom will be $N - 1$, or 29; reading down the proper column, you should find a value of 2.045. Thus, the critical region for this test will begin at t(critical) $= \pm 2.045$.

Take a moment to notice some additional features of the t distribution. First, note that the t(critical) we found above is larger in value than the comparable Z(critical), which for a two-tailed test at an alpha of 0.05 would be ± 1.96. This relationship reflects the fact that the t distribution is flatter than the Z distribution (see Figure 7.6). When you use the t distribution, the critical regions will begin farther away from the mean of the sampling distribution and, therefore, the null hypothesis will be harder to reject. Furthermore, the smaller the sample size (the lower the degrees of freedom), the larger the value of t(obtained) necessary for a rejection of the H_0.

Second, scan the column for an alpha of 0.05, two-tailed test. Note that, for one degree of freedom, the t(critical) is ± 12.706 and that the value of t(critical) decreases as degrees of freedom increase. For degrees of freedom greater than 120, the value of t(critical) is the same as the comparable value of Z(critical), or ± 1.96. As sample size increases, the t distribution comes to resemble the Z distribution more and more until, with sample sizes greater than 120, the two distributions are essentially identical.[3]

A Test of Hypothesis Using Student's t. To demonstrate the uses of the t distribution in more detail, we will work through an example problem. Note that, in terms of the five-step model, the changes required by using t scores occur mostly in steps 3 and 4. In step 3, the sampling distribution will be the t distribution, and degrees of freedom (df) must be computed before locating the critical region as marked by t(critical). In step 4, we will use a slightly different formula for computing the test statistic, **t(obtained).** As compared with the formula for Z(obtained), s will replace σ and $N - 1$ will replace N.

Specifically,

FORMULA 7.2
$$t(\text{obtained}) = \frac{\bar{X} - \mu}{s/\sqrt{N - 1}}$$

[3] Appendix B abbreviates the t distribution by presenting a limited number of critical t scores for degrees of freedom between 31 and 120. If the degrees of freedom for a specific problem equal 77 and alpha equals 0.05, two-tailed, we have a choice between a t(critical) of ± 2.000 (df = 60) and a t(critical) of ± 1.980 (df = 120). In situations such as these, take the larger table value as t(critical). This will make rejection of the H_0 less likely and is therefore the more conservative course of action.

A researcher wonders if commuter students are different from the general student body in terms of academic achievement. She has gathered a random sample of 30 commuter students and has learned from the registrar that the mean grade-point average for all students is 2.50 ($\mu = 2.50$), but the standard deviation of the population (σ) is not available. Sample data are reported below. Does the sample come from a population that has a mean of 2.50?

Student Body	Commuter Students
$\mu = 2.50 \, (= \mu_{\bar{X}})$	$\bar{X} = 2.78$
$\sigma = ?$	$s = 1.23$
	$N = 30$

Step 1. Making Assumptions and Meeting Test Requirements.

Model: Random sampling
Level of measurement is interval-ratio
Sampling distribution is normal

Step 2. Stating the Null Hypothesis.

$$H_0: \mu = 2.50$$
$$(H_1: \mu \neq 2.50)$$

You can see from the research hypothesis that the researcher has not predicted a direction for the difference. This will be a two-tailed test.

Step 3. Selecting the Sampling Distribution and Establishing the Critical Region. Because σ is unknown and sample size is small, we will use the t distribution to find the critical region. Alpha will be set at 0.01.

Sampling distribution = t distribution
$\alpha = 0.01$, two-tailed test
df = $(N - 1) = 29$
t(critical) = ± 2.756

Step 4. Computing the Test Statistic.

$$t(\text{obtained}) = \frac{\bar{X} - \mu}{s/\sqrt{N - 1}}$$

$$t(\text{obtained}) = \frac{2.78 - 2.50}{1.23\sqrt{29}}$$

$$t(\text{obtained}) = \frac{0.28}{0.23}$$

$$t(\text{obtained}) = +1.22$$

FIGURE 7.8   SAMPLING DISTRIBUTION SHOWING t (OBTAINED) VERSUS t (CRITICAL) (α = 0.05, two-tailed test, df = 29)

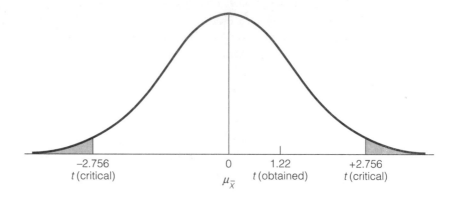

−2.756	+2.756
t (critical)	t (critical)

0   1.22
$\mu_{\overline{X}}$ t (obtained)

Step 5. **Making a Decision and Interpreting Test Results**. The test statistic does not fall into the critical region. Therefore, the researcher fails to reject the H_0. The difference between the sample mean (2.78) and the population mean (2.50) is no greater than what we would expect if only random chance were operating. Figure 7.8 displays the test statistic and critical regions.

To summarize, when testing single-sample means, we must decide which theoretical distribution we will use to establish the critical region. The choice is

---

STEP BY STEP	**Testing the Significance of the Difference Between a Sample Mean and a Population Mean (Small Samples): Computing t(obtained) and Interpreting Results**

When sample size is small, we use Formula 7.2 to convert the sample mean into a t score called t(obtained). Then, in step 5, we compare t(obtained) with t(critical). If the test statistic is in the critical region, we reject the null hypothesis Otherwise, we fail to reject the null hypothesis.

To solve Formula 7.2: $t(\text{obtained}) = \dfrac{\overline{X} - \mu}{s/\sqrt{N-1}}$

Step 1: Find the square root of $N - 1$.

Step 2: Divide the quantity you found in step 1 into the sample standard deviation (s).

Step 3: Subtract the population mean from the sample mean.

Step 4: Divide the quantity you found in step 3 by the quantity you found in step 2. This is t(obtained).

To decide whether or not to reject the null hypothesis:

Step 5: Compare t(obtained) to t(critical). If t(obtained) is *in* the critical region, reject the null hypothesis. If t(obtained) is *not* in the critical region, fail to reject the null hypothesis.

Step 6: Interpret the decision to reject or fail to reject the null hypothesis in terms of the original question. For example, our conclusion for the example problem used in Section 7.6 was "There is no significant difference between the average GPAs of commuter students and the student body as a whole."

straightforward. If the population standard deviation (σ) is known or sample size is large, we will use the Z distribution (summarized in Appendix A). If σ is unknown and the sample is small, we will use the t distribution (summarized in Appendix B). *(For practice in using the* t *distribution in a test of hypothesis, see problems 7.8 to 7.10 and 7.17.)*

7.7 TESTS OF HYPOTHESES FOR SINGLE-SAMPLE PROPORTIONS (LARGE SAMPLES)

Introduction.   In many cases, the variables in which we are interested will not be measured in a way that justifies the assumption of interval-ratio level of measurement. One alternative in this situation is to use a sample proportion (P_s) rather than a sample mean as the test statistic. As we shall see below, the overall procedures for testing single-sample proportions are the same as those for testing means. The central question is still "Does the population from which the sample was drawn have a certain characteristic?" We still conduct the test based on the assumption that the null hypothesis is true, and we still evaluate the probability of the obtained sample outcome against a sampling distribution of all possible sample outcomes. Our decision at the end of the test is also the same. If the obtained test statistic falls into the critical region (is unlikely, given the assumption that the H_0 is true), we reject the H_0.

Having stressed the continuity in procedures and logic, I must hastily point out the important differences as well. These differences are best related in terms of the five-step model for hypothesis testing. In step 1, when working with sample proportions, we assume that the variable is measured at the nominal level of measurement. In step 2, the symbols used to state the null hypothesis are different even though the null is still a statement of "no difference."

In step 3, we will use only the standardized normal curve (the Z distribution) to find areas under the sampling distribution and locate the critical region. This will be appropriate as long as sample size is large. We will not consider small-sample tests of hypothesis for proportions in this text.

In step 4, computing the test statistic, the format of the formula remains the same. That is, the test statistic, Z(obtained), equals the sample statistic minus the mean of the sampling distribution, divided by the standard deviation of the sampling distribution. However, the symbols will change because we are basing the tests on sample proportions. The formula can be stated as:

FORMULA 7.3
$$Z(\text{obtained}) = \frac{P_s - P_u}{\sqrt{P_u(1 - P_u)/N}}$$

Step 5 is exactly the same as before. If the test statistic, Z(obtained), falls into the critical region, as marked by Z(critical), reject the H_0.

A Test of Hypothesis Using Sample Proportions.   An example should clarify these procedures. A random sample of 122 households in a low-income neighborhood revealed that 53 (or a proportion of 0.43) of the households were headed by females. In the city as a whole, the proportion of female-headed households is 0.39. Are households in the lower-income neighborhood significantly different from households in the city as a whole in terms of this characteristic?

Step 1. Making Assumptions and Meeting Test Requirements.

> Model: Random sampling
> Level of measurement is nominal
> Sampling distribution is normal in shape

Step 2. Stating the Null Hypothesis. The research question, as stated above, asks only if the sample proportion is different from the population proportion. Because we have not predicted any direction for the difference, we will use a two-tailed test.

$$H_0: P_u = 0.39$$
$$(H_1: P_u \neq 0.39)$$

Step 3. Selecting the Sampling Distribution and Establishing the Critical Region.

> Sampling distribution = Z distribution
> $\alpha = 0.10$, two-tailed test
> $Z(\text{critical}) = \pm 1.65$

Step 4. Computing the Test Statistic.

$$Z(\text{obtained}) = \frac{P_s - P_u}{\sqrt{P_u(1 - P_u)/N}}$$

$$Z(\text{obtained}) = \frac{0.43 - 0.39}{\sqrt{0.39(0.61)/122}}$$

$$Z(\text{obtained}) = +0.91$$

Step 5. Making a Decision and Interpreting Test Results. The test statistic, Z(obtained), does not fall into the critical region. Therefore, we fail to reject the H_0. There is no statistically significant difference between the low-income community and the city as a whole in terms of the proportion of households headed by females. Figure 7.9 displays the sampling distribution, the critical region, and the Z(obtained). *(For practice in tests of significance using sample proportions, see problems 7.1c, 7.11 to 7.14, 7.15a to d, and 7.16.)*

FIGURE 7.9   SAMPLING DISTRIBUTION SHOWING Z (OBTAINED) VERSUS Z (CRITICAL) ($\alpha = 0.10$, two-tailed test)

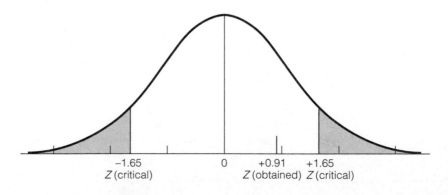

STEP BY STEP	**Testing the Significance of the Difference Between a Sample Proportion and a Population Proportion (Large Samples): Computing Z(Obtained) and Interpreting Results**

In step 4 of the five-step model, we use Formula 7.3 to convert the sample proportion into a Z score called Z(obtained). Then, in step 5, we compare Z(obtained) with Z(critical). If the test statistic is in the critical region, we reject the null hypothesis. Otherwise, we fail to reject the null hypothesis.

To solve Formula 7.3: $Z(\text{obtained}) = \dfrac{P_s - P_u}{\sqrt{P_u(1 - P_u)/N}}$

Step 1: Start with the denominator and substitute the proper value for P_u. The other term in the numerator will be equal to $1 - P_u$. Multiply these terms together.

Step 2: Divide the quantity you found in step 1 by N.

Step 3: Take the square root of the value you found in step 2.

Step 4: Subtract the value of P_u from P_s.

Step 5: Divide the quantity you found in step 4 by the quantity you found in step 3. The result is the value of Z(obtained).

To decide whether or not to reject the null hypothesis:

Step 6: Compare Z(obtained) to Z(critical). If Z(obtained) is *in* the critical region, reject the null hypothesis. If Z(obtained) is *not* in the critical region, fail to reject the null hypothesis.

Step 7: Interpret the decision to reject or fail to reject the null hypothesis in terms of the original question. For example, our conclusion for the example problem used in the Section 7.7 was "There is no significant difference in the proportion of households headed by females between the low-income community and the city as a whole."

Application 7.2

In a random sample ($N = 103$) drawn from the most affluent neighborhood in a community, 76% of the respondents reported that they had voted Republican in the most recent presidential election. For the community as a whole, 66% of the electorate voted Republican. Was the affluent neighborhood significantly more likely to have voted Republican?

Step 1. Making Assumptions and Meeting Test Requirements.

> Model: Random sampling
> Level of measurement is nominal
> Sampling distribution is normal

This is a large sample, so we may assume a normal sampling distribution. The variable, percent Republican, is only nominal in level of measurement.

Step 2. Stating the Null Hypothesis (H_0). The null hypothesis says that the affluent neighborhood is not different from the community as a whole.

$$H_0: P_u = 0.66$$

The original question ("Was the affluent neighborhood *more* likely to vote Republican?") suggests a one-tailed research hypothesis:

$$(H_1: P_u > .66)$$

Step 3. Selecting the Sampling Distribution and Establishing the Critical Region.

Sampling distribution = Z distribution
$\alpha = 0.05$
$Z(\text{critical}) = +1.65$

(continued next page)

Application 7.2 *(continued)*

The research hypothesis says that we will be concerned only with outcomes in which the neighborhood is more likely to vote Republican or with sample outcomes in the upper tail of the sampling distribution.

Step 4. Computing the Test Statistic. The information necessary for a test of the null hypothesis, expressed in the form of proportions, is:

Neighborhood	Community
$P_s = 0.76$	$P_u = 0.66$
$N = 103$	

The test statistic, Z(obtained), would be:

$$Z(\text{obtained}) = \frac{P_s - P_u}{\sqrt{P_u(1 - P_u)/N}}$$

$$Z(\text{obtained}) = \frac{0.76 - 0.66}{\sqrt{(0.66)(1 - 0.66)/103}}$$

$$Z(\text{obtained}) = \frac{0.10}{\sqrt{(0.224)/103}}$$

$$Z(\text{obtained}) = \frac{0.100}{0.047}$$

$$Z(\text{obtained}) = 2.13$$

Step 5. Making a Decision and Interpreting Test Results. With alpha set at 0.05, one-tailed, the critical region would begin at Z(critical) $= +1.65$. With an obtained Z score of 2.13, the null hypothesis is rejected. The difference between the affluent neighborhood and the community as a whole is statistically significant and in the predicted direction. Residents of the affluent neighborhood were significantly more likely to have voted Republican in the last presidential election.

SUMMARY

1. This chapter presented all the basic concepts and techniques for testing hypotheses. We saw how to test the null hypothesis of "no difference" for single-sample means and proportions. In both cases, the central question is whether the population represented by the sample has a certain characteristic.

2. All tests of a hypothesis involve finding the probability of the observed sample outcome, given that the null hypothesis is true. If the outcome has a low probability, we reject the null hypothesis. In the usual research situation, we will want to reject the null hypothesis and thereby support the research hypothesis.

3. The five-step model will be our framework for decision making throughout the hypothesis-testing chapters. We will always (1) make assumptions, (2) state the null hypothesis, (3) select a sampling distribution, specify alpha, and find the critical region, (4) compute a test statistic, and (5) make a decision and interpret the results of the test. What we do during each step, however, will vary, depending on the specific test being conducted.

4. If we can predict a direction for the difference in stating the research hypothesis, a one-tailed test

is called for. If no direction can be predicted, a two-tailed test is appropriate. There are two kinds of errors in hypothesis testing. Type I, or alpha, error is rejecting a true null; Type II, or beta, error is failing to reject a false null. The probabilities of committing these two types of error are inversely related and cannot be simultaneously minimized in the same test. By selecting an alpha level, we try to balance the probability of these two kinds of error.

5. When testing sample means, use the t distribution to find the critical region when the population standard deviation is unknown and sample size is small.

6. Sample proportions can also be tested for significance. Tests are conducted using the five-step model. Compared to the test for the sample mean, the major differences lie in the level-of-measurement assumption (step 1), the statement of the null (step 2), and the computation of the test statistic (step 4).

7. If you are still confused about the uses of inferential statistics described in this chapter, don't be alarmed or discouraged. A sizable volume of rather complex material has been presented and

only rarely will a beginning student fully comprehend the unique logic of hypothesis testing on the first exposure. After all, it is not every day

that you learn how to test a s believe (the null hypothesis tion that doesn't exist (the s

SUMMARY OF FORMULAS

Single-sample means, large samples

7.1

$$Z(\text{obtained}) = \frac{\bar{X} - \mu}{\sigma/\sqrt{N}}$$

Single-sample means when samples are small and population standard deviation is unknown

7.2

$$t(\text{obtained}) = \frac{\bar{X} - \mu}{s/\sqrt{N-1}}$$

Single-sample proportions, large samples

7.3

$$Z(\text{obtained}) = \frac{P_s - P_u}{\sqrt{P_u(1 - P_u)/N}}$$

GLOSSARY

Alpha level (α). The proportion of area under the sampling distribution that contains unlikely sample outcomes, given that the null hypothesis is true. Also, the probability of Type I error.

Critical region (region of rejection). The area under the sampling distribution that, in advance of the test itself, is defined as including unlikely sample outcomes, given that the null hypothesis is true.

Five-step model. A step-by-step guideline for conducting tests of hypotheses. A framework that organizes decisions and computations for all tests of significance.

Hypothesis testing. Statistical tests that estimate the probability of sample outcomes if assumptions about the population (the null hypothesis) are true.

Null hypothesis (H_0). A statement of "no difference." For single-sample tests of significance, the population from which the sample was drawn is assumed to have a certain characteristic or value.

One-tailed test. A type of hypothesis test used when (1) the direction of the difference can be predicted or (2) concern focuses on outcomes in only one tail of the sampling distribution.

Research hypothesis (H_1). A statement that contradicts the null hypothesis. For single-sample tests of significance, the research hypothesis says

that the population from which the sample was drawn does not have a certain characteristic or value.

Significance testing. See Hypothesis testing.

Student's t distribution. A distribution used to find the critical region for tests of sample means when σ is unknown and sample size is small.

t(critical). The t score that marks the beginning of the critical region of a t distribution.

t(obtained). The test statistic computed in step 4 of the five-step model. The sample outcome expressed as a t score.

Test statistic. The value computed in step 4 of the five-step model that converts the sample outcome into either a t score or a Z score.

Two-tailed test. A type of hypothesis test used when (1) the direction of the difference cannot be predicted or (2) concern focuses on outcomes in both tails of the sampling distribution.

Type I error (alpha error). The probability of rejecting a null hypothesis that is, in fact, true.

Type II error (beta error). The probability of failing to reject a null hypothesis that is, in fact, false.

Z(critical). The Z score that marks the beginnings of the critical region on a Z distribution.

Z(obtained). The test statistic computed in step 4 of the five-step model. The sample outcomes expressed as a Z score.

MEDIA RESOURCES

The Wadsworth Sociology Resource Center: Virtual Society at
http://www.thomsonedu.com/sociology

Visit the companion website for the seventh edition of *Statistics: A Tool for Social Research* to access a wide range of student resources. Begin by clicking on the Student Resources section of the book's website to access the following study tools:

- Basic math review
- Flash cards

- Additional chapter problems
- Internet links
- Table of random numbers
- MicroCase and SPSS examples and exercises
- Hypothesis testing for variables measured at the ordinal level

PROBLEMS

7.1 a. For each situation, find Z(critical).

Alpha	Form	Z(Critical)
.05	One-tailed	
.10	Two-tailed	
.06	Two-tailed	
.01	One-tailed	
.02	Two-tailed	

b. For each situation, find the critical t score.

Alpha	Form	N	t(Critical)
.10	Two-tailed	31	
.02	Two-tailed	24	
.01	Two-tailed	121	
.01	One-tailed	31	
.05	One-tailed	61	

c. Compute the appropriate test statistic (Z or t) for each situation:

1. $\mu = 2.40$ $\bar{X} = 2.20$
    $\alpha = 0.75$ $N = 200$

2. $\mu = 17.1$ $\bar{X} = 16.8$
            $s = 0.9$
            $N = 45$

3. $\mu = 10.2$ $\bar{X} = 9.4$
            $s = 1.7$
            $N = 150$

4. $P_u = .57$ $P_s = 0.60$
            $N = 117$

5. $P_u = 0.32$ $P_s = 0.30$
            $N = 322$

7.2 SOC **a.** The student body at St. Algebra College attends an average of 3.3 parties per month. A random sample of 117 sociology majors averages 3.8 parties per month with a standard deviation of 0.53. Are sociology majors significantly different from the student body as a whole? (*HINT: The wording of the research question suggests a two-tailed test. This means that the alternative or research hypothesis in step 2 will be stated as* H_1: $\mu \neq 3.3$ *and that the critical region will be split between the upper and lower tails of the sampling distribution. See Table 7.2 for values of Z(critical) for various alpha levels.*)

b. What if the research question were changed to "Do sociology majors attend a significantly *greater* number of parties"? How would the test conducted in 7.2a change? (*HINT: This wording implies a one-tailed test of significance. How would the research hypothesis change? For the alpha you used in problem 7.2a, what would the value of Z(critical) be?*)

7.3 SW **a.** Nationally, social workers average 10.2 years of experience. In a random sample, 203 social workers in greater metropolitan Shinbone average only 8.7 years with a standard deviation of 0.52. Are social workers in Shinbone significantly less experienced? (*Note the wording of the research hypothesis. This situation may justify one-tailed tests of significance. If you chose a one-tailed test, what form would the research hypothesis take, and where would the critical region begin?*)

b. The same sample of social workers reports an average annual salary of $35,782 with a standard deviation of $622. Is this figure significantly higher than the national average of $34,509? *(The wording of the research hypotheses suggests a one-tailed test. What form would the research hypothesis take, and where would the critical region begin?)*

7.4 SOC Nationally, the average score on the college entrance exams (verbal test) is 453 with a standard deviation of 95. A random sample of 152 freshmen entering St. Algebra College shows a mean score of 502. Is there a significant difference?

7.5 SOC A random sample of 423 Chinese Americans has finished an average of 12.7 years of formal education with a standard deviation of 1.7. Is this significantly different from the national average of 12.2 years?

7.6 SOC A sample of 105 workers in the Overkill Division of the Machismo Toy Factory earns an average of $24,375 per year. The average salary for all workers is $24,230 with a standard deviation of $523. Are workers in the Overkill Division overpaid? Conduct both one- and two-tailed tests.

7.7 GER **a.** Nationally, the population as a whole watches 6.2 hours of TV per day. A random sample of 1,017 senior citizens reports watching an average of 5.9 hours per day with a standard deviation of 0.7. Is the difference significant?
b. The same sample of senior citizens reports that they belong to an average of 2.1 voluntary organizations and clubs with a standard deviation of 0.5. Nationally, the average is 1.7. Is the difference significant?

7.8 SOC A school system has assigned several hundred "chronic and severe underachievers" to an alternative educational experience. To assess the program, a random sample of 35 has been selected for comparison with all students in the system.
a. In terms of GPA, did the program work?

Systemwide GPA	Program GPA
$\mu = 2.47$	$\bar{X} = 2.55$
	$s = 0.70$
	$N = 35$

b. In terms of absenteeism (number of days missed per year), what can you say about the success of the program?

Systemwide	Program
$\mu = 6.13$	$\bar{X} = 4.78$
	$s = 1.11$
	$N = 35$

c. In terms of standardized test scores in math and reading, was the program a success?

Math Test: Systemwide	Math Test: Program
$\mu = 103$	$\bar{X} = 106$
	$s = 2.0$
	$N = 35$

Reading Test: Systemwide	Reading Test: Program
$\mu = 110$	$\bar{X} = 113$
	$s = 2.0$
	$N = 35$

(HINT: Note the wording of the research questions. Is a one-tailed test justified? Is the program a success if the students in the program are no different from students systemwide? What if the program students were performing at lower levels? If a one-tailed test is used, what form should the research hypothesis take? Where will the critical region begin?)

7.9 SOC A random sample of 26 local sociology graduates scored an average of 458 on the GRE advanced sociology test with a standard deviation of 20. Is this significantly different from the national average ($\mu = 440$)?

7.10 PA Nationally, the per capita property tax is $130. A random sample of 36 southeastern cities average $98 with a standard deviation of $5. Is the difference significant? Summarize your conclusions in a sentence or two.

7.11 GER/CJ A survey shows that 10% of the population is victimized by property crime each year. A random sample of 527 older citizens (65 years or more of age) shows a victimization rate of 14%. Are older people more likely to be victimized? Conduct both one- and two-tailed tests of significance.

7.12 [CJ] A random sample of 113 convicted rapists in a state prison system completed a program designed to change their attitudes toward women, sex, and violence before being released on parole. Fifty-eight eventually became repeat sex offenders. Is this recidivism rate significantly different from the rate for all offenders (57%) in that state? Summarize your conclusions in a sentence or two. *(HINT: You must use the information given in the problem to compute a sample proportion. Remember to convert the population percentage to a proportion.)*

7.13 [PS] In a recent statewide election, 55% of the voters rejected a proposal to institute a state lottery. In a random sample of 150 urban precincts, 49% of the voters rejected the proposal. Is the difference significant? Summarize your conclusions in a sentence or two.

7.14 [CJ] Statewide, the police clear by arrest 35% of the robberies and 42% of the aggravated assaults reported to them. A researcher takes a random sample of all the robberies ($N = 207$) and aggravated assaults ($N = 178$) reported to a metropolitan police department in one year and finds that 83 of the robberies and 80 of the assaults were cleared by arrest. Are the local arrest rates significantly different from the statewide rate? Write a sentence or two interpreting your decision.

7.15 [SOC/SW] A researcher has compiled a file of information on a random sample of 317 families in a city that has chronic, long-term patterns of child abuse. Some characteristics of the sample along with values for the city as a whole are reported here. For each trait, test the null hypothesis of "no difference" and summarize your findings.

a. Mothers' educational level (proportion completing high school):

City	Sample
$P_u = 0.63$	$P_s = 0.61$

b. Family size (proportion of families with four or more children):

City	Sample
$P_u = 0.21$	$P_s = 0.26$

c. Mothers' work status (proportion of mothers with jobs outside the home):

City	Sample
$P_u = 0.51$	$P_s = 0.27$

d. Relations with kin (proportion of families that have contact with kin at least once a week):

City	Sample
$P_u = 0.82$	$P_s = 0.43$

e. Fathers' educational achievement (average years of formal schooling):

City	Sample
$\mu = 12.3$	$\bar{X} = 12.5$
	$s = 1.7$

f. Fathers' occupational stability (average years in present job):

City	Sample
$\mu = 5.2$	$\bar{X} = 3.7$
	$s = 0.5$

7.16 [SW] You are the head of an agency seeking funding for a program to reduce unemployment among teenage males. Nationally, the unemployment rate for this group is 18%. A random sample of 323 teenage males in your area reveals an unemployment rate of 21.7%. Is the difference significant? Can you demonstrate a need for the program? Should you use a one-tailed test in this situation? Why? Explain the result of your test of significance as you would to a funding agency.

7.17 [PA] The city manager of Shinbone has received a complaint from the local union of firefighters that they are underpaid. Not having much time, the city manager gathers the records of a random sample of 27 firefighters and finds that their average salary is $38,073 with a standard deviation of $575. If she knows that the average salary nationally is $38,202, how can she respond to the complaint? Should she use a one-tailed test in this situation? Why? What would she say in a memo to the union that would respond to the complaint?

7.18 The following essay questions review the basic principles and concepts of inferential statistics. The order of the questions roughly follows the five-step model.

a. Researchers can conduct hypothesis testing or significance testing only with a random sample. Why?

b. Under what specific conditions can we assume that the sampling distribution is normal in shape?

c. Explain the role of the sampling distribution in a test of hypothesis.

d. The null hypothesis is an assumption about reality that makes it possible to test sample outcomes for their significance. Explain.

e. What is the critical region? How is the size of the critical region determined?

f. Describe a research situation in which a one-tailed test of hypothesis would be appropriate.

g. Consider the shape of the sampling distribution. Why does use of the t distribution (as opposed to the Z distribution) make it more difficult to reject the null hypothesis?

h. What exactly can we conclude in the one-sample case when the test statistic falls into the critical region?

8

Hypothesis Testing II
The Two-Sample Case

By the end of this chapter, you will be able to:

1. Identify and cite examples of situations in which the two-sample test of hypothesis is appropriate.
2. Explain the logic of hypothesis testing as applied to the two-sample case.
3. Explain what an independent random sample is.
4. Perform a test of hypothesis for two-sample means or two-sample proportions following the five-step model and correctly interpret the results.
5. List and explain each of the factors (especially sample size) that affects the probability of rejecting the null hypothesis. Explain the differences between statistical significance and importance.

8.1 INTRODUCTION

In Chapter 7, we dealt with hypothesis testing in the one-sample case. In that situation, our concern was with the significance of the difference between a sample value and a population value. In this chapter, we will consider research situations where we will be concerned with the significance of the difference between two separate populations. For example, do men and women in the United States vary in their support for gun control? Obviously, we cannot ask all adult males and females for their opinions on this issue. Instead, we must draw random samples of both groups and use the information gathered from these samples to infer population patterns.

The central question asked in hypothesis testing in the two-sample case is "Is the difference between the samples large enough to allow us to conclude (with a known probability of error) that the populations represented by the samples are different?" Thus, if we find a large enough difference in support for gun control between random samples of men and women, we can argue that the difference between the samples did not occur by simple random chance but, rather, represents a real difference between all men and all women in the United States (the population).

In this chapter, we will consider tests for the significance of the difference between sample means and sample proportions. In both tests, the five-step model will serve as a framework for organizing our decision making. The general flow of the hypothesis-testing process is very similar to the one followed in the one-sample case, but we will also need to consider some important differences.

8.2 HYPOTHESIS TESTING WITH SAMPLE MEANS (LARGE SAMPLES)

Introduction: Two-Sample Versus One-Sample Tests. There are several important differences between the two-sample tests covered in this chapter and the one-sample tests explained in Chapter 7. The first occurs in step 1 of the five-step model. The one-sample case requires that the sample be selected following the principle of EPSEM (each case in the population must have an equal chance of being selected for the sample). The two-sample situation requires that the samples be selected independently as well as randomly. This requirement is met when the selection of a case for one sample has no effect on the probability that any particular case will be included in the other sample. In our example, this would mean that the selection of a specific male for the sample would have no effect on the probability of selecting any particular female. This new requirement will be stated as **independent random sampling** in step 1.

We can satisfy the requirement of independent random sampling by drawing EPSEM samples from separate lists (for example, one for females and one for males). It is usually more convenient, however, to draw a single EPSEM sample from a single list of the population and then subdivide the cases into separate groups (males and females, for example). As long as the original sample is selected randomly, any subsamples created by the researcher will meet the assumption of independent random samples.

The second important difference in the five-step model for the two-sample case is in the form of the null hypothesis. The null is still a statement of "no difference." Now, however, instead of saying that the population from which the sample is drawn has a certain characteristic, it will say that the two populations are not different—for example, "There is no significant difference between men and women in their support of gun control." If the test statistic falls in the critical region, we can reject the null hypothesis of no difference between the populations, and the argument that the populations are different on the trait of interest will be supported.

A third important new element concerns the sampling distribution: the distribution of all possible sample outcomes. In Chapter 7, the sample outcome was either a mean or a proportion. Now, we are dealing with two samples (e.g., samples of men and women) and the sample outcome is the *difference between* the sample statistics. In terms of our example, the sampling distribution would include all possible differences in sample means between men and women regarding support for gun control. If the null hypothesis is true and men and women do *not* have different views about gun control, the difference between the population means will be zero, the mean of the sampling distribution will be zero, and the huge majority of differences between sample means will be zero (or, at any rate, very small in value). The greater the differences between the sample means, the further the sample outcome (the *difference* between the two sample means) will be from the mean of the sampling distribution (zero), and the more likely that the difference reflects a real difference between the populations represented by the samples.

A Test of Hypothesis for Two-Sample Means.   To illustrate the procedure for testing sample means, assume that a researcher has access to a nationally representative random sample and that the individuals in the sample have responded to a scale that measures attitudes toward gun control. The sample is divided by sex, and sample statistics are computed for males and females separately. If the scale yields interval-ratio-level data, then we can conduct a test for the significance of the difference in sample means.

As long as sample size is large (that is, as long as the combined number of cases in the two samples exceeds 100), the sampling distribution of the differences in sample means will be normal, and we can use the normal curve (Appendix A) to establish the critical regions. The test statistic, Z(obtained), will be computed by the usual formula: sample outcome (the difference between the sample means) minus the mean of the sampling distribution, divided by the standard deviation of the sampling distribution. (See Formula 8.1.) Note that numerical subscripts identify the samples and the two populations they represent. The subscript attached to σ ($\sigma_{\bar{X}-\bar{X}}$) indicates that we are dealing with the sampling distribution of the *differences* in sample means.

FORMULA 8.1
$$Z(\text{obtained}) = \frac{(\bar{X}_1 - \bar{X}_2) - (\mu_1 - \mu_2)}{\sigma_{\bar{X}-\bar{X}}}$$

where $(\bar{X}_1 - \bar{X}_2)$ = the difference in the sample means

$(\mu_1 - \mu_2)$ = the difference in the population means

$\sigma_{\bar{X}-\bar{X}}$ = the standard deviation of the sampling distribution of the differences in sample means

The second term in the numerator, $(\mu_1 - \mu_2)$, reduces to zero because we assume that the null hypothesis (which will be stated as H_0: $\mu_1 = \mu_2$) is true. Recall that we always base tests of significance on the assumption that the null hypothesis is true. If the means of the two populations are equal, then the term $(\mu_1 - \mu_2)$ will be zero and can be dropped from the equation. In effect, then, the formula we will actually use to compute the test statistic in step 4 will be:

FORMULA 8.2
$$Z(\text{obtained}) = \frac{(\bar{X}_1 - \bar{X}_2)}{\sigma_{\bar{X}-\bar{X}}}$$

For large samples, the standard deviation of the sampling distribution of the difference in sample means is defined as:

FORMULA 8.3
$$\sigma_{\bar{X}-\bar{X}} = \sqrt{\frac{\sigma_1^2}{N_1} + \frac{\sigma_2^2}{N_2}}$$

Because we will rarely, if ever, know the values of the population standard deviations (σ_1 and σ_2), we must use the sample standard deviations, suitably corrected for bias, to estimate them. Formula 8.4 displays the equation for estimating the standard deviation of the sampling distribution in this situation. This is called a **pooled estimate** because it combines information from both samples.

FORMULA 8.4
$$\sigma_{\bar{X}-\bar{X}} = \sqrt{\frac{s_1^2}{N_1 - 1} + \frac{s_2^2}{N_2 - 1}}$$

The sample outcomes for support of gun control are reported below; now we can test for the significance of the difference.

Sample 1 (Men)	Sample 2 (Women)
$\overline{X} = 6.2$	$\overline{X} = 6.5$
$s_1 = 1.3$	$s_2 = 1.4$
$N_1 = 324$	$N_2 = 317$

We see from the sample statistics that men have a lower average score on the Support for Gun Control Scale and are less supportive of gun control. The test of hypothesis will tell us if this difference is large enough to justify the conclusion that it did not occur by random chance alone but rather reflects an actual difference between the populations of men and women on this issue.

Step 1. Making Assumptions and Meeting Test Requirements. Note that, although we now assume that the random samples are independent, the rest of the model is the same as in the one-sample case:

> Model: Independent random samples
> Level of measurement is interval-ratio
> Sampling distribution is normal

Step 2. Stating the Null Hypothesis. The null hypothesis states that the *populations* represented by the samples are not different on this variable. Because no direction for the difference has been predicted, a two-tailed test is called for, as the research hypothesis indicates:

$$H_0: \mu_1 = \mu_2$$
$$(H_1: \mu_1 \neq \mu_2)$$

Step 3. Selecting the Sampling Distribution and Establishing the Critical Region. For large samples, we can use the Z distribution to find areas under the sampling distribution and establish the critical region. Alpha will be set at 0.05:

> Sampling distribution = Z distribution
> Alpha = 0.05
> Z(critical) = ± 1.96

Step 4. Computing the Test Statistic. The population standard deviations are unknown, so we will use Formula 8.4 to estimate the standard deviation of the sampling distribution. We will then substitute this value into Formula 8.2 to compute Z(obtained):

$$\sigma_{\overline{X} - \overline{x}} = \sqrt{\frac{s_1^2}{N_1 - 1} + \frac{s_2^2}{N_2 - 1}} = \sqrt{\frac{(1.3)^2}{324 - 1} + \frac{(1.4)^2}{317 - 1}} = \sqrt{(0.0052) + (0.0062)}$$

$$= \sqrt{0.0114} = 0.107$$

$$Z\text{(obtained)} = \frac{(\overline{X}_1 - \overline{X}_2)}{\sigma_{\overline{X} - \overline{x}}} = \frac{6.2 - 6.5}{0.107} = \frac{-0.300}{0.107} = -2.80$$

STEP BY STEP	**Testing the Difference in Sample Means for Significance (Large Samples): Computing Z(obtained) and Interpreting Results**

In step 4 of the five-step model, we compute the test statistic and use Formula 8.2 to convert the difference between the sample means into a Z score called Z(obtained). Then, in step 5, we compare Z(obtained) with Z(critical) and decide whether to reject or fail to reject the null hypothesis

To solve Formula 8.2, we must first solve Formula 8.4:

$$\sigma_{\bar{X}-\bar{X}} = \sqrt{\frac{s_1^2}{N_1 - 1} + \frac{s_2^2}{N_2 - 1}}$$

Step 1: Subtract 1 from N_1.

Step 2: Square the value of the standard deviation for the first sample (s_1^2).

Step 3: Divide the quantity you found in step 2 by the quantity you found in step 1.

Step 4: Subtract 1 from N_2.

Step 5: Square the value of the standard deviation for the second sample (s_2^2).

Step 6: Divide the quantity you found in step 5 by the quantity you found in step 4.

Step 7: Add the quantity you found in step 6 to the quantity you found in step 3.

Step 8: Take the square root of the quantity you found in step 7.

To solve Formula 8.2: Z(obtained) $= \dfrac{(\bar{X}_1 - \bar{X}_2)}{\sigma_{\bar{X}-\bar{X}}}$

Step 9: Subtract $\bar{X}_2$ from $\bar{X}_1$.

Step 10: Divide the quantity you found in step 9 by the quantity you found in step 8 ($\sigma_{\bar{X}-\bar{X}}$).

Step 11: Compare Z(obtained) to Z(critical). If Z(obtained) is in the critical region, reject the null hypothesis. If Z(obtained) is not in the critical region, fail to reject the null hypothesis.

Step 12: Interpret the decision to reject or fail to reject the null hypothesis in terms of the original question. For example, our conclusion for the example problem used in Section 8.2 was "There is a significant difference between men and women in their support for gun control."

Step 5. Making a Decision and Interpreting the Results of the Test. Comparing the test statistic with the critical region:

$$Z(\text{obtained}) = -2.80$$
$$Z(\text{critical}) = \pm 1.96$$

we see that the Z score clearly falls into the critical region. This outcome indicates that a difference as large as -0.3 ($6.2 - 6.5$) between the sample means is unlikely if the null hypothesis is true. The null hypothesis of no difference can be rejected, and the notion that men and women are different in terms of their support of gun control is supported. The decision to reject the null hypothesis has only a 0.05 probability (the alpha level) of being incorrect.

Note that the value for Z(obtained) is negative, indicating that men have significantly lower scores than women for support for gun control. The sign of the test statistics reflects our arbitrary decision to label men sample 1 and women sample 2. If we had reversed the labels and called women sample 1 and men sample 2, the sign of the Z(obtained) would have been positive, but its value (2.80) would have been exactly the same, as would our decision in step 5. *(For practice in testing the significance of the difference between sample means for large samples, see problems 8.1 to 8.6, 8.9, and 8.15d to f.)*

Application 8.1

Researchers have administered an attitude scale measuring satisfaction with family life to a sample of married respondents. On this scale, higher scores indicate greater satisfaction. The sample has been divided into respondents with no children and respondents with at least one child, and means and standard deviations have been computed for both groups. Is there a significant difference in satisfaction with family life between these two groups? The sample information is:

Sample 1 (No Children)	Sample 2 (At Least One Child)
$\overline{X}_1 = 11.3$	$\overline{X}_2 = 10.8$
$s_1 = 0.6$	$s_2 = 0.5$
$N_1 = 78$	$N_2 = 93$

We can see from the sample results that respondents with no children are more satisfied. The significance of this difference will be tested by following the five-step model.

Step 1. Making Assumptions and Meeting Test Requirements.

Model: Independent random samples
Level of measurement is interval-ratio
Sampling distribution is normal

Step 2. Stating the Null Hypothesis.

$$H_0: \mu_1 = \mu_2$$
$$(H_1: \mu_1 \neq \mu_2)$$

Step 3. Selecting the Sampling Distribution and Establishing the Critical Region.

Sampling distribution = Z distribution
Alpha = 0.05, two-tailed
Z(critical) = ± 1.96

Step 4. Computing the Test Statistic.

$$\sigma_{\overline{X}-\overline{X}} = \sqrt{\frac{s_1^2}{N_1 - 1} + \frac{s_2^2}{N_2 - 1}} = \sqrt{\frac{(0.6)^2}{78 - 1} + \frac{(0.5)^2}{93 - 1}}$$
$$= \sqrt{0.008} = 0.09$$

$$Z\text{(obtained)} \frac{(\overline{X}_1 - \overline{X}_2)}{\sigma_{\overline{X}-\overline{X}}} = \frac{11.3 - 10.8}{0.09} = \frac{0.50}{0.09} = 5.56$$

Step 5. Making a Decision.
Comparing the test statistic with the critical region,

$$Z\text{(obtained)} = 5.56$$
$$Z\text{(critical)} = \pm 1.96$$

we would reject the null hypothesis. This test supports the conclusion that parents and childless couples are different with respect to satisfaction with family life. Given the direction of the difference, we also note that childless couples are significantly happier.

STATISTICS IN EVERYDAY LIFE: Culture Wars

According to many observers, the United States is experiencing an increasingly bitter "culture war" over the morality of abortion, gay marriage, prayer in schools, evolution, and a host of other issues. How deeply is American society divided? How wide is the chasm that separates "traditionalists" and conservatives from "progressives" and liberals?

We can address some of these questions by observing the size and statistical significance of the differences between the opposing sides. We will focus on a single item from the 2004 General Social Survey (GSS) that asked people about their support for gay marriage. Recall that the GSS is administered to a nationally representative sample of about 3,000 adult Americans, so our conclusion will be relevant for the entire society. The item asked people for their level of support or opposition to the statement "Homosexuals should have the right to marry." As you are no doubt aware, the issue of gay marriage has been at the forefront of the culture

(continued)

STATISTICS IN EVERYDAY LIFE *(continued)*

wars. Gay marriage has been legalized (at least briefly) in some localities, but many people bitterly oppose the concept and are working to pass an amendment to the U.S. Constitution that would define a legal marriage as consisting *only* of a man and a woman.

Let's begin by looking at the overall level of support for gay marriage in the United States. The table shows the percentage of the GSS sample selecting each of the five degrees of approval and disapproval for gay marriage. Recall from Chapter 6 that we can be 95% confident that these sample percentages closely approximate (within ±3%) the attitudes of the entire U.S. adult population.

Support for Marriage for Homosexuals, 2004 General Social Survey

	Frequency	Percent
1. *Strongly Agree*	136	11.4
2. *Agree*	216	18.2
3. *Neither Agree nor Disagree*	173	14.6
4. *Disagree*	240	20.3
5. *Strongly Disagree*	420	35.5
Totals	1,185	100.00

While there is a fair amount of variation, a majority of respondents (55.8% combining "disagree" with "strongly disagree") opposes marriage for homosexuals. Furthermore, over a third of the sample chose "disagree strongly," the single most common (modal) score.

What are the correlates of support and opposition for gay marriage? The opposing sides in the culture war are often said to be defined by politics

and religion, and the table below reports the results of some *t* tests using these factors as independent variables. I also included a measure of education to see if this had any impact on the debate.

The average level of support for marriage for homosexuals was calculated from the scores stated in the preceding table. The average score for the entire sample was 3.5, which is higher than the neutral point (3) on the scale. This result is consistent with our previous conclusion that the sample as a whole disapproves of marriage for homosexuals. The *t* tests will tell us if the differences in subcategory means are significant—that is, which independent variables are associated with real differences in the population of all adult U.S. citizens.

The first test compares people with a high school education or less with people who have had at least some college. This test indicates that, to some extent, the culture divide is defined by education with the more educated tending toward a higher level of approval of gay marriage (as indicated by their lower score—see the scale in the previous table).

Second, we compare people who define themselves as "strong" in religious intensity with people who have no religious affiliation. Respondents who indicated that they were "moderately" religious were omitted from this test so we are looking at the extremes of religiosity in this test. As expected, people with a strong religious commitment are significantly more disapproving (have a significantly higher score on the scale).

The final two tests compared the political correlates of the debate. First, all people who identified themselves as supporters of the Democratic Party are compared with all supporters of the

t Tests for the Significance of the Differences in Sample Means

Variable	Categories	Average Score	t Score	Number of Cases	Significant at 0.05?
Education	H. S. or less	3.7	4.99	718	Yes
	Some college	3.3		466	
Religious intensity	Strong	3.9	8.32	462	Yes
	No religion	3.3		168	
Political party	All Democrats	3.0	−10.74	510	Yes
	All Republicans	3.8		462	
Political party	Strong Democrats	2.9	−10.12	206	Yes
	Strong Republicans	4.2		178	

(continued)

STATISTICS IN EVERYDAY LIFE *(continued)*

Republican Party. As expected, the Republicans are significantly more opposed. Finally, we compare people who identified themselves as "strong" Democrats with those who labeled themselves "strong" Republicans, a test which will give us an indication of how far apart the extreme wings of the parties are. We can see that strong Republicans have the highest average score among the subgroups and are significantly more likely than strong Democrats to disapprove.

What do these test show? The size of the *t* scores verifies that there are deep divides in values and outlooks in American society. Opposition to gay marriage is particularly intense among strong Republicans and people who classify themselves as strongly religious. Support for gay marriage is highest among strong Democrats. Judging from the size of the *t* scores, it may also be that the political divides are greater and deeper than those formed around religion and education.

Issues such as gay marriage may tend to pull us apart, but do other values and beliefs tend to bind us together? The bitterness and intensity of the culture wars may be somewhat mitigated if opposing sides subscribe to a common code of values and beliefs. For example, if Americans maintain a firm commitment to civil liberties such as freedom of speech, it seems less likely that the culture wars will escalate to extremes and truly fracture U.S. society. How strong is the commitment to civil liberties? Are they endorsed by both sides of the debate?

To investigate, we will look at two items in the GSS that measure support for free speech for gays. The first item asks if the respondent would allow a homosexual to speak in their community and the other asks if they would allow books written by homosexuals in their community library. I added scores on these two items together to generate an overall measure of support for free speech for gays. The scale has three scores, as listed in the table below. The table also shows the percentage of the respondents at each score of the scale

Support for Civil Liberties for Homosexuals, 2004 General Social Survey

	Frequency	Percent
1. *Would allow both speech and book*	613	69.6
2. *Would either ban speech or ban book*	168	19.1
3. *Would ban both speech and book*	100	11.4
Totals	881	100.1

The vast majority of respondents (almost 70%) support civil liberties, but about 20% have mixed feelings and, perhaps disturbingly, a little over 10% of the Americans would not extend the right of free speech to homosexuals. How does support for civil liberties vary by subgroup?

The divisions around education and religion remain significant and the less educated and the strongly religious are less supportive of civil liberties for gays (have significantly higher scale scores). On the other hand, there are no significant differences along political lines. So, while the divisions over issues such as gay marriage may be deep, they are somewhat counterbalanced by a common commitment to core values, at least along political lines.

t Tests for the Significance of the Differences in Sample Means

Variable	Categories	Average Score	t Score	Number of Cases	Significant at 0.05?
Education	H. S. or less	2.5	6.03	569	Yes
	Some college	2.2		311	
Religious intensity	Strong	2.6	5.54	339	Yes
	No religion	2.2		128	
Political party	All Democrats	2.4	0.74	393	No
	All Republicans	2.4		329	
Political party	Strong Democrats	2.5	0.43	130	No
	Strong Republicans	2.5		116	

8.3 HYPOTHESIS TESTING WITH SAMPLE MEANS (SMALL SAMPLES)

Introduction. As with single-sample means, when the population standard deviation is unknown and sample size is small (combined *N*s of less than 100), we can no longer use the *Z* distribution to find areas under the sampling distribution. Instead, we will use the *t* distribution to find the critical region and thus to identify unlikely sample outcomes. To use the *t* distribution for testing two-sample means, we need to perform one additional calculation and make one additional assumption. The calculation is for degrees of freedom, a quantity required for proper use of the *t* table (Appendix B). In the two-sample case, degrees of freedom are equal to $N_1 + N_2 - 2$.

The additional assumption is a more complex matter. When samples are small, we must assume that the variances of the populations of interest are equal to justify the assumption of a normal sampling distribution and to form a pooled estimate of the standard deviation of the sampling distribution. The assumption of equal variance in the population can be tested but, for our purposes here, we will simply assume equal population variances without formal testing. This assumption is safe as long as sample sizes are approximately equal.

A Test of Hypothesis for Two-Sample Means: Small Samples. To illustrate this procedure, assume that a researcher believes that center-city families are significantly larger than suburban families, as measured by number of children. Random samples from both areas are gathered and sample statistics computed.

Sample 1 (Suburban)	Sample 2 (Center City)
$\overline{X}_1 = 2.37$	$\overline{X}_2 = 2.78$
$s_1 = 0.63$	$s_2 = 0.95$
$N_1 = 42$	$N_2 = 37$

The sample data reveal a difference in the predicted direction. We can test the significance of this observed difference with the five-step model.

Step 1. Making Assumptions and Meeting Test Requirements. Sample size is small, and the population standard deviation is unknown. Hence, we must assume equal population variances in the model:

> Model: Independent random samples
> Level of measurement is interval-ratio
> → Population variances are equal ($\sigma_1^2 = \sigma_2^2$)
> Sampling distribution is normal

Step 2. Stating the Null Hypothesis. A direction has been predicted (center-city families are larger), so we will use a one-tailed test. The research hypothesis is stated in accordance with this decision:

$$H_0: \mu_1 = \mu_2$$
$$(H_1: \mu_1 < \mu_2)$$

Step 3. Selecting the Sampling Distribution and Establishing the Critical Region. With small samples, we use the *t* distribution to establish the critical region. Alpha will be set at 0.05, and a one-tailed test will be used.

Sampling distribution = t distribution
Alpha = 0.05, one-tailed
Degrees of freedom = $N_1 + N_2 - 2 = 42 + 37 - 2 = 77$
t(critical) = -1.671

Note that the critical region is placed in the lower tail of the sampling distribution in accordance with the direction specified in H_1.

Step 4. Computing the Test Statistic. With small samples, we use a different formula (Formula 8.5) for the pooled estimate of the standard deviation of the sampling distribution. We then substitute this value directly into the denominator of the formula for t(obtained) in Formula 8.6:

FORMULA 8.5
$$\sigma_{\bar{X}-\bar{X}} = \sqrt{\frac{N_1 s_1^2 + N_2 s_2^2}{N_1 + N_2 - 2}} \sqrt{\frac{N_1 + N_2}{N_1 N_2}}$$

$$\sigma_{\bar{X}-\bar{X}} = \sqrt{\frac{(42)(.63)^2 + (37)(.95)^2}{42 + 37 - 2}} \sqrt{\frac{42 + 37}{(42)(37)}}$$

$$\sigma_{\bar{X}-\bar{X}} = \sqrt{\frac{50.06}{77}} \sqrt{\frac{79}{1554}}$$

$$\sigma_{\bar{X}-\bar{X}} = (.81)(.23)$$

$$\sigma_{\bar{X}-\bar{X}} = 0.19$$

FORMULA 8.6
$$t(\text{obtained}) = \frac{(\bar{X}_1 - X_2) - 0}{\sigma_{\bar{X}-\bar{X}}}$$

$$t(\text{obtained}) = \frac{2.37 - 2.78}{0.19} = \frac{-0.41}{0.19} = -2.16$$

Step 5. Making a Decision and Interpreting Test Results. Comparing the test statistic with the critical region,

$$t(\text{obtained}) = -2.16$$
$$t(\text{critical}) = -1.671$$

we can see that the test statistic falls into the critical region. If the null ($\mu_1 = \mu_2$) were true, this would be a very unlikely outcome, so the null can be rejected. There is a statistically significant difference (a difference so large that it is unlikely to be due to random chance) in the sizes of center-city and suburban families. Furthermore, center-city families are significantly larger in size. Figure 8.1

FIGURE 8.1   THE SAMPLING DISTRIBUTION WITH CRITICAL REGION AND TEST STATISTIC DISPLAYED

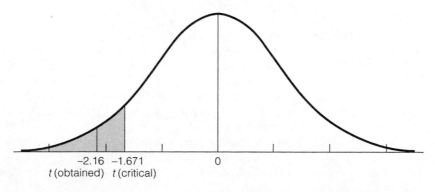

-2.16 -1.671 0
t(obtained) t(critical)

STEP BY STEP	**Testing the Difference in Sample Means for Significance (Small Samples): Computing t(obtained) and Interpreting Results**

In step 4 of the five-step model, we compute the test statistic and use Formula 8.6 to convert the difference between the sample means into a t score called t(obtained). Then, in step 5, we compare t(obtained) with t(critical) and decide to reject or fail to reject the null hypothesis.

To solve Formula 8.6, we must first solve Formula 8.5:

$$\sigma_{\bar{X}-\bar{X}} = \sqrt{\frac{N_1 s_1^2 + N_2 s_2^2}{N_1 + N_2 - 2}} \sqrt{\frac{N_1 + N_2}{N_1 N_2}}$$

Step 1: Add N_1 and N_2 and then subtract 2 from this total.

Step 2: Square the standard deviation for the first sample (s_1^2).

Step 3: Multiply s_1^2 by N_1.

Step 4: Square the standard deviation for the second sample (s_2^2).

Step 5: Multiply s_2^2 by N_2.

Step 6: Add the quantities you found in step 3 and step 5.

Step 7: Divide the quantity you found in step 6 by the quantity you found in step 1.

Step 8: Take the square root of the quantity you found in step 7.

Step 9: Multiply N_1 and N_2.

Step 10: Add N_1 and N_2.

Step 11: Divide the quantity you found in step 10 by the quantity you found in step 9.

Step 12: Take the square root of the quantity you found in step 11.

Step 13: Multiply the quantity you found in step 12 by the quantity you found in step 8.

To solve Formula 8.6:

$$t(\text{obtained}) = \frac{(\bar{X}_1 - \bar{X}_2)}{\sigma_{\bar{X}-\bar{X}}}$$

Step 14: Subtract the mean of sample 2 from the mean of sample 1.

Step 15: Divide the quantity you found in step 14 by the quantity you found in step 13.

Step 16: Compare t(obtained) to t(critical). If t(obtained) is *in* the critical region, reject the null hypothesis. If t(obtained) is *not* in the critical region, fail to reject the null hypothesis.

Step 17: Interpret the decision to reject or fail to reject the null hypothesis in terms of the original question. For example, our conclusion for the example problem used in Section 8.3 was "There is a significant difference between the average size of center-city and suburban families."

depicts the test statistic and sampling distribution. *(For practice in testing the significance of the difference between sample means for small samples, see problems 8.7 and 8.8.)*

8.4 HYPOTHESIS TESTING WITH SAMPLE PROPORTIONS (LARGE SAMPLES)

Introduction.   Testing for the significance of the difference between two sample proportions is analogous to testing sample means. The null hypothesis states that no difference exists between the populations from which the samples are drawn on the trait being tested. The sample proportions form the basis of the test statistic computed in step 4, which is then compared with the critical region. When sample sizes are large (combined Ns of more than 100), we can use the Z distribution to find the critical region. We will not consider tests of significance for proportions based on small samples in this text.

To find the value of the test statistic, we must solve several preliminary equations. Formula 8.7 uses the values of the two-sample proportions (P_s) to give us an estimate of the population proportion (P_u), the proportion of cases in the population that have the trait under consideration assuming the null hypothesis is true:

FORMULA 8.7

$$P_u = \frac{N_1 P_{s1} + N_2 P_{s2}}{N_1 + N_2}$$

We then use the estimated value of P_u to determine a value for the standard deviation of the sampling distribution of the difference in sample proportions in Formula 8.8:

FORMULA 8.8

$$\sigma_{p-p} = \sqrt{P_u(1 - P_u)}\sqrt{\frac{N_1 + N_2}{N_1 N_2}}$$

Next, we substitute this value into the formula for computing the test statistic, as Formula 8.9 shows:

FORMULA 8.9

$$Z(\text{obtained}) = \frac{(P_{s1} - P_{s2}) - (P_{u1} - P_{u2})}{\sigma_{p-p}}$$

where $(P_{s1} - P_{s2})$ = the difference between the sample proportions

$(P_{u1} - P_{u2})$ = the difference between the population proportions

σ_{p-p} = the standard deviation of the sampling distribution of the difference between sample proportions

As was the case with sample means, we assume the second term in the numerator is zero by the null hypothesis. Therefore, the formula reduces to:

FORMULA 8.10

$$Z(\text{obtained}) = \frac{(P_{s1} - P_{s2})}{\sigma_{p-p}}$$

Remember to solve these equations in order, starting with Formula 8.7 (and skipping Formula 8.9).

A Test of Hypothesis for Two-Sample Proportions.   An example will make these procedures clearer. Assume that random samples of black and white senior citizens have been selected, and each respondent has been classified as high or low in terms of the number of memberships he or she holds in voluntary associations. Is there a statistically significant difference in the participation patterns of the elderly by race? The proportion of each group classified as "high" in participation and sample size for both groups is reported here:

Sample 1 (Black Senior Citizens)	Sample 2 (White Senior Citizens)
$P_{s1} = 0.34$	$P_{s1} = 0.25$
$N_1 = 83$	$N_2 = 103$

Step 1. Making Assumptions and Meeting Test Requirements.

Model: Independent random samples
Level of measurement is nominal
Sampling distribution is normal

Step 2. Stating the Null Hypothesis.   Because no direction has been predicted, this will be a two-tailed test:

$$H_0: P_{u1} = P_{u2}$$
$$(H_1: P_{u1} \neq P_{u2})$$

Step 3. Selecting the Sampling Distribution and Establishing the Critical Region. The sample size is large, so we will use the Z distribution to establish the critical region. Setting alpha at 0.05, we have:

$$\text{Sampling distribution} = Z \text{ distribution}$$
$$\text{Alpha} = 0.05, \text{ two-tailed}$$
$$Z(\text{critical}) = \pm 1.96$$

Step 4. Computing the Test Statistic.   Begin with the formula for estimating P_u (Formula 8.7), substitute the resultant value into Formula 8.8, and then solve for Z(obtained) with Formula 8.10.

$$P_u = \frac{N_1 P_{s1} + N_2 P_{s2}}{N_1 + N_2} = \frac{(83)(0.34) + (103)(0.25)}{83 + 103} = 0.29$$

$$\sigma_{p-p} = \sqrt{P_u(1 - P_u)}\sqrt{\frac{N_1 + N_2}{N_1 N_2}} = \sqrt{(0.29)(0.71)}\sqrt{\frac{83 + 103}{(83)(103)}}$$

$$= (0.45)(0.15) = 0.07$$

$$Z(\text{obtained}) = \frac{(P_{s1} - P_{s2})}{\sigma_{p-p}} = \frac{0.34 - 0.25}{0.07} = 1.29$$

Step 5. Making a Decision and Interpreting the Results of the Test.   Because the test statistic, Z(obtained) $= 1.29$, does not fall into the critical region as marked by the Z(critical) of ± 1.96, we fail to reject the null hypothesis. The difference between the sample proportions is no greater than what would be expected if the null hypothesis were true and only random chance were operating. Black and white senior citizens are not significantly different in terms of participation patterns as measured in this test. *(For practice in testing the significance of the difference between sample proportions, see problems 8.10 to 8.14 and 8.15a to c.)*

8.5 THE LIMITATIONS OF HYPOTHESIS TESTING: SIGNIFICANCE VERSUS IMPORTANCE

Given that we are usually interested in rejecting the null hypothesis, we should take a moment to consider systematically the factors that affect our decision in step 5. Generally speaking, the probability of rejecting the null hypothesis is a function of four independent factors:

1. The size of the observed difference(s)
2. The alpha level
3. The use of one- or two-tailed tests
4. The size of the sample

STEP BY STEP	**Testing the Difference in Sample Proportions for Significance (Large Samples): Computing Z(obtained) and Interpreting Results**

In step 4 of the five-step model, we compute the test statistic and use Formula 8.10 to convert the difference between the sample proportions into a Z score called Z(obtained). Then, in step 5, we compare Z(obtained) with Z(critical) and decide whether to reject or fail to reject the null hypothesis.

To test the difference between sample proportions, we must first solve Formula 8.7:

$$P_u = \frac{N_1 P_{s1} + N_2 P_{s2}}{N_1 + N_2}$$

Step 1: Add N_1 and N_2.

Step 2: Multiply P_{s1} by N_1.

Step 3: Multiply P_{s2} by N_2.

Step 4: Add the quantity you found in step 3 to the quantity you found in step 2.

Step 5: Divide the quantity you found in step 4 by the quantity you found in step 1.

Next, we solve Formula 8.8:

$$\sigma_{p-p} = \sqrt{P_u(1 - P_u)}\sqrt{\frac{N_1 + N_2}{N_1 N_2}}$$

Step 6: Multiply P_u (the quantity you found in Step 5) by $(1 - P_u)$.

Step 7: Take the square root of the quantity you found in step 6.

Step 8: Multiply N_1 and N_2.

Step 9: Add N_1 and N_2. (See step 1.)

Step 10: Divide the quantity you found in step 9 by the quantity you found in step 8.

Step 11: Take the square root of the quantity you found in step 10.

Step 12: Multiply the quantity you found in step 11 by the quantity you found in step 7.

To solve Formula 8.10:

$$Z(\text{obtained}) = \frac{(P_{s1} + P_{s2})}{\sigma_{p-p}}$$

Step 13: Subtract P_{s2} from P_{s1}.

Step 14: Divide the quantity you found in step 13 by the quantity you found in step 12.

Step 15: Compare Z(obtained) to Z(critical). If Z(obtained) is *in* the critical region, reject the null hypothesis. If Z(obtained) is *not* in the critical region, fail to reject the null hypothesis.

Step 16: Interpret the decision to reject or fail to reject the null hypothesis in terms of the original question. For example, our conclusion for the example problem used in Section 8.4 was "There is no significant difference between the participation patterns of black and white senior citizens."

Only the first of these four is not under the direct control of the researcher. The size of the difference (either between the sample outcome and the population value or between two sample outcomes) is partly a function of the testing procedures (that is, how variables are measured) but should generally reflect the underlying realities we are trying to probe.

The relationship between alpha level and the probability of rejection is straightforward. The higher the alpha level, the larger the critical region, the higher the percentage of all possible sample outcomes that fall in the critical region, and the greater the probability of rejection. Thus, it is easier to reject the H_0 at the 0.05 level than at the 0.01 level, and easier still at the 0.10 level. The danger here, of course, is that higher alpha levels will lead to more frequent Type I errors, and we might find ourselves declaring small differences to be statistically significant. In similar fashion, using a one-tailed test will increase the probability of rejection (assuming that the proper direction has been predicted).

Application 8.2

Do attitudes toward sex vary by gender? The respondents in a national survey have been asked if they think that premarital sex is "always wrong" or only "sometimes wrong." The proportion of each sex that feels that premarital sex is always wrong follows:

Females	Males
$P_{s1} = 0.35$	$P_{s2} = 0.32$
$N_1 = 450$	$N_2 = 417$

This is all the information we will need to conduct a test of the null hypothesis following the familiar five-step model with alpha set at .05, two-tailed test.

Step 1. Making Assumptions and Meeting Test Requirements.

> Model: Independent random samples
> Level of measurement is nominal
> Sampling distribution is normal

Step 2. Stating the Null Hypothesis.

$$H_0: P_{u1} = P_{u2}$$
$$(H_1: P_{u1} \neq P_{u2})$$

Step 3. Selecting the Sampling Distribution and Establishing the Critical Region.

> Sampling distribution = Z distribution
> Alpha = 0.05, two-tailed
> $Z(\text{critical}) = \pm 1.96$

Step 4. Computing the Test Statistic. Remember to start with Formula 8.7, substitute the value for P_u into Formula 8.8, and then substitute that value into Formula 8.10 to solve for Z(obtained).

$$P_u = \frac{N_1 P_{s1} + N_2 P_{s2}}{N_1 + N_2} = \frac{(450)(0.35) + (417)(0.32)}{450 + 417}$$

$$= \frac{290.94}{867} = 0.34$$

$$\sigma_{p-p} = \sqrt{P_u(1 - P_u)}\sqrt{\frac{N_1 + N_2}{N_1 N_2}}$$

$$= \sqrt{(0.34)(0.66)}\sqrt{\frac{450 + 417}{(450)(417)}}$$

$$= \sqrt{0.2244}\ \sqrt{0.0046}$$

$$= (0.47)(0.068) = 0.032$$

$$Z(\text{obtained}) = \frac{(P_{s1} - P_{s2})}{\sigma_{p-p}} = \frac{0.35 - 0.32}{0.032}$$

$$= \frac{0.030}{0.032} = 0.94$$

Step 5. Making a Decision. With an obtained Z score of 0.94, we would fail to reject the null hypothesis. There is no statistically significant difference between males and females on attitudes toward premarital sex.

The final factor is sample size: with all other factors constant, the probability of rejecting H_0 increases with sample size. In other words, the larger the sample, the more likely we are to reject the null hypothesis. With very large samples (say those with thousands of cases), we may declare small, unimportant differences to be statistically significant.

This relationship may appear to be surprising, but we can appreciate the reasons for it by briefly considering the formulas used to compute test statistics in step 4. In all these formulas, for all tests of significance, sample size (N) is in the "denominator of the denominator." Algebraically, this is equivalent to being in the numerator of the formula and means that the value of the test statistic is directly proportional to N and that the two will increase together. To illustrate, consider Table 8.1, which shows the value of the test statistic for single-sample means from samples of various sizes. The value of the test statistic, Z(obtained), increases as N increases even though none of the other terms in the formula changes. This pattern of higher probabilities for rejecting H_0 with larger samples holds for all tests of significance.

TABLE 8.1   TEST STATISTICS FOR SINGLE-SAMPLE MEANS COMPUTED FROM SAMPLES OF VARIOUS SIZES ($\bar{X} = 80$, $\mu = 77$, $s = 5$ throughout)

Sample Size	Test Statistic, Z(obtained)
100	1.99
200	2.82
500	4.47

On one hand, the relationship between sample size and the probability of rejecting the null should not alarm us unduly. Larger samples are, after all, better approximations of the populations they represent. Thus, we can trust decisions based on larger samples more than decisions based on small samples.

On the other hand, this relationship clearly underlines what is perhaps the most significant limitation of hypothesis testing. Simply because a difference is statistically significant does not guarantee that it is important in any other sense. Particularly with very large samples, relatively small differences may be statistically significant. Even with small samples, of course, differences that are otherwise trivial or uninteresting may be statistically significant. The crucial point is that statistical significance and theoretical or practical importance can be two very different things. Statistical significance is a necessary but not sufficient condition for theoretical or practical importance. A difference that is not statistically significant is almost certainly unimportant in every other sense. However, significance by itself does not guarantee importance. Even when research results are clearly not produced by random chance, the researcher must still assess their importance. Do they firmly support a theory or hypothesis? Are they clearly consistent with a prediction or analysis? Do they strongly indicate a line of action in solving some problem? These are the kinds of questions a researcher must ask when assessing the importance of the results of a statistical test.

Also, note that researchers have access to some very powerful ways of analyzing the importance (versus the statistical significance) of research results. Parts III and IV of this text will introduce these statistics, including bivariate measures of association and multivariate statistical techniques.

SUMMARY

1. A common research situation is to test for the significance of the difference between two populations. Sample statistics are calculated for random samples of each population, and then we test for the significance of the difference between the samples as a way of inferring differences between the specified populations.

2. When sample information is summarized in the form of sample means, and N is large, we use the Z distribution to find the critical region. When N is small, we use the t distribution to establish the critical region. In the latter circumstance, we must also assume equal population variances before forming a pooled estimate of the standard deviation of the sampling distribution.

3. Differences in sample proportions may also be tested for significance. For large samples, we use the Z distribution to find the critical region.

4. In all tests of hypothesis, numerous factors affect the probability of rejecting the null: the size of the difference, the alpha level, the use of one-versus two-tailed tests, and sample size. Statistical significance is not the same thing as theoretical or practical importance. Even after a difference is found to be statistically significant, the researcher must still demonstrate the relevance or importance of his or her findings. The statistics presented in Parts III and IV will give us the tools we need to deal directly with issues beyond statistical significance.

SUMMARY OF FORMULAS

Test statistic for two-sample means, large samples	8.1	$Z(\text{obtained}) = \dfrac{(\bar{X}_1 - \bar{X}_2) - (\mu_1 - \mu_2)}{\sigma_{\bar{X}-\bar{X}}}$
Test statistic for two-sample means, large samples (simplified formula)	8.2	$Z(\text{obtained}) = \dfrac{(\bar{X}_1 - \bar{X}_2)}{\sigma_{\bar{X}-\bar{X}}}$
Standard deviation of the sampling distribution of the difference in sample means, large samples	8.3	$\sigma_{\bar{X}-\bar{X}} = \sqrt{\dfrac{\sigma_1^{\,2}}{N_1} + \dfrac{\sigma_2^{\,2}}{N_2}}$
Pooled estimate of the standard deviation of the sampling distribution of the difference in sample means, large samples	8.4	$\sigma_{\bar{X}-\bar{X}} = \sqrt{\dfrac{s_1^{\,2}}{N_1 - 1} + \dfrac{s_2^{\,2}}{N_2 - 1}}$
Pooled estimate of the standard deviation of the sampling distribution of the difference in sample means, small samples	8.5	$\sigma_{\bar{X}-\bar{X}} = \sqrt{\dfrac{N_1 s_1^{\,2} + N_2 s_2^{\,2}}{N_1 + N_2 - 2}} \sqrt{\dfrac{N_1 + N_2}{N_1 N_2}}$
Test statistic for two-sample means, small samples	8.6	$t(\text{obtained}) = \dfrac{(\bar{X}_1 - \bar{X}_2)}{\sigma_{\bar{X}-\bar{X}}}$
Pooled estimate of population proportion, large samples	8.7	$P_u = \dfrac{N_1 P_{s1} + N_2 P_{s2}}{N_1 + N_2}$
Standard deviation of the sampling distribution of the difference in sample proportions, large samples	8.8	$\sigma_{p-p} = \sqrt{P_u(1 - P_u)} \sqrt{\dfrac{N_1 + N_2}{N_1 N_2}}$
Test statistic for two-sample proportions, large samples	8.9	$Z(\text{obtained}) = \dfrac{(P_{s1} - P_{s2}) - (P_{u1} - P_{u2})}{\sigma_{p-p}}$
Test statistic for two-sample proportions, large samples (simplified formula)	8.10	$Z(\text{obtained}) = \dfrac{(P_{s1} - P_{s2})}{\sigma_{p-p}}$

GLOSSARY

Independent random samples. Random samples gathered in such a way that the selection of a particular case for one sample has no effect on the probability that any other particular case will be selected for the other samples.

Pooled estimate. An estimate of the standard deviation of the sampling distribution of the difference in sample means based on the standard deviations of both samples.

σ_{p-p} Symbol for the standard deviation of the sampling distribution of the differences in sample proportions.

$\sigma_{\bar{x}-\bar{x}}$ Symbol for the standard deviation of the sampling distribution of the differences in sample means.

MULTIMEDIA RESOURCES

The Wadsworth Sociology Resource Center: Virtual Society at
http://www.thomsonedu.com/sociology

Visit the companion website for the seventh edition of *Statistics: A Tool for Social Research* to access a wide range of student resources. Begin by clicking on the Student Resources section of the book's website to access the following study tools:

- Basic math review
- Flash cards

- Additional chapter problems
- Internet links
- Table of random numbers
- MicroCase and SPSS examples and exercises
- Hypothesis testing for variables measured at the ordinal level

PROBLEMS

8.1 For each problem below, test for the significance of the difference in sample statistics using the five-step model. (*HINT: Remember to solve Formula 8.4 for before attempting to solve Formula 8.2. Also, in Formula 8.4, perform the mathematical operations in the proper sequence. First square each sample standard deviation, then divide by N − 1, add the resultant values, and then find the square root of the sum.*)

a.

Sample 1	Sample 2
$\overline{X}_1 = 72.5$	$\overline{X}_2 = 76.0$
$s_1 = 14.3$	$s_2 = 10.2$
$N_1 = 136$	$N_2 = 257$

b.

Sample 1	Sample 2
$\overline{X}_1 = 107$	$\overline{X}_2 = 103$
$s_1 = 14$	$s_2 = 17$
$N_1 = 175$	$N_2 = 200$

8.2 [SOC] My colleague John Gessner and I administered questionnaires to samples of undergraduates. Among other things, the questionnaires contained a scale that measured attitudes toward interpersonal violence (higher scores indicate greater approval of interpersonal violence). Test the results as reported below for sexual, racial, and social-class differences.

a.

Sample 1 (Males)	Sample 2 (Females)
$\overline{X}_1 = 2.99$	$\overline{X}_2 = 2.29$
$s_1 = 0.88$	$s_2 = 0.91$
$N_1 = 122$	$N_2 = 251$

b.

Sample 1 (Blacks)	Sample 2 (Whites)
$\overline{X}_1 = 2.76$	$\overline{X}_2 = 2.49$
$s_1 = 0.68$	$s_2 = 0.91$
$N_1 = 43$	$N_2 = 304$

c.

Sample 1 (White Collar)	Sample 2 (Blue Collar)
$\overline{X}_1 = 2.46$	$\overline{X}_2 = 2.67$
$s_1 = 0.91$	$s_2 = 0.87$
$N_1 = 249$	$N_2 = 97$

d. Summarize your results in terms of the significance and the direction of the differences. Which of these three factors seems to make the biggest difference in attitudes toward interpersonal violence?

8.3 [SOC] Do athletes in different sports vary in terms of intelligence? College Board scores of random samples of college basketball and football players are reported here. Is there a significant difference? Write a sentence or two explaining the difference.

a.

Sample 1 (Basketball Players)	Sample 2 (Football Players)
$\overline{X}_1 = 460$	$\overline{X}_2 = 442$
$s_1 = 92$	$s_2 = 57$
$N_1 = 102$	$N_2 = 11$

b. What about male and female college athletes?

Sample 1 (Males)	Sample 2 (Females)
$\overline{X}_1 = 452$	$\overline{X}_2 = 480$
$s_1 = 88$	$s_2 = 75$
$N_1 = 107$	$N_2 = 105$

8.4 PA A number of years ago, the fire department in Shinbone, Kansas, began recruiting minority group members through an affirmative action program. In terms of efficiency ratings as compiled by their superiors, how do the affirmative action employees rate? The ratings of random samples of both groups were collected, and the results are reported below (higher ratings indicate greater efficiency).

Sample 1 (Affirmative Action)	Sample 2 (Regular)
$\overline{X}_1 = 15.2$	$\overline{X}_2 = 15.5$
$s_1 = 3.9$	$s_2 = 2.0$
$N_1 = 97$	$N_2 = 100$

Write a sentence or two of interpretation.

8.5 SOC Are middle-class families more likely than working-class families to maintain contact with kin? Write a paragraph summarizing the results of the following tests.

a. A sample of middle-class families reported an average of 7.3 visits per year with close kin while a sample of working-class families averaged 8.2 visits. Is the difference significant?

Visits

Sample 1 (Middle Class)	Sample 2 (Working Class)
$\overline{X}_1 = 7.3$	$\overline{X}_2 = 8.2$
$s_1 = 0.3$	$s_2 = 0.5$
$N_1 = 89$	$N_2 = 55$

b. The middle-class families averaged 2.3 phone calls and 8.7 e-mail messages per month with close kin. The working-class families averaged 2.7 calls and 5.7 e-mail messages per month. Are these differences significant?

Phone Calls

Sample 1 (Middle Class)	Sample 2 (Working Class)
$\overline{X}_1 = 2.3$	$\overline{X}_2 = 2.7$
$s_1 = 0.5$	$s_2 = 0.8$
$N_1 = 89$	$N_2 = 55$

E-Mail Messages

Sample 1 (Middle Class)	Sample 2 (Working Class)
$\overline{X}_1 = 8.7$	$\overline{X}_2 = 5.7$
$s_1 = 0.3$	$s_2 = 1.1$
$N_1 = 89$	$N_2 = 55$

8.6 SOC Are college students who live in dormitories significantly more involved in campus life than students who commute to campus? The data below report the average number of hours per week students devote to extracurricular activities. Is the difference between these randomly selected samples of commuter and residential students significant?

Sample 1 (Residential)	Sample 2 (Commuter)
$\overline{X}_1 = 12.4$	$\overline{X}_2 = 10.2$
$s_1 = 2.0$	$s_2 = 1.9$
$N_1 = 158$	$N_2 = 173$

8.7 SOC Are senior citizens who live in retirement communities more socially active than those who live in age-integrated communities? Write a sentence or two explaining the results of these tests. (*HINT: Remember to use the proper formulas for small sample sizes.*)

a. In a random sample, senior citizens living in a retirement village reported that they had an average of 1.42 face-to-face interactions per day with their neighbors. A random sample of those living in age-integrated communities reported 1.58 interactions. Is the difference significant?

Sample 1 (Retirement Community)	Sample 2 (Age-Integrated Neighborhood)
$\overline{X}_1 = 1.42$	$\overline{X}_2 = 1.58$
$s_1 = 0.10$	$s_2 = 0.78$
$N_1 = 43$	$N_2 = 37$

b. Senior citizens living in the retirement village reported that they had 7.43 telephone calls with friends and relatives each week while those in the age-integrated communities reported 5.50 calls. Is the difference significant?

Sample 1 (Retirement Community)	Sample 2 (Age-Integrated Neighborhood)
$\bar{X}_1 = 7.43$	$\bar{X}_2 = 5.50$
$s_1 = 0.75$	$s_2 = 0.25$
$N_1 = 43$	$N_2 = 37$

8.8 [SOC] As the director of the local Boys Club, you have claimed for years that membership in your club reduces juvenile delinquency. Now, a cynical member of your funding agency has demanded proof of your claim. Fortunately, your local sociology department is on your side and springs to your aid with student assistants, computers, and hand calculators at the ready. Random samples of members and nonmembers are gathered and interviewed with respect to their involvement in delinquent activities. Each respondent is asked how many delinquent acts he has engaged in over the past year. The results are in and reported below (the average number of admitted acts of delinquency). What can you tell the funding agency?

Sample 1 (Members)	Sample 2 (Nonmembers)
$\bar{X}_1 = 10.3$	$\bar{X}_2 = 12.3$
$s_1 = 2.7$	$s_2 = 4.2$
$N_1 = 40$	$N_2 = 55$

8.9 [SOC] A survey has been administered to random samples of respondents in each of five nations. For each nation, are men and women significantly different in terms of their reported levels of satisfaction? Respondents were asked: "How satisfied are you with your life as a whole?" Responses varied from 1 (very dissatisfied) to 10 (very satisfied). Conduct a test for the significance of the difference in mean scores for each nation.

France

Males	Females
$\bar{X}_1 = 7.4$	$\bar{X}_2 = 7.7$
$s_1 = .20$	$s_2 = .25$
$N_1 = 1,005$	$N_2 = 1,234$

Nigeria

Males	Females
$\bar{X}_1 = 6.7$	$\bar{X}_2 = 7.8$
$s_1 = .16$	$s_2 = .23$
$N_1 = 1,825$	$N_2 = 1,256$

China

Males	Females
$\bar{X}_1 = 7.6$	$\bar{X}_2 = 7.1$
$s_1 = .21$	$s_2 = .11$
$N_1 = 1,400$	$N_2 = 1,200$

Mexico

Males	Females
$\bar{X}_1 = 8.3$	$\bar{X}_2 = 9.1$
$s_1 = .29$	$s_2 = .30$
$N_1 = 1,645$	$N_2 = 1,432$

Japan

Males	Females
$\bar{X}_1 = 8.8$	$\bar{X}_2 = 9.3$
$s_1 = .34$	$s_2 = .32$
$N_1 = 1,621$	$N_2 = 1,683$

8.10 For each problem, test the sample statistics for the significance of the difference. *(HINT: In testing proportions, remember to begin with Formula 8.7, then solve Formulas 8.8 and 8.10.)*

a.

Sample 1	Sample 2
$P_{s1} = 0.17$	$P_{s2} = 0.20$
$N_1 = 101$	$N_2 = 114$

b.

Sample 1	Sample 2
$P_{s1} = 0.62$	$P_{s2} = 0.60$
$N_1 = 532$	$N_2 = 478$

8.11 [CJ] About half of the police officers in Shinbone, Kansas, have completed a special course in investigative procedures. Has the course increased their efficiency in clearing crimes by arrest? The proportions of cases cleared by arrest for random samples of trained and untrained officers are reported below.

Sample 1 (Trained)	Sample 2 (Untrained)
$P_{s1} = 0.47$	$P_{s2} = 0.43$
$N_1 = 157$	$N_2 = 113$

8.12 [SW] A large counseling center needs to evaluate several experimental programs. Write a paragraph summarizing the results of these tests. Did the new programs work?

a. One program is designed for divorce counseling; the key feature of the program is its counselors, who are married couples working in teams. About half of all clients have been randomly assigned to this special program and half to the regular program, and the proportion of cases that eventually ended in divorce was recorded for both. The results for random samples of couples from both programs are reported below. In terms of preventing divorce, did the new program work?

Sample 1 (Special Program)	Sample 2 (Regular Program)
$P_{s1} = 0.53$	$P_{s2} = 0.59$
$N_1 = 78$	$N_2 = 82$

b. The agency is also experimenting with peer counseling for depressed children. About half of all clients were randomly assigned to peer counseling. After the program had run for a year, a random sample of children from the new program was compared with a random sample of children who did not receive peer counseling. In terms of the percentage judged to be "much improved," did the new program work?

Sample 1 (Peer Counseling)	Sample 2 (No Peer Counseling)
$P_{s1} = 0.10$	$P_{s2} = 0.15$
$N_1 = 52$	$N_2 = 56$

8.13 SOC At St. Algebra College, the sociology and psychology departments have been feuding for years about the respective quality of their programs. In an attempt to resolve the dispute, you have gathered data about the graduate school experience of random samples of both groups of majors. The results are presented below: the proportion of majors who applied to graduate schools, the proportion of majors accepted into their preferred programs, and the proportion of these who completed their programs. As measured by these data, is there a significant difference in program quality?

a. Proportion of majors who applied to graduate school:

Sample 1 (Sociology)	Sample 2 (Psychology)
$P_{s1} = 0.53$	$P_{s2} = 0.40$
$N_1 = 150$	$N_2 = 175$

b. Proportion accepted by program of first choice:

Sample 1 (Sociology)	Sample 2 (Psychology)
$P_{s1} = 0.75$	$P_{s2} = 0.85$
$N_1 = 80$	$N_2 = 70$

c. Proportion completing the programs:

Sample 1 (Sociology)	Sample 2 (Psychology)
$P_{s1} = 0.75$	$P_{s2} = 0.69$
$N_1 = 60$	$N_2 = 60$

8.14 CJ The local police chief started a "crimeline" program some years ago and wonders if it's really working. The program publicizes unsolved violent crimes in the local media and offers cash rewards for information leading to arrests. Are "featured" crimes more likely to be cleared by arrest than other violent crimes? Results from random samples of both types of crimes are reported as follows:

Sample 1 (Crimeline Crimes Cleared by Arrest)	Sample 2 (Noncrimeline Crimes Cleared by Arrest)
$P_{s1} = 0.35$	$P_{s2} = 0.25$
$N_1 = 178$	$N_2 = 212$

8.15 SOC Some results from the 2004 General Social Survey are reported here in terms of differences by sex. The actual questions appear in Appendix G. Which of these differences, if any, are significant? Write a sentence or two of interpretation for each test. (Note: The results presented below are for the entire 2004 GSS sample, not for the smaller GSS data set used in the SPSS exercises in this text.)

a. Proportion favoring the legalization of marijuana (GRASS):

Sample 1 (Males)	Sample 2 (Females)
$P_{s1} = 0.39$	$P_{s2} = 0.34$
$N_1 = 379$	$N_2 = 423$

b. Proportion "definitely agreeing that human beings developed from earlier species of animals" (SCITESTY):

Sample 1 (Males)	Sample 2 (Females)
$P_{s1} = 0.19$	$P_{s2} = 0.12$
$N_1 = 642$	$N_2 = 800$

c. Proportion voting for President Bush in 2000 (PRES00):

Sample 1 (Males)	Sample 2 (Females)
$P_{s1} = 0.59$	$P_{s2} = 0.51$
$N_1 = 744$	$N_2 = 944$

d. Average hours spent on the Internet each week, not counting e-mail (WWWHR):

Sample 1 (Males)	Sample 2 (Females)
$\overline{X}_1 = 8.79$	$\overline{X}_2 = 6.30$
$s_1 = 11.04$	$s_2 = 9.92$
$N_1 = 793$	$N_2 = 908$

e. Average rate of church attendance (ATTEND):

Sample 1 (Males)	Sample 2 (Females)
$\overline{X}_1 = 3.40$	$\overline{X}_2 = 4.14$
$s_1 = 2.67$	$s_2 = 2.67$
$N_1 = 1,276$	$N_2 = 1,525$

f. Number of children (CHILDS):

Sample 1 (Males)	Sample 2 (Females)
$\overline{X}_1 = 1.65$	$\overline{X}_2 = 1.96$
$s_1 = 1.61$	$s_2 = 1.61$
$N_1 = 1,279$	$N_2 = 1,529$

SPSS for Windows

Using *SPSS for Windows* to Test the Significance of the Difference Between Two Means

SPSS DEMONSTRATION 8.1 Do Men or Women Watch More TV?

SPSS for Windows includes several tests for the significance of the difference between means. In this demonstration, we'll use the **Independent-Samples T Test**, the test we covered in Sections 8.2 and 8.3, to test for the significance of the difference between men and women in average hours of reported TV watching. If there is a statistically significant difference between the sample means for men and women, we can conclude that the populations (all U.S. adult men and women) are different on this variable.

Start *SPSS for Windows* and load the 2004 GSS database. From the main menu bar, click **Analyze**, then **Compare Means**, and then **Independent-Samples T Test**. The **Independent-Samples T Test** dialog box will open with the usual list of variables on the left. Find and move the cursor over *tvhours* (the variable label is HOURS PER DAY WATCHING TV) and click the top arrow in the middle of the window to move *tvhours* to the **Test Variable(s)** box. Next, find and highlight *sex* (the label is RESPONDENTS' SEX) and click the bottom arrow in the middle of the window to move *sex* to the **Grouping Variable** box. Two question marks will appear in the **Grouping Variable** box, and the **Define Groups** button will become active. SPSS needs to know which cases go in which groups and, in the case at hand, the instructions we need to supply are straightforward. Males (indicated by a score of 1 on *sex*) go into group 1 and females (a score of 2) go into group 2.

Click the **Define Groups** button, and the **Define Groups** window will appear. The cursor will be blinking in the box beside Group 1—SPSS is asking for the score that will determine which cases go into this group. Type a 1 in this box (for males) and then click the box next to Group 2 and type a 2 (for females). Click **Continue** to return to the **Independent-Samples T Test** window and click **OK** and the output shown below will be produced. (The 95% confidence interval automatically produced by this program has been deleted to conserve space.)

In the first block of output are some descriptive statistics. The 195 males in the sample reported that they watched TV an average of 2.68 hours a day with a standard deviation of 2.388. The 227 females watched an average of 3.21 hours per day with a standard deviation of 3.360. We can see from this output that the sample means are different and that, on the average, females watch more TV. Is the difference between the sample means significant?

The results of the test for significance are reported in the next block of output. *SPSS for Windows* does a separate test for each assumption about the population variance (see Sections 8.2 and 8.3), but we will look only at the "Equal variances assumed" reported in the top row. This is basically the same model used in this chapter.

Skip over the first columns of the output block (reporting the results of a test for equality of the population variances). In the top row, *SPSS for Windows* reports a *t* value (−1.837), the degrees of freedom (df = 420), and the "Sig. (2-tailed)" (.067). This last piece of information is an alpha level except that it is the *exact* probability of getting the observed difference in sample means if only chance is operating. Thus, there is no need to look up the test statistic in a *t* or *Z* table. This value is slightly greater than .05, our usual indicator of significance. We will fail to reject the null hypothesis and conclude that the difference is not statistically significant. There is no difference between the average number of hours spent

Group Statistics

	RESPONDENTS' SEX	N	Mean	Std. Deviation	Std. Error Mean
HOURS PER DAY	MALE	195	2.68	2.388	.171
WATCHING TV	FEMALE	227	3.21	3.360	.223

Independent Samples Test

		Levene's Test for Equality of Variances		T Test for Equality of Means				
		F	Sig.	t	Df	Sig. (2-tailed)	Mean Difference	Std. Error Difference
HOURS PER DAY WATCHING TV	Equal variances assumed	11.136	.001	−1.837	420	.067	−.529	.037
	Equal variances not assumed			−1.884	406.266	.060	−.529	.281

viewing TV between men and women. (Note that, at the 0.10 level, we would have found the difference to be significant.)

SPSS DEMONSTRATION 8.2 Using the Compute Command to Test for Gender Differences in Attitudes About Abortion

The **Compute** command was introduced in Demonstration 4.2. To refresh your memory, I used **Compute** to create a summary scale (*abscale*) for attitudes toward abortion by adding the scores on the two constituent items (*abhlth* and *abany*). Remember that, once created, a computed variable is added to the active file and can be used like any of the variables actually recorded in the file. If you did not save the data file with *abscale* included, you can quickly recreate the variable by following the instructions in Demonstration 4.2.

Here, we will test *abscale* for the significance of the difference by gender. Our question is: Do men and women have different attitudes toward abortion? If the difference in the sample means is large enough, we can reject the null hypothesis and conclude that the populations are different. Before we conduct the test, I should point out that the abortion scale used in this test is only ordinal in level of measurement. Scales like this are often treated as interval-ratio variables, but we should still be cautious in interpreting our results.

Follow the instructions in Demonstration 8.1 for using the **Independent-Samples T Test** command, and move *abscale* rather than *tvhours* to the **Test Variable(s)** box. If necessary, repeat the procedure for making *sex* the grouping variable. Your output will be as shown below.

The sample means are quite close in value (2.70 versus 2.74) and the test statistic ($t = -.607$) is not significant, as indicated by the "Sig. (2-tailed)" value of .544. There is no significant difference in support for abortion by gender.

As a final point, let me direct your attention to the *N* column. This test was based on 195 men and 226 women, for a total of 421 people. The original sample included almost 1,500 people. What happened to all those cases?

Group Statistics

RESPONDENTS' SEX		N	Mean	Std. Deviation	Std. Error Mean
abscale	MALE	195	2.7026	.68294	.04898
	FEMALE	226	2.7434	.68998	.04590

Independent Samples Test

		Levene's Test for Equality of Variances		T Test for Equality of Means				
		F	Sig.	t	Df	Sig. (2-tailed)	Mean Difference	Std. Error Difference
HOURS PER DAY WATCHING TV	Equal variances assumed	.033	.857	−.607	419	.544	−.0408	.06717
	Equal variances not assumed			−.608	411.029	.544	−.0408	.06712

Recall from Appendix F that relatively few items on the GSS are given to the entire sample. Furthermore, as I mentioned at the end of Chapter 4, the **Compute** command is designed to drop a case from the computations if it is missing a score on any of the constituent variables. So, most of the "missing cases" were not asked about their attitudes about abortion, and the rest did not answer one or both of the constituent abortion items. This phenomenon of diminishing sample size is a common problem in survey research that, at some point, may jeopardize the integrity of the inquiry.

Exercises

8.1 Are men significantly different from women in occupational prestige or education? Using Demonstration 8.1 as a guide, substitute *prestg80* and *educ* for *tvhours*. Write a sentence or two summarizing the results of this test.

8.2 Using Demonstration 8.2 as a guide, construct a summary scale for attitudes about traditional gender roles using *fepresch* and *fefam* in place of *abhlth* and *abany*. Test for the significance of the difference using *sex* as the independent variable. Write a sentence or two summarizing the test results.

8.3 What other independent variables might explain differences in opinions about abortion? Select several more independent variables besides *sex* and conduct additional *t* tests with *abscale* as the dependent variable. Remember that the *t* test requires the independent variable to have *only* two categories. For variables with more than two categories (*relig* or *attend*, for example), you can meet this requirement in one of two ways. First, you can use the **Define Groups** button in the **Grouping Variables** box to select specific categories of a variable. You could, for example, compare Protestants and Catholics by choosing scores of 1 (Protestants) and 2 (Catholics).

Second, you can collapse the scores of variables with more than two categories, such as *attend,* by using the **Recode** command. Use Demonstration 2.2 as a guide and collapse your selected independent variables at the median or some other logical point. Don't forget to choose the "recode into different variable" option. Summarize your results in a sentence or two.

9

Hypothesis Testing III
The Analysis of Variance

By the end of this chapter, you will be able to:

1. Identify and cite examples of situations in which ANOVA is appropriate.
2. Explain the logic of hypothesis testing as applied to ANOVA.
3. Perform the ANOVA test, using the five-step model as a guide, and correctly interpret the results.
4. Define and explain the concepts of population variance, total sum of squares, the sum of squares between, the sum of squares within, and mean square estimates.
5. Explain the difference between the statistical significance and the importance of relationships between variables.

9.1 INTRODUCTION

In this chapter, we will examine a very flexible and widely used test of significance called the **analysis of variance** (often abbreviated as **ANOVA**). This test is designed for use with interval-ratio-level dependent variables and is a powerful tool for analyzing the most sophisticated and precise measurements you are likely to encounter.

It is perhaps easiest to think of ANOVA as an extension of the *t* test for the significance of the difference between two-sample means, which Chapter 8 discussed. We can use the *t* test only in situations in which our independent variable has exactly two categories (e.g., Protestants and Catholics). The analysis of variance, on the other hand, is appropriate for independent variables with more than two categories (e.g., Protestants, Catholics, Jews, people with no religious affiliation, and so forth).

To illustrate, suppose we were interested in examining the social basis of support for capital punishment. Why does support for the death penalty vary from person to person? Could there be a relationship between religion (the independent variable) and support for capital punishment (the dependent variable)? Opinion about the death penalty has an obvious moral dimension and may well be affected by a person's religious background.

Suppose we administered a scale that measures support for capital punishment at the interval-ratio level to a randomly selected sample that includes Protestants, Catholics, Jews, people with no religious affiliation ("None"), and people from other religions ("Other"). We want to see if there is significant variation in support for the death penalty among these five religious affiliations. We will also want to answer other questions such as: Which religion shows the least or most support for capital punishment? Are Protestants significantly more supportive than Catholics or Jews? How do people with no religious affiliation compare to people in the other categories? The analysis of variance provides a very useful statistical context for addressing such questions.

**9.2 THE LOGIC OF THE
ANALYSIS OF VARIANCE**

For ANOVA, the null hypothesis is that the populations from which the samples are drawn are equal on the characteristic of interest. As applied to our problem, the null hypothesis could be phrased as "People from different religious denominations do not vary in their support for the death penalty" or symbolically as $\mu_1 = \mu_2 = \mu_3 = \ldots = \mu_k$. (Note that this is an extended version of the null hypothesis for the two-sample t test.) As usual, the researcher will normally be interested in rejecting the null hypothesis and, in this case, showing that support is related to religion.

If the null hypothesis of "no difference" among the various religious populations (*all* Catholics, *all* Protestants, and so forth) is true, then any means calculated from randomly selected samples should be roughly equal in value. If the populations are truly the same, the average score for the Protestant sample should be about the same as the average score for the Catholic sample, the Jewish sample, and so forth. Note that the averages are unlikely to be exactly the same value even if the null hypothesis really is true because we will always encounter some error or chance fluctuations in the measurement process. We are *not* asking: "Are there differences among the samples or categories of the independent variable (or, in our example, the religions)?" Rather, we are asking: "Are the differences among the samples large enough to reject the null hypothesis and justify the conclusion that the populations represented by the samples are different?"

Now consider what kinds of outcomes we might encounter if we actually administered a "Support of Capital Punishment Scale" and organized the scores by religion. Of the infinite variety of possibilities, let's focus on two extreme outcomes as exemplified by Tables 9.1 and 9.2. In the first set of hypothetical results (Table 9.1), we see that the means and standard deviations of the groups are similar. The average scores are about the same for every religious group, and all five groups exhibit about the same dispersion. These results would be consistent with the null hypothesis of no difference. Neither the average score nor the dispersion of the scores changes in any important way by religion.

Now consider another set of fictitious results in Table 9.2. Here we see substantial differences in average score from category to category, with Jews showing the lowest support and Protestants showing the highest. Also, the standard deviations are low and similar from category to category, indicating that there is not much variation within the religions. Table 9.2 shows marked differences *among* religions combined with homogeneity *within* religions, as indicated by the low values of the standard deviations. These results would contradict the null hypothesis and support the notion that support for the death penalty does vary by religion.

The ANOVA test is based on the kinds of comparisons just outlined. The test compares the amount of variation between categories (e.g., from Protestant

TABLE 9.1   SUPPORT FOR CAPITAL PUNISHMENT BY RELIGION: MEANS ARE SIMILAR (fictitious data)

	Protestant	Catholic	Jewish	None	Other
Mean	10.3	11.0	10.1	9.9	10.5
Standard Deviation	2.4	1.9	2.2	1.7	2.0

TABLE 9.2 SUPPORT FOR CAPITAL PUNISHMENT BY RELIGION: MEANS ARE NOT SIMILAR (fictitious data)

	Protestant	Catholic	Jewish	None	Other
Mean	14.7	11.3	5.7	8.3	7.1
Standard Deviation	2.4	1.9	2.2	1.7	2.0

to Catholic to Jewish to "None" to "Other") with the amount of variation within categories (among Protestants, among Catholics, and so forth). The greater the differences *between* categories, relative to the differences *within* categories, the more likely that the null hypothesis of "no difference" is false and can be rejected. If support for capital punishment truly varies by religion, then the sample mean for each religion should be quite different from the others and dispersion within the categories should be relatively low.

9.3 THE COMPUTATION OF ANOVA

Even though we have been thinking of ANOVA as a test for the significance of the difference between sample means, the computational routine actually involves developing two separate estimates of the population variance, σ^2 (hence the name "analysis of variance"). Recall from Chapter 4 that the variance and standard deviation both measure dispersion and that the variance is simply the standard deviation squared. One estimate of the population variance is based on the amount of variation *within* each of the categories of the independent variable, and the other is based on the amount of variation *between* categories.

Before constructing these estimates, we need to introduce some new concepts and statistics. The first new concept is the total variation of the scores, which we measure by a quantity called the **total sum of squares,** or **SST.**

FORMULA 9.1
$$\text{SST} = \Sigma(X_i - \overline{X})^2$$

To find this quantity, we would take each score, subtract the mean, square the difference, and then add to get the total of the squared differences. If this formula seems vaguely familiar, it's because the same expression appears in the numerator of the formula for the sample variance and the standard deviation (see Chapter 4). This redundancy is not surprising, considering that the SST is also a way of measuring dispersion.

To construct the two separate estimates of the population variance, we divide the total variation (SST) into two components. One of these reflects the pattern of variation within the categories and is called the **sum of squares within (SSW).** The other component is based on the variation between categories and is called the **sum of squares between (SSB).** SSW and SSB are components of SST, as Formula 9.2 indicates:

FORMULA 9.2
$$\text{SST} = \text{SSB} + \text{SSW}$$

The sum of squares within is defined as:

FORMULA 9.3
$$\text{SSW} = \Sigma(X_i - \overline{X}_k)^2$$

where SSW = the sum of squares within the categories
$\overline{X}_k$ = the mean of a category

This formula directs us to take each score, subtract the mean of the category from the score, square the result, and then sum the squared differences. We will do this for each category separately and then sum the squared differences for all categories to find SSW.

The sum of squares between (SSB) reflects the variation between the samples, with the category means serving as summary statistics for each category. Each category mean will be treated as a "case" for purposes of this estimate. The formula for the sum of squares between (SSB) is:

FORMULA 9.4
$$\text{SSB} = \Sigma N_k(\overline{X}_k - \overline{X})^2$$

where SSB = the sum of squares between the categories
N_k = the number of cases in a category
$\overline{X}_k$ = the mean of a category

To find SSB, subtract the overall mean of all scores from each category mean, square the difference, multiply by the number of cases in the category, and add the results across all the categories.

Let's pause to remember what we are after here. If the null hypothesis is true, then there should not be much variation from category to category, relative to the variation within categories, and the two estimates to the population variance based on SSW and SSB should be roughly equal. The larger the difference between the two estimates, the more likely we will be to reject the null hypothesis. If the category means are about the same value, the differences will not be significant. The larger the differences between category means and the more homogeneous the categories, the more likely that the differences are statistically significant.

The next step in the computational routine is to construct the estimates of the population variance. To do this, we will divide each sum of squares by its respective degrees of freedom. To find the degrees of freedom associated with SSW, subtract the number of categories (k) from the number of cases (N). The degrees of freedom associated with SSB are the number of categories minus one. In summary,

FORMULA 9.5
$$\text{dfw} = N - k$$

where dfw = degrees of freedom associated with SSW
N = total number of cases
k = number of categories

FORMULA 9.6
$$\text{dfb} = k - 1$$

where dfb = degrees of freedom associated with SSB
k = number of categories

The actual estimates of the population variance, called the **mean square estimates,** are calculated by dividing each sum of squares by its respective degrees of freedom:

FORMULA 9.7
$$\text{Mean square within} = \text{SSW/dfw}$$

FORMULA 9.8
$$\text{Mean square between} = \text{SSB/dfb}$$

The test statistic calculated in step 4 of the five-step model is called the ***F* ratio,** and its value is determined by the following formula:

FORMULA 9.9
$$F = \text{Mean square between/Mean square within}$$

As you can see, the value of the F ratio will be a function of the amount of variation between categories compared to the amount of variation within the categories. The greater the variation between the categories relative to the variation within, the higher the value of the F ratio and the more likely we will reject the null hypothesis.

9.4 A COMPUTATIONAL SHORTCUT

The computational routine for ANOVA, as summarized in the previous section, requires a number of separate steps and different formulas and will surely seem complicated the first time you see it. As is almost always the case, if you proceed systematically from formula to formula (in the correct order, of course), you will see that the computations aren't nearly as formidable as they appear initially. Unfortunately, they will still be lengthy and time consuming. So, at the risk of stretching your patience, let me introduce a way to save some time and computational effort. This will require the introduction of even more formulas, but the eventual savings in time will be worth the effort. Formula 9.10 shows a quicker, more convenient way to calculate SST:

FORMULA 9.10

$$SST = \Sigma X^2 - N\bar{X}^2$$

To solve this formula, first find the sum of the squared scores (in other words, square each score and then add up the squared scores). Next, square the overall mean, multiply that value by the total number of cases in the sample (N), and subtract that quantity from the sum of the squared scores.

Once we have the values of SST and SSB, we can find SSW by manipulating Formula 9.2 and doing some simple subtraction:

FORMULA 9.11

$$SSW = SST - SSB$$

STEP BY STEP	**Computing ANOVA**

Step 1: Find SST by Formula 9.10: $SST = \Sigma X^2 - N\bar{X}^2$
  a. To find ΣX^2, square each score and then add the squared scores together.
  b. To find $N\bar{X}^2$, square the value of the mean of all scores and then multiply the result by N.
  c. Subtract the quantity you found in step b from the quantity you found in step a.

Step 2: Find SSB by Formula 9.4:
$$SSB = \Sigma N_k(\bar{X}_k - \bar{X})^2$$
  a. Subtract the mean of all scores ($\bar{X}$) from the mean of each category ($\bar{X}_k$) and then square each difference.
  b. Multiply each of the squared differences you found in step a by the number of cases in the category (N_k).
  c. Add the quantities you found in step b together.

Step 3: Find SSW by Formula 9.11: $SSW = SST - SSB$. Subtract the value of SSB (see step 2) from the value of SST (see step 1).

Step 4: Calculate degrees of freedom:
  a. For dfw, use Formulas 9.5: $dfw = N - k$. Subtract the number of categories (k) from the number of cases (N).
  b. For dfb, use Formula 9.6: $dfb = k - 1$. Subtract 1 from the number of categories (k).

Step 5: Construct the two mean square estimates to the population variance:
  a. To find the mean square within (MSW), use Formula 9.7: $MSW = SSW/dfw$. Divide SSW (see step 3) by dfw (see step 4a).
  b. For the mean square between (MSB), use Formula 9.8: $MSB = SSB/dfb$. Divide SSB (see step 2) by dfb (see step 4b).

Step 6: Find the obtained F ratio by Formula 9.9: $F = MSB/MSW$. Divide the MSB estimate (see step 5b) by the MSW estimate (see step 5a).

The revised computational routine for ANOVA is summarized in the preceding step-by-step box. These computations and the actual test of significance will be illustrated in the next sections.

9.5 A COMPUTATIONAL EXAMPLE

Assume that we have administered our "Support for Capital Punishment Scale" to a sample of 20 individuals who are equally divided into the five religious categories. (Obviously, this sample is much too small for any serious research and is intended solely for purposes of illustration.) Table 9.3 reports all scores, along with the squared scores, the category means, and the overall mean.

TABLE 9.3   SUPPORT FOR CAPITAL PUNISHMENT BY RELIGION FOR 20 SUBJECTS (fictitious data)

Protestant		Catholic		Jewish		None		Other	
X	X^2	X	X^2	X	X^2	X	X^2	X	X^2
8	64	12	144	12	144	15	225	10	100
12	144	20	400	13	169	16	256	18	324
13	169	25	625	18	324	23	529	12	144
17	289	27	729	21	441	28	784	12	144
50	666	84	1,898	64	1,078	82	1,794	52	712
$\bar{X}_k = 12.5$		$\bar{X}_k = 21.0$		$\bar{X}_k = 16.0$		$\bar{X}_k = 20.5$		$\bar{X}_k = 13.0$	
				$\bar{X}_k = 16.6$					

To organize our computations, we'll follow the routine summarized in the step-by-step box. These steps are presented in Table 9.4.

TABLE 9.4   COMPUTING ANOVA: SUPPORT FOR CAPITAL PUNISHMENT BY RELIGION

Step	Quantity	Formula	Solution
1	SST	9.10 SST = $\Sigma X^2 - N\bar{X}^2$	SST = 6,148 − (20)(16.6)^2 = **636.8**
2	SSB	9.4 SSB = $\Sigma N_k(\bar{X}_K - \bar{X})^2$	SSB = 4(12.5 − 16.6)^2 + 4(21.0 − 16.6)^2 + 4(16.0 − 16.6)^2 + 4(20.5 − 16.6)^2 + 4(13.0 − 16.6)^2 = 67.24 + 77.44 + 1.44 + 60.84 + 51.84 = **258.80**
3	SSW	9.11 SSW = SST − SSB	SSW = 636.8 − 258.8 = **378.00**
4	dfw	9.5 dfw = $N - k$	dfw = 20 − 5 = **15**
	dfb	9.6 dfb = $k - 1$	dfb = 5 − 1 = **4**
5	MSW	9.7 MSW = SSW/dfw	MSW = 378.00/15 = **25.20**
	MSB	9.8 MSB = SSB/dfb	MSB = 258.80/4 = **64.70**
6	F ratio	9.9 F = MSW/MSB	F = 64.70/25.20 = **2.57**

The F ratio computed in step 6 of the step-by-step box must still be evaluated for its significance. (*Solve any of the end-of-chapter problems to practice computing these quantities and solving these formulas.*)

9.6 A TEST OF SIGNIFICANCE FOR ANOVA

In this section, we will see how to test an F ratio for significance and also look at some of the assumptions underlying the ANOVA test. As usual, we will follow the five-step model as a convenient way of organizing our decisions.

Step 1. Making Assumptions and Meeting Test Requirements.

> Model: Independent random samples
> Level of measurement is interval-ratio
> Populations are normally distributed
> Population variances are equal

The model assumptions are stringent and underscore the fact that we should use ANOVA only with dependent variables that have been carefully and precisely measured. However, as long as sample sizes are equal (or nearly so), ANOVA can tolerate some violation of the model assumptions. In situations where you are uncertain or have samples of very different size, it is probably advisable to use an alternative test. (Chi square in Chapter 10 is one option.)

Step 2. Stating the Null Hypothesis. For ANOVA, the null hypothesis always states that the means of the populations from which the samples were drawn are equal. For our example problem, we are concerned with five different populations or categories, so our null hypothesis would be:

$$H_0: \mu_1 = \mu_2 = \mu_3 = \mu_4 = \mu_5$$

where μ_1 represents the mean for Protestants, μ_2, the mean for Catholics, and so forth.

The alternative hypothesis states simply that at least one of the population means is different. The wording here is important. If we reject the null, ANOVA does not identify which mean or means are significantly different, but we can usually identify the most important differences by inspecting the sample means.

(H_1: At least one of the population means is different.)

Step 3. Selecting the Sampling Distribution and Establishing the Critical Region. The sampling distribution for ANOVA is the F distribution, which Appendix D summarizes. Note that there are separate tables for alphas of .05 and .01, respectively. As with the t table, the value of the critical F score will vary by degrees of freedom. For ANOVA, there are two separate degrees of freedom, one for each estimate of the population variance. The numbers across the top of the table are the degrees of freedom associated with the between estimate (dfb), and the numbers down the side of the table are those associated with the within estimate (dfw). In our example, dfb is $(k - 1)$, or 4, and dfw is $(N - k)$, or 15 (see Formulas 9.5 and 9.6). So, if we set alpha at .05, our critical F score will be 3.06.

Summarizing these considerations:

> Sampling distribution $= F$ distribution
> Alpha $= .05$
> Degrees of freedom (within) $= (N - k) = 15$
> Degrees of freedom (between) $= (k - 1) = 4$
> F(critical) $= 3.06$

Taking a moment to inspect the two F tables, you will notice that all the values are greater than 1.00. This is because ANOVA is a one-tailed test, and we are concerned only with outcomes in which there is more variance between categories than within categories. F values of less than 1.00 would indicate that the between estimate was lower in value than the within estimate and, because we would always fail to reject the null in such cases, we simply ignore this class of outcomes.

Step 4. Computing the Test Statistic. We computed the test statistics in the previous section, where we found an obtained F ratio of 2.57.

Step 5. Making a Decision and Interpreting the Results of the Test. Compare the test statistic with the critical value:

$$F\text{(critical)} = 3.06$$
$$F\text{(obtained)} = 2.57$$

Because the test statistic does not fall into the critical region, our decision would be to fail to reject the null. Support for capital punishment does not differ significantly by religion, and the variation we observed in the sample means is unimportant.

Application 9.1

A large university recently conducted an experiment in teaching introductory biology. One section was taught by the traditional lecture-lab method, a second was taught by an all-lab/demonstration approach with no lectures, and a third was taught entirely by a series of videotaped lectures and demonstrations that the students were free to view at any time and as often as they wanted. Students were randomly assigned to each of the three sections and, at the end of the semester, random samples of final exam scores were collected from each section. Is there a significant difference in student performance by teaching method?

FINAL EXAM SCORES BY TEACHING METHOD

Lecture		Demonstration		Videotape	
X	X^2	X	X^2	X	X^2
55	3,025	56	3,136	50	2,500
57	3,249	60	3,600	52	2,704
60	3,600	62	3,844	60	3,600
63	3,969	67	4,489	61	3,721
72	5,184	70	4,900	63	3,969
73	5,329	71	5,041	69	4,761
79	6,241	82	6,724	71	5,041
85	7,225	88	7,744	80	6,400
92	8,464	95	9,025	82	6,724
$\Sigma X = 636$	$\Sigma X^2 = 46{,}286$	$\Sigma X = 651$	$\Sigma X^2 = 48{,}503$	$\Sigma X = 588$	$\Sigma X^2 = 39{,}420$
$\bar{X}_k = 70.67$		$\bar{X}_k = 72.33$		$\bar{X}_k = 65.33$	

$$\bar{X} = 1{,}875/27 = 69.44$$

We can see by inspection that the "Videotape" group had the lowest average score and that the "Demonstration" group had the highest average score. The ANOVA test will tell us if these differences are large enough to justify the conclusion that they did not occur by chance alone. The table below follows the computational routine established at the end of Section 9.4:

(continued next page)

Application 9.1 (continued)

COMPUTING ANOVA

Step	Quantity	Formula	Solution
1	SST	9.10 SST $= \Sigma X^2 - N\bar{X}^2$	SST $= (46{,}286 + 48{,}503 + 39{,}420) -$ $27(69.44)^2 = \mathbf{4{,}017.33}$
2	SSB	9.4 SSB $= \Sigma N_k(\bar{X}_k - \bar{X})^2$	SSB $= (9)(70.67 - 69.44)^2 - (9)(72.33 -$ $69.44)^2 + (9)(65.33 - 69.44)^2 =$ $13.62 + 75.17 + 152.03 = \mathbf{240.82}$
3	SSW	9.11 SSW $=$ SST $-$ SSB	SSW $= 4{,}017.33 - 240.82 = \mathbf{3{,}776.51}$
4	dfw	9.5 dfw $= N - k$	dfw $= 27 - 3 = \mathbf{24}$
	dfb	9.6 dfb $= k - 1$	dfb $= 3 - 1 = \mathbf{2}$
5	MSW	9.7 MSW $=$ SSW/dfw	MSW $= 3776.51/24 = \mathbf{157.36}$
	MSB	9.8 MSB $=$ SSB/dfb	MSB $= 240.82/2 = \mathbf{120.41}$
6	F ratio	9.9 F $=$ MSW/MSB	F $= 120.41/157.36 = \mathbf{0.77}$

We can now conduct the test of significance.

Step 1. Making Assumptions and Meeting Test Requirements.

> Model: Independent random samples
> Level of measurement is interval-ratio
> Populations are normally distributed
> Population variances are equal

Step 2. Stating the Null Hypothesis.

$$H_0: \mu_1 = \mu_2 = \mu_3$$

(H_1: At least one of the population means is different.)

Step 3. Selecting the Sampling Distribution and Establishing the Critical Region.

> Sampling distribution $= F$ distribution
> Alpha $= .05$

Degrees of freedom (within) $= (N - k)$
$$= (27 - 3) = 24$$
Degrees of freedom (between) $= (k - 1)$
$$= (3 - 1) = 2$$
$$F(\text{critical}) = 3.40$$

Step 4. Computing the Test Statistic. We found an obtained F ratio of 0.77.

Step 5. Making a Decision and Interpreting the Results of the Test. Compare the test statistic with the critical value:

$$F(\text{critical}) = 3.40$$
$$F(\text{obtained}) = 0.77$$

We would clearly fail to reject the null hypothesis ("the population means are equal") and would conclude that the observed differences among the category means were the results of random chance. Student performance in this course does not vary significantly by teaching method.

9.7 AN ADDITIONAL EXAMPLE FOR COMPUTING AND TESTING THE ANALYSIS OF VARIANCE

In this section, we will work through an additional example of the computation and interpretation of the ANOVA test. We will first review matters of computation, find the obtained F ratio, and then test the statistic for its significance. In the computational section, we will follow the step-by-step guidelines presented at the end of Section 9.4.

A researcher is evaluating the efficiency with which each of three social service agencies is administering a particular program. One area of concern is the speed of the agencies in processing paperwork and determining the eligibility of potential clients. The researcher has gathered information on the number of days required for processing a random sample of 10 cases in each agency. Is there a significant difference? The data appear in Table 9.5, which also includes some additional information we will need to complete our calculations.

Table 9.6 gives the actual computations for ANOVA, once again following the computational routine introduced at the end of Section 9.4.

TABLE 9.5 NUMBER OF DAYS REQUIRED TO PROCESS CASES FOR THREE AGENCIES (fictitious data)

Client	Agency A		Agency B		Agency C	
	X	X^2	X	X^2	X	X^2
1	5	25	12	144	9	81
2	7	49	10	100	8	64
3	8	64	19	361	12	144
4	10	100	20	400	15	225
5	4	16	12	144	20	400
6	9	81	11	121	21	441
7	6	36	13	169	20	400
8	9	81	14	196	19	361
9	6	36	10	100	15	225
10	6	36	9	81	11	121
	$\Sigma X = 70$	$\Sigma X^2 = 524$	$\Sigma X = 130$	$\Sigma X^2 = 1,816$	$\Sigma X = 150$	$\Sigma X^2 = 2,462$
		$\bar{X}_k = 7.0$		$\bar{X}_k = 13.0$		$\bar{X}_k = 15.0$

$$\bar{X}_k = 350/30 = 11.67$$

TABLE 9.6 COMPUTING ANOVA: PROCESSING SPEED BY AGENCY

Step	Quantity	Formula	Solution
1	SST	9.10 SST$= \Sigma X^2 - N\bar{X}^2$	SST $= (524 + 1,816 + 2,462) - 30(11.67)^2 = 4,802 - 30(136.19) = 4,802 - 4,085.7 = $ **716.30**
2	SSB	9.4 SSB $= \Sigma N_k(\bar{X}_K - \bar{X})^2$	SSB $= (10)(7.0\text{-}11.67)^2 + (10)(13.0 - 11.67)^2 + (10)(15.0 - 11.67)^2 = (10)(21.81) + (10)(1.77) + (10)(11.09) = 218.09 + 17.70 + 110.89 = $ **346.68**
3	SSW	9.11 SSW $=$ SST $-$ SSB	SSW $= 716.30 - 346.68 = $ **369.62**
4	dfw	9.5 dfw $= N - k$	dfw $= 30 - 3 = $ **27**
	dfb	9.6 dfb $= k - 1$	dfb $= 3 - 1 = $ **2**
5	MSW	9.7 MSW $=$ SSW/dfw	MSW $= 369.62/27 = $ **13.69**
	MSB	9.8 MSB $=$ SSB/dfb	MSB $= 346.68/2 = $ **173.34**
6	F ratio	9.9 $F =$ MSW/MSB	$F = 173.35/13.69 = $ **12.66**

We can now test the F ratio for its significance.

Step 1. **Making Assumptions and Meeting Test Requirements.**

Model: Independent random samples
Level of measurement is interval-ratio
Populations are normally distributed
Population variances are equal

The researcher will always be in a position to judge the adequacy of the first two assumptions in the model. The second two assumptions are more problematical, but remember that ANOVA will tolerate some deviation from its assumptions as long as sample sizes are roughly equal.

Step 2. Stating the Null Hypothesis.

$$H_0: \mu_1 = \mu_2 = \mu_3$$

(H_1: At least one of the population means is different.)

Step 3. Selecting the Sampling Distribution and Establishing the Critical Region.

Sampling distribution = F distribution
Alpha = .05
Degrees of freedom (within) = $(N - k) = (30 - 3) = 27$
Degrees of freedom (between) = $(k - 1) = (3 - 1) = 2$
F(critical) = 3.35

Step 4. Computing the Test Statistic. We found an obtained F ratio of 12.66.

Step 5. Making a Decision and Interpreting the Results of the Test. Compare the test statistic with the critical value:

$$F(\text{critical}) = 3.35$$
$$F(\text{obtained}) = 12.66$$

The test statistic is in the critical region, and we would reject the null of no difference. The differences between the three agencies are very unlikely to have occurred by chance alone. The agencies are significantly different in the speed with which they process paperwork and determine eligibility. *(For practice in conducting the ANOVA test, see problems 9.2 to 9.9. Begin with the lower-numbered problems; they have smaller data sets, fewer categories, and, therefore, the simplest calculations.)*

STATISTICS IN EVERYDAY LIFE: Who Gets Rich?

Does social inequality reproduce itself from generation to generation? How much of an advantage do children of well-to-do parents have over children from families of more modest circumstances? Are high levels of education and income reserved for the children of the affluent or can anyone achieve success through his or her individual effort and hard work?

Although we cannot offer a definitive answer to these questions, we can use ANOVA to investigate the importance of the social class of a person's family of origin as an explanation for individual success. If social inequality does reproduce itself across the generations, there should be large,

statistically significant differences between people born to privilege and people born to humble backgrounds. If, on the other hand, success is simply a matter of hard work and effort—as many Americans argue—we should find that the parents' social class has little effect on (or "no significant relationship with") the success of their children.

To assess the sources of individual mobility and success in American life, I used the 3,000 respondents to the 2004 General Social Survey. As you recall, the GSS respondents are an EPSEM sample, and any conclusions we reach about the sample will apply to the entire adult population of the United States. To conduct the test, I divided the

(continued)

STATISTICS IN EVERYDAY LIFE *(continued)*

sample into five categories based on the educational level of the respondents' parents. The least advantaged respondents were those who came from families in which neither parent had finished high school ("Neither HS"), and the most privileged respondents came from families in which both parents held college degrees ("Both BA"). To avoid the confounding effects of race, the test was restricted to whites only.

The table below reports the results. For each category of parents' education, the table shows the mean years of respondent's education ("R's Educ.") and income ("R's Inc."). The table shows that respondents with the least educated parents averaged 12.26 years of education and had an average income of $31,166. In contrast, respondents with the most educated parents averaged 16.19 years of education (nearly four years more than the least advantaged respondents) and had an average income of $50,189. Note also that the sample means increase in regular fashion across the levels of

parents' social class. For both education and income, people from the lowest social classes had the lowest levels of success, and the levels increase step by step for each social class.

I conducted two separate analysis of variance tests, one for each dependent variable. The table also shows the *F* ratio and sample size for both tests. Both *F* ratios were statistically significant at the 0.05 level. In other words, there was a significant difference in average years of education and average income by the level of parents' education. These results strongly suggest that social inequality does in fact reproduce itself. Of course, there is variation within each category (as indicated by the standard deviations), and a person's social class at birth does not simply determine his or her success in life. Still, these results demonstrate that people who have the good sense to be born to affluence have a strong advantage in the competition for success in American society.

Respondent's Success by Parents' Education

	Parents' Education							
R's Educ.	*Neither HS*	*One HS*	*Both HS*	*One BA*	*Both BA*	**F ratio**	**Significant at 0.05?**	**N**
$\bar{X}$ s =	12.26	13.56	14.13	15.34	16.19	*116.951*	Yes	**2,062**
s =	3.024	2.559	2.352	2.425	2.306			
N	*458*	*384*	*695*	*308*	*218*			
R's Inc.*								
$\bar{X}$ s =	$31,166	$36,306	$41,124	$44,192	$50,189	*10.068*	Yes	1,275
s =	24,968	29,58	31,284	34,997	36,060			
N	*209*	*222*	*482*	*209*	*154*			

* The GSS has no variable that measures income in dollars. Instead, I used a variable that measured income in broad categories ($20,000 to $24,999, for example) and took the midpoint of each interval to represent income ($22,500 for the $20,000 to $24,999 interval, for example).

9.8 THE LIMITATIONS OF THE TEST

ANOVA is appropriate whenever you want to test differences between the means of an interval-ratio-level variable across three or more categories of an independent variable. This application is called **one-way analysis of variance** because it involves the effect of a single variable (for example, religion) on another (for example, support for capital punishment). This is the simplest application of ANOVA, and you should be aware that the technique has numerous more advanced and complex forms. For example, you may encounter research projects that study the effects of two separate variables (for example, religion and gender) on some third variable.

One important limitation of ANOVA is that it requires interval-ratio measurement of the dependent variable and roughly equal numbers of cases in each of the categories of the independent variable. The former condition may be difficult to meet with complete confidence for many variables of interest to the social sciences. The latter condition may create problems when the research hypothesis calls for comparisons between groups that are, by their nature, unequal in numbers (for example, white versus black Americans) and may call for some unusual sampling schemes in the data-gathering phase of a research project. Because ANOVA can tolerate some deviation from its model assumptions, neither limitation should be crippling, but you should be aware of these limitations in planning your own research and in judging the adequacy of research conducted by others.

A second limitation of ANOVA actually applies to all forms of significance testing and was introduced in Section 8.5. These tests are designed to detect nonrandom differences: differences so large that they are very unlikely to be produced by random chance alone. The problem is that differences that are statistically significant are not necessarily important in any other sense. Statistical techniques that can assess the importance of results directly appear in Parts III and IV of this text.

A final limitation of ANOVA relates to the research hypothesis. As you recall, when the null hypothesis is rejected, the alternative hypothesis is supported. The limitation is that the alternative hypothesis is not specific: It simply asserts that at least one of the population means is different from the others. Obviously, we would like to know which differences are significant. We can sometimes determine this by simple inspection. In our problem involving social service agencies, for example, Table 9.5 indicates that Agency A is the source of most of the differences. This informal, "eyeball" method can be misleading, however, and you should exercise caution in making conclusions about which means are significantly different.

SUMMARY

1. One-way analysis of variance is a powerful test of significance that researchers commonly use when comparisons across more than two categories or samples are of interest. It is perhaps easiest to conceptualize ANOVA as an extension of the test for the difference in sample means.

2. ANOVA compares the amount of variation within the categories to the amount of variation between categories. If the null hypothesis of no difference is false, there should be relatively great variation between categories and relatively little variation within categories. The greater the differences from category to category relative to the differences within the categories, the more likely we will be able to reject the null.

3. The computational routine for even simple applications of ANOVA can quickly become complex. The basic process is to construct separate estimates of the population variance based on the variation within the categories and the variation between the categories. The test statistic is the F ratio, which is based on a comparison of these two estimates. The step-by-step box summarizes the basic computational routine. In addition, statistical packages such as SPSS are widely available to perform complex calculations such as these accurately and quickly. If you haven't yet learned how to use such programs, ANOVA may provide you with the necessary incentive.

4. The ANOVA test can be organized into the familiar five-step model for testing the significance of

sample outcomes. Although the model assumptions (step 1) require high-quality data, the test can tolerate some deviation as long as sample sizes are roughly equal. The null takes the familiar form of stating that there is no difference of any importance among the population values, while the alternative hypothesis asserts that at least one population mean is different. The sampling distribution is the F distribution, and the test is always one tailed. The decision to reject or to fail to reject the null is based on a comparison of the obtained F ratio with the critical F ratio as determined for a given alpha level and degrees of freedom. The decision to reject the null indicates only that one or more of the population means is different from the others. We can often determine which sample mean(s) account for the difference by inspecting the sample data, but this informal method should be used with caution.

SUMMARY OF FORMULAS

Total sum of squares	9.1	$SST = \Sigma(X_i - \overline{X})^2$
The two components of the total sum of squares	9.2	$SST = SSB + SSW$
Sum of squares within	9.3	$SSW = \Sigma(X_i - \overline{X}_k)^2$
Sum of squares between	9.4	$SSB = \Sigma N_k(\overline{X}_k - \overline{X})^2$
Degrees of freedom for SSW	9.5	$dfw = N - k$
Degrees of freedom for SSB	9.6	$dfb = k - 1$
Mean square within	9.7	Mean square within = SSW/dfw
Mean square between	9.8	Mean square between = SSB/dfb
F ratio	9.9	F = Mean square between/Mean square within
Computational formula for SST	9.10	$SST = \Sigma X^2 - N\overline{X}^2$
Finding SSW by subtraction	9.11	$SSW = SST - SSB$

GLOSSARY

Analysis of variance. A test of significance appropriate for situations in which we are concerned with the differences among more than two sample means.

ANOVA. See Analysis of variance.

F ratio. The test statistic computed in step 4 of the ANOVA test.

Mean square estimate. An estimate of the variance calculated by dividing the sum of squares within (SSW) or the sum of squares between (SSB) by the proper degrees of freedom.

One-way analysis of variance. Applications of ANOVA in which the effect of a single independent variable on a dependent variable is observed.

Sum of squares between (SSB). The sum of the squared deviations of the sample means from the overall mean, weighted by sample size.

Sum of squares within (SSW). The sum of the squared deviations of scores from the category means.

Total sum of squares (SST). The sum of the squared deviations of the scores from the overall mean.

MULTIMEDIA RESOURCES

 The Wadsworth Sociology Resource Center: Virtual Society at
http://www.thomsonedu.com/sociology

Visit the companion website for the seventh edition of *Statistics: A Tool for Social Research* to access a wide range of student resources. Begin by clicking on the Student Resources section of the book's website to access the following study tools:

- Basic math review
- Flash cards

- Additional chapter problems
- Internet links
- Table of random numbers
- MicroCase and SPSS examples and exercises
- Hypothesis testing for variables measured at the ordinal level

PROBLEMS

(*NOTE:* The number of cases in these problems is very low—a fraction of the sample size necessary for any serious research—to simplify computations.)

9.1 Conduct the ANOVA test for each set of scores below. (*HINT: Follow the computational shortcut outlined in Section 9.4 and keep track of all sums and means by constructing computational tables like Table 9.3 or 9.4.)*

a.

	Category	
A	B	C
5	10	12
7	12	16
8	14	18
9	15	20

b.

	Category	
A	B	C
1	2	3
10	12	10
9	2	7
20	3	14
8	1	1

c.

A	B	C	D
13	45	23	10
15	40	78	20
10	47	80	25
11	50	34	27
10	45	30	20

9.2 SOC   What type of person is most involved in the neighborhood and community? Who is more likely to volunteer for organizations such as PTA, scouts, or Little League? In a random sample, 15 people have been asked for their number of memberships in community voluntary organizations and some other information. Which differences are significant?

a.

Membership by Education		
Less Than High School	High School	College
0	1	0
1	3	3
2	3	4
3	4	4
4	5	4

b.

Membership by Length of Residence in Present Community

Less Than 2 Years	2–5 Years	More Than 5 Years
0	0	1
1	2	3
3	3	3
4	4	4
4	5	4

c.

Membership by Extent of Television Watching

Little or None	Moderate	High
0	3	4
0	3	4
1	3	4
1	3	4
2	4	5

d.

Membership by Number of Children

None	One Child	More Than One Child
0	2	0
1	3	3
1	4	4
3	4	4
3	4	5

9.3 SOC In a local community, a random sample of 18 couples has been assessed on a scale that measures the extent to which power and decision making are shared (lower scores) or monopolized by one partner (higher scores) and on marital happiness (higher scores indicate higher levels of happiness). The couples were also classified by type of relationship: traditional (only the husband works outside the home), dual career (both partners work), and cohabitational (partners living together but not legally married, regardless of work patterns). Does decision making or happiness vary significantly by type of relationship?

a.

Decision Making

Traditional	Dual Career	Cohabitational
7	8	2
8	5	1
2	4	3
5	4	4
7	5	1
6	5	2

b.

Happiness

Traditional	Dual Career	Cohabitational
10	12	12
14	12	14
20	12	15
22	14	17
23	15	18
24	20	22

9.4 CJ The city of Shinbone has implemented two separate crime-reduction programs. One involves a neighborhood watch program with citizens actively involved in crime prevention. The second involves officers patrolling the neighborhoods on foot rather than in patrol cars. In terms of the percentage reduction in crimes reported to the police over a one-year period, were the programs successful? The results are for random samples of 18 neighborhoods drawn from the entire city.

Neighborhood Watch	Foot Patrol	No Program
−10	−21	+30
−20	−15	−10
+10	−80	+14
+20	−10	+80
+70	−50	+50
+10	−10	−20

9.5 SOC Are sexually active teenagers any better informed about AIDS and other potential health problems related to sex than teenagers who are sexually inactive? A 15-item test of general knowledge about sex and health was administered to random samples of teens who are sexually inactive, teens who are sexually active but with only a single partner ("going steady"), and teens who are sexually active with more than one partner. Is there any significant difference in the test scores?

Inactive	Active— One Partner	Active—More Than One Partner
10	11	12
12	11	12
8	6	10
10	5	4
8	15	3
5	10	15

9.6 SOC Does the rate of voter turnout vary significantly by the type of election? A random sample of voting precincts displays the following pattern of voter turnout by election type. Assess the results for significance.

Local Only	State	National
33	35	42
78	56	40
32	35	52
28	40	66
10	45	78

Local Only	State	National
12	42	62
61	65	57
28	62	75
29	25	72
45	47	51
44	52	69
41	55	59

9.7 GER Do older citizens lose interest in politics and current affairs? A brief quiz on recent headline stories was administered to random samples of respondents from each of four different age groups. Is there a significant difference? The data below represent numbers of correct responses.

High School (15–18)	Young Adult (21–30)	Middle Aged (30–55)	Retired (65+)
0	0	2	5
1	0	3	6
1	2	3	6
2	2	4	6
2	4	4	7
2	4	5	7
3	4	6	8
5	6	7	10
5	7	7	10
7	7	8	10
7	7	8	10
9	10	10	10

9.8 SOC A small random sample of respondents has been selected from the General Social Survey database. Each respondent has been classified as a city dweller, a suburbanite, or a rural dweller. Are there statistically significant differences by place of residence for any of the variables listed below?

a.

Occupational Prestige (PRESTG80)

Urban	Suburban	Rural
32	40	30
45	48	40
42	50	40
47	55	45
48	55	45
50	60	50
51	65	52
55	70	55
60	75	55
65	75	60

b.

Number of Children (CHILDS)

Urban	Suburban	Rural
1	0	1
1	1	4
0	0	2
2	0	3
1	2	3
0	2	2
2	3	5
2	2	0
1	2	4
0	1	6

c.

Family Income (INCOME98)

Urban	Suburban	Rural
5	6	5
7	8	5
8	11	11
11	12	10
8	12	9
9	11	6
8	11	10
3	9	7
9	10	9
10	12	8

d.

Church Attendance (ATTEND)

Urban	Suburban	Rural
0	0	1
7	0	5
0	2	4
4	5	4
5	8	0
8	5	4
7	8	8
5	7	8
7	2	8
4	6	5

e.

Hours of TV Watching per Day (TVHOURS)

Urban	Suburban	Rural
5	5	3
3	7	7
12	10	5
2	2	0
0	3	1
2	0	8
3	1	5
4	3	10
5	4	3
9	1	1

9.9 SOC Does support for suicide by the terminally ill ("death with dignity") vary by social class? Is this relationship different in different nations? Small samples in three nations were asked if it is ever justified for a person with an incurable disease to take his or her own life. Respondents answered in terms of a 10-point scale on which 10 was "always justified" (the strongest support for "death with dignity") and 1 was "never justified" (the lowest level of support). Results are reported below.

CANADA

Lower Class	Working Class	Middle Class	Upper Class
7	5	1	5
7	6	3	7
6	7	4	8
4	8	5	9
7	8	7	10
8	9	8	10
9	5	8	8
9	6	9	5
6	7	9	8
5	8	5	9

MEXICO

Lower Class	Working Class	Middle Class	Upper Class
5	2	1	2
2	2	1	4
4	1	3	5
5	1	4	7
4	6	1	8
2	5	2	10
3	7	1	10
1	2	5	9
1	3	1	8
3	1	1	8

UNITED STATES

Lower Class	Working Class	Middle Class	Upper Class
4	4	4	1
5	5	6	5
6	1	7	8
1	4	5	9
3	3	8	9
3	3	9	9
3	4	9	8
5	2	8	6
3	1	7	9
6	1	2	9

SSPS for Windows

Using *SPSS for Windows* to Conduct Analysis of Variance

SPSS DEMONSTRATION 9.1 Does Political Conservatism Increase with Age?

SPSS provides several different ways of conducting the analysis of variance test. The procedure summarized below is the most accessible of these, but it still incorporates options and capabilities that we have not covered in this chapter. If you want to explore these possibilities, please consult the SPSS manual or use the online Help facility.

In designing examples to demonstrate the ANOVA procedure, my choices are constrained by the scarcity of interval-ratio variables in the GSS data set. Some variables that "should be" interval ratio are actually measured at the ordinal level (e.g., income), while others are unsuitable dependent variables on logical grounds (e.g., age). To have something to discuss, I have taken some liberties with respect to level-of-measurement criteria in several of the examples that follow (a practice that is, in fact, common in social science research).

Let's begin by exploring the idea that political ideology is linked to age in U.S. society. People are often said to become more conservative about a wide range of issues as they age, and this might result in greater political conservatism in older age groups. For a measure of ideology, we will use *polviews*. Higher scores on this variable indicate higher levels of conservatism. We need to collapse *age* into a few categories to fit the ANOVA design. To find reasonable cutting points for age, I ran the **Frequencies** command to find the ages that divided the sample into three groups of roughly equal size.

Use these scores to recode *age:*

Interval	% of Sample
18–36	33.6%
37–52	33.5%
53–89	33.9%

To use the ANOVA procedure, click **Analyze, Compare Means,** and then **One-way ANOVA.** The **One-way ANOVA** window appears. Find *polviews* (the label for this variable is 'THINK OF SELF AS LIBERAL OR CONSERVATIVE') in the variable list on the left and click the arrow to move the variable name into the **Dependent List** box. Note that you can request more than one dependent variable at a time. Next, find the name of the recoded *age* variable (*ager?*) and click the arrow to move the variable name into the **Factor** box.

Click **Options** and then click the box next to **Descriptive** in the **Statistics** box to request means and standard deviations along with the analysis of variance. Click **Continue** and then click **OK,** and the following output will be produced:

Descriptives

THINK OF SELF AS LIBERAL OR CONSERVATIVE

					95% Confidence Interval for Mean			
	N	Mean	Std. Deviation	Std. Error	Lower Bound	Upper Bound	Minimum	Maximum
1.00	204	4.12	1.418	.099	3.93	4.32	1	7
2.00	209	4.27	1.506	.104	4.07	4.48	1	7
3.00	219	4.27	1.373	.093	4.09	4.45	1	7
Total	632	4.22	1.432	.057	4.11	4.33	1	7

ANOVA

THINK OF SELF AS LIBERAL OR CONSERVATIVE

	Sum of Squares	df	Mean Square	*F*	Sig.
Between Groups	3.047	2	1.523	.743	.476
Within Groups	1290.496	629	2.052		
Total	1293.543	631			

The output box labeled ANOVA includes the various degrees of freedom, all of the sums of squares, the Mean Square estimates, the *F* ratio (.743), and, at the far right, the exact probability ("Sig.") of getting these results if the null hypothesis is true. This is reported as .476, which is much higher than our usual alpha level of .05. The differences in *polviews* for the various age groups are *not* statistically significant.

The report also displays some summary statistics. An inspection of the means shows that the average score for the entire sample was 4.22. The youngest age group is the least conservative (remember that lower scores indicate greater liberalism) but the two older groups have identical means. These results indicate that conservatism does not increase with age (at least for the older age groups).

SPSS DEMONSTRATION 9.2 Does Political Ideology Vary by Social Class?

Let's continue our analysis of *polviews* to see if there are any significant differences in political ideology by social class. The GSS includes several variables that might indicate class, including *degree, income, class,* and *prestg80.* We should probably investigate all of these, but—to conserve space—we'll confine our attention to *class* (the label for this variable is SUBJECTIVE CLASS IDENTIFICATION). Follow the instructions for **One-way ANOVA** in Demonstration 9.1 and specify *polviews* again as the dependent variable and *class* as the factor, or independent variable. The output will look like this:

Descriptives

THINK OF SELF AS LIBERAL OR CONSERVATIVE

	N	Mean	Std. Deviation	Std. Error	95% Confidence Interval for Mean Lower Bound	95% Confidence Interval for Mean Upper Bound	Minimum	Maximum
LOWER CLASS	46	4.33	1.578	.233	3.86	4.79	1	7
WORKING CLASS	276	4.20	1.329	.080	4.04	4.35	1	7
MIDDLE CLASS	290	4.27	1.486	.087	4.09	4.44	1	7
UPPER CLASS	22	3.86	1.642	.350	3.14	4.59	1	7
Total	634	4.23	1.431	.057	4.11	4.34	1	7

ANOVA

THINK OF SELF AS LIBERAL OR CONSERVATIVE

	Sum of Squares	df	Mean Square	F	Sig.
Between Groups	4.056	3	1.352	.659	.578
Within Groups	1292.690	630	2.052		
Total	1296.746	633			

An inspection of the group means shows that upper-class respondents were most liberal (had the lowest average score on *polviews*) followed by working-class respondents. Lower-class and upper-class respondents had nearly identical scores. The *F* ratio is quite low (.659), and the Sig. (.578) reported at the far right of the ANOVA output box shows that these differences are not statistically significant. According to these results, there is no significant difference in political ideology by social class.

SPSS DEMONSTRATION 9.3 Another Test for Differences in Attitudes on Abortion

In Demonstration 8.2, we found that *sex* did not have a statistically significant impact on *abscale.* In this demonstration, we will investigate the impact of religious affiliation (*relig*). Remember that higher scores on *abscale* indicate greater opposition to legalized abortion. Click **Analyze, Compare Means,** and **One-way ANOVA** and name *abscale* as the dependent variable and *relig* as the factor. If necessary, see Demonstration 8.2 for instructions on computing *abscale.* The output will look like this:

Descriptives

abscale

	N	Mean	Std. Deviation	Std. Error	95% Confidence Interval for Mean		Minimum	Maximum
					Lower Bound	Upper Bound		
PROTESTANT	231	2.8009	.70043	.04609	2.7101	2.8917	2.00	4.00
CATHOLIC	100	2.8200	.65721	.06572	2.6896	2.9504	2.00	4.00
JEWISH	4	2.0000	.00000	.00000	2.0000	2.0000	2.00	2.00
NONE	59	2.4576	.62483	.08135	2.2948	2.6205	2.00	4.00
OTHER (SPECIFY)	2	2.0000	.00000	.00000	2.0000	2.0000	2.00	2.00
Total	396	2.7424	.68939	.03464	2.6743	2.8105	2.00	4.00

ANOVA

abscale

	Sum of Squares	df	Mean Square	F	Sig.
Between Groups	9.483	4	2.371	5.201	.000
Within Groups	178.244	391	.456		
Total	187.727	395			

The *F* ratio (5.201) and Sig. (.000) indicate a significant difference at the .05 level. Protestant and Catholic respondents were the most opposed and Jewish and "other" respondents were the most supportive of the right to a legal abortion.

Exercises

9.1 As a follow-up on Demonstration 9.2, test *income98, prestg80,* and *papres80* as independent variables (factors) against *polviews.* Do these measures of social class display the same type of relationship with *polviews* as *class?* First, recode each of these independents into three categories. Run **Frequencies** for each and find cutting points that divide the sample into three groups of roughly equal size.

9.2 What other variable might have a significant relationship with *abscale?* Pick three potential independent variables (factors) and test their relationships with *abscale.* Some possible independent variables would be *race, attend,* or any of the measures of social class. If necessary, recode the independent variables into a few (three or four) categories.

10

Hypothesis Testing IV
Chi Square

LEARNING OBJECTIVES

By the end of this chapter, you will be able to:

1. Identify and cite examples of situations in which the chi square test is appropriate.
2. Explain the structure of a bivariate table and the concept of independence as applied to expected and observed frequencies in a bivariate table.
3. Explain the logic of hypothesis testing as applied to a bivariate table.
4. Perform the chi square test using the five-step model and correctly interpret the results.
5. Explain the limitations of the chi square test and, especially, the difference between statistical significance and importance.

10.1 INTRODUCTION

The **chi square (χ^2) test** has probably been the most frequently used test of hypothesis in the social sciences, a popularity that is due largely to the fact that the assumptions and requirements in step 1 of the five-step model are easy to satisfy. Specifically, the chi square test has no restrictions in terms of level of measurement and can be conducted with variables measured at the nominal level (the lowest level of measurement). Also, the test is **nonparametric** or "distribution free,"[1] which means that it requires no assumption at all about the shape of the population or sampling distribution.

These easily satisfied assumptions are an advantage because the decision to reject the null hypothesis (step 5) is not specific: It indicates only that one statement in the model (step 1) *or* the null hypothesis (step 2) is wrong. Usually, of course, we single out the null hypothesis for rejection. The more certain we are of the model, the greater our confidence that the null hypothesis is the faulty assumption. A "weak" or easily satisfied model means that we can decide to reject the null hypothesis with even greater certainty.

Chi square has also been popular for its flexibility. Unlike the tests of significance presented in Chapter 8, we can use chi square with variables that have more than two categories or scores. For example, in Chapter 8, we tested the significance of the difference in the proportions of black and white citizens who were "highly participatory" in voluntary associations. What if the researcher wished to expand the test to include Americans of Hispanic and Asian descent? The two-sample test would no longer be applicable, but chi square handles the more complex variable easily. Also, unlike the ANOVA test we covered in Chapter 9, the chi square can be conducted with variables at all levels of measurement.

---

[1]See the website for this text for other nonparametric tests of significance.

10.2 BIVARIATE TABLES

Chi square is computed from **bivariate tables,** so called because they display the scores of cases on two different variables at the same time. Bivariate tables are used to ascertain the significance of the relationship between the variables and for other purposes that we will investigate in later chapters. In fact, these tables are very commonly used in research, so let's examine them in detail.

First of all, bivariate tables have, as their name indicates, two dimensions. We refer to the horizontal (across) dimension as **rows** and call the vertical dimension (up and down) **columns.** Each column or row represents a score on a variable, and the intersections of the rows and columns **(cells)** represent the various combined scores on both variables.

Let's use an example to clarify. Suppose a researcher is interested in the relationship between racial group membership and participation in voluntary groups, community-service organizations, and so forth. Do blacks and whites vary in their level of involvement in volunteer groups? We have two variables (race and number of memberships) and, for the sake of simplicity, assume that both are simple dichotomies. That is, people have been classified as either black or white and as either high or low in their level of involvement in voluntary associations.

By convention, we place the independent variable (the variable assumed to be the cause) in the columns and the dependent variable in the rows. In the example at hand, race is the causal variable (the question was "Is membership *affected by* race?"), and each column will represent a score on this variable. Each row, on the other hand, will represent a score on level of membership (high or low). Table 10.1 displays the outline of the bivariate table for a sample of 100 people.

Note some details of the table. First, subtotals have been added to each column and row. These are called the row or column **marginals,** and in this case, they tell us that 50 members of the sample were black and 50 were white (the column marginals) and 50 were rated as high in participation and 50 were rated low (the row marginals). Second, the total number of cases in the sample ($N = 100$) is reported at the intersection of the row and column marginals. Finally, take careful note of the labeling of the table. Each row and column is identified and the table has a descriptive title that includes the names of the variables with the dependent variable listed first. Remember to use clear, complete labels and concise titles in *all* tables, graphs, and charts.

As you have noticed, Table 10.1 lacks one piece of crucial information: the numbers of each racial group that rated high or low on the dependent variable. To finish the table, we need to classify all members of the sample in terms of both their race and their level of participation, count how often each

TABLE 10.1   RATES OF PARTICIPATION IN VOLUNTARY ASSOCIATIONS BY RACIAL GROUP FOR 100 SENIOR CITIZENS

Participation Rates	Racial Group		
	Black	White	
High			50
Low			50
	50	50	100

combination of scores occurs, and record these numbers in the appropriate cell of the table. Because each of our variables (race and participation rates) has two scores, there are four possible combinations of scores, each corresponding to a cell in the table. For example, blacks with high levels of participation would be counted in the upper-left-hand cell, whites with low levels of participation would be counted in the lower-right-hand cell, and so forth. When we are finished counting, each cell will display the number of times each combination of scores occurred.

Finally, note how we could expand the bivariate table to accommodate variables with more scores. If we wished to include more groups in the test (e.g., Asian Americans or Hispanic Americans), then we would simply add columns to the table. More elaborate dependent variables could also be easily accommodated. If we had measured participation rates with three categories (e.g., high, moderate, and low) rather than two, we would simply add another row to the table.

10.3 THE LOGIC OF CHI SQUARE

Chi square is a test for the independence of the relationship between the variables. We have encountered the term *independence* in connection with the requirements for the two-sample case (Chapter 8) and for the ANOVA test (Chapter 9). In those situations, we noted that independent random samples are gathered such that the selection of a particular case for one sample has no effect on the probability that any particular case will be selected for the other sample.

In the context of chi square, the concept of **independence** takes on a slightly different meaning because it refers to the relationship between the variables, not the samples. Two variables are independent if the classification of a case into a particular category of one variable has no effect on the probability that the case will fall into any particular category of the second variable. For example, race and participation in voluntary associations would be independent of each other if the classification of a person as black or white has no effect on his or her classification as high or low on participation. In other words, the variables would be independent if level of participation and race were completely unrelated to each other.

Consider Table 10.1 again. If these two variables are truly independent, the cell frequencies will be determined solely by random chance and we would find that, just as an honest coin will show heads about 50% of the time when flipped, about half of the black respondents will rank high on participation and half will rank low. The same pattern would hold for the 50 white respondents and, therefore, each of the four cells would have about 25 cases in it, as Table 10.2 shows.

TABLE 10.2 THE EXPECTED CELL FREQUENCIES IF RATES OF PARTICIPATION AND RACIAL GROUP WERE INDEPENDENT

| | Racial Group | | |
Participation Rates	Black	White	
High	25	25	50
Low	25	25	50
	50	50	100

This pattern of cell frequencies indicates that the racial classification of the subjects has no effect on the probability that they would be either high or low in participation. The probability of being classified as high or low would be 0.5 for both blacks and whites, and the variables would therefore be independent.

The null hypothesis for chi square is that the variables are independent. Under the assumption that the null hypothesis is true, the cell frequencies we would expect to find if only random chance were operating are computed. These frequencies, called **expected frequencies** (symbolized f_e), are then compared, cell by cell, with the frequencies actually observed in the table **(observed frequencies,** symbolized f_o**).** If the null hypothesis is true and the variables are independent, then there should be little difference between the expected and observed frequencies. If the null is false, however, there should be large differences between the two. The greater the differences between expected (f_e) and observed (f_o) frequencies, the less likely that the variables are independent and the more likely that we will be able to reject the null hypothesis.

10.4 THE COMPUTATION OF CHI SQUARE

As with all tests of hypothesis, with chi square we compute a test statistic, χ^2**(obtained),** from the sample data and then place that value on the sampling distribution of all possible sample outcomes. Specifically, we compare the χ^2(obtained) with the value of χ^2**(critical),** which we will determine by consulting a chi square table (Appendix C) for a particular alpha level and degrees of freedom. Prior to conducting the formal test of hypothesis, let's take a moment to consider the calculation of chi square, as defined by Formula 10.1:

FORMULA 10.1

$$\chi^2(\text{obtained}) = \sum \frac{(f_o - f_e)^2}{f_e}$$

where f_o = the cell frequencies observed in the bivariate table
f_e = the cell frequencies that would be expected if the variables were independent

We must work on a cell-by-cell basis to solve this formula. The formula tells us to subtract the expected frequency from the observed frequency for each cell, square the result, divide by the expected frequency for that cell, and then sum the resultant values for all cells.

This formula requires an expected frequency for each cell in the table. In Table 10.2, the marginals are the same value for all rows and columns, and the expected frequencies are obvious by intuition: $f_e = 25$ for all four cells. In the more usual case, the expected frequencies will not be obvious, marginals will be unequal, and we must use Formula 10.2 to find the expected frequency for each cell:

FORMULA 10.2

$$f_e = (\text{Row marginal} \times \text{Column marginal})/N$$

That is, the expected frequency for any cell is equal to the total number of cases in the row in which the cell is located (the row marginal) times the total number of cases in the column in which the cell is located (the column marginal) divided by the total number of cases in the table (N).

An example using Table 10.3 should clarify these procedures. A random sample of 100 social work majors has been classified in terms of whether the Council

TABLE 10.3 EMPLOYMENT OF 100 SOCIAL WORK MAJORS BY ACCREDITATION STATUS OF UNDERGRADUATE PROGRAM

	Accreditation Status		
Employment Status	Accredited	Not Accredited	Totals
Working as a social worker	30	10	40
Not working as a social worker	25	35	60
Totals	55	45	100

TABLE 10.4 EXPECTED FREQUENCIES FOR TABLE 10.3

	Accreditation Status		
Employment Status	Accredited	Not Accredited	Totals
Working as a social worker	22	18	40
Not working as a social worker	33	27	60
Totals	55	45	100

TABLE 10.5 COMPUTATIONAL TABLE FOR TABLE 10.3

(1)	(2)	(3)	(4)	(5)
f_o	f_e	$f_o - f_e$	$(f_o - f_e)^2$	$(f_o - f_e)^2/f_e$
30	22	8	64	2.91
10	18	−8	64	3.56
25	33	−8	64	1.94
35	27	8	64	2.37
$N = 100$	$N = 100$	0		χ^2(obtained) = 10.78

on Social Work Education has accredited their undergraduate programs (the column or independent variable) and whether they were hired in social work positions within three months of graduation (the row or dependent variable).

Beginning with the upper-left-hand cell (graduates of accredited programs who are working as social workers), the expected frequency for this cell, using Formula 10.2, is $(40 \times 55)/100$, or 22. For the other cell in this row (graduates of unaccredited programs who are working as social workers), the expected frequency is $(40 \times 45)/100$, or 18. For the two cells in the bottom row, the expected frequencies are $(60 \times 55)/100$, or 33, and $(60 \times 45)/100$, or 27, respectively. Table 10.4 displays the expected frequencies for all four cells.

Now we can find the value for chi square for these data by solving Formula 10.1. It will be helpful to use a computing table, such as Table 10.5, to organize the several steps required to compute chi square. The table lists the observed frequencies (f_o) in column 1 in order from the upper-left-hand cell to the lower-right- hand cell, moving left to right across the table and top to bottom. Column 2 lists the expected frequencies (f_e) in exactly the same order.

STEP BY STEP	Computing Chi Square

Step 1: Prepare a computational table similar to Table 10.5. List the observed frequencies (f_o) in column 1. The total of column 1 is the number of cases (N).

Step 2: To find the expected frequencies for each cell in the table, use Formula 10.2:

f_e = (Row marginal × Column marginal)/N

a. Start with the upper-left-hand cell and multiply the row marginal by the column marginal.

b. Divide the quantity you found in step a by N. The result is the expected frequency (f_e) for that cell. Record this value in the proper row of your computational table.

c. Repeat steps a and b for each cell in the table. Double-check to make sure that you are using the correct row and column marginals. Record each f_e in column 2 of the computational table.

d. Find the total of column 2 of the computational table. If this total does not equal the total of column 1 (N), you have made a mistake and need to check your computations.

Step 3: Subtract the expected frequency (f_e) from the observed frequency (f_o) for each cell and list these values in column 3 of the computational table. Find the total of column 3. If this total does not equal zero, you have made an error and need to check your computations.

Step 4: Square each value in column 3 and record the result in column 4 of the computational table.

Step 5: Divide each value in column 4 by the expected frequency for that cell and record the result in column 5 of the computational table.

Step 6: Add up column 5. The sum of column 5 is χ^2(obtained).

Double-check to make sure that you have listed the cell frequencies in the same order for both of these columns. The complete procedure for computing χ^2 appears in the step-by-step box above.

$$\chi^2\text{(obtained)} = 10.78$$

Note that the totals for columns 1 and 2 (f_o and f_e) are exactly the same. This will always be the case. If the totals do not match, you have made a computational error (probably in the calculation of the expected frequencies). Also note that the sum of column 3 will always be zero, another convenient way to check your math to this point.

This sample value for chi square must still be tested for its significance. *(For practice in computing chi square, see problem 10.1.)*

10.5 THE CHI SQUARE TEST FOR INDEPENDENCE

As always, the five-step model for significance testing will provide the framework for organizing our decision making. The data presented in Table 10.3 will serve as our example.

Step 1. Making Assumptions and Meeting Test Requirements. Note that we make no assumptions at all about the shape of the sampling distribution.

Model: Independent random samples
Level of measurement is nominal

Step 2. Stating the Null Hypothesis. As noted previously, the null hypothesis in the case of chi square states that the two variables are independent. If the null is true, the differences between the observed and expected frequencies will be small. As usual, the research hypothesis directly contradicts the null. Thus, if we reject H_0, the research hypothesis will be supported.

$$H_0: \text{The two variables are independent}$$
$$(H_1: \text{The two variables are dependent})$$

Step 3. Selecting the Sampling Distribution and Establishing the Critical Region. The sampling distribution of sample chi squares, unlike the Z and t distributions, is positively skewed, with higher values of sample chi squares in the upper tail of the distribution (to the right). Thus, with the chi square test, the critical region is established in the upper tail of the sampling distribution.

Appendix C gives values for χ^2(critical). This table is similar to the t table, with alpha levels arrayed across the top and degrees of freedom down the side. A major difference, however, is that we use the following formula to find degrees of freedom (df) for chi square:

FORMULA 10.3
$$df = (r - 1)(c - 1)$$

A table with two rows and two columns (a 2 × 2 table) has one degree of freedom regardless of the number of cases in the sample.[2] A table with two rows and three columns would have $(2 - 1)(3 - 1)$, or two degrees of freedom. Our sample problem involves a 2 × 2 table with df = 1, so if we set alpha at 0.05, the critical chi square score would be 3.841. Any value for the sample statistic—χ^2(obtained)—greater than 3.841 would cause us to reject the null hypothesis. Summarizing these decisions, we have:

$$\text{Sampling distribution} = \chi^2 \text{ distribution}$$
$$\text{Alpha} = .05$$
$$\text{Degrees of freedom} = 1$$
$$\chi^2(\text{critical}) = 3.841$$

Step 4. Computing the Test Statistic. Section 10.4 introduced the mechanics of these computations. As you recall, we had:

$$\chi^2(\text{obtained}) = \sum \frac{(f_o - f_e)^2}{f_e}$$

$$\chi^2(\text{obtained}) = 10.78$$

[2]Degrees of freedom are the number of values in a distribution that are free to vary for any particular statistic. A 2 × 2 table has one degree of freedom because, for a given set of marginals, once one cell frequency is determined, all other cell frequencies are fixed (that is, they are no longer free to vary). In Table 10.3, for example, if any cell frequency is known, all others are determined. If the upper-left-hand cell is known to be 30, the remaining cell in that row must be 10, since there are 40 cases total in the row and 40 − 30 = 10. Once we establish the frequencies of the cells in the top row, we determine cell frequencies for the bottom row by subtraction from the column marginals. Incidentally, this relationship can be used to good advantage when computing expected frequencies. For example, in a 2 × 2 table, only one expected frequency needs to be computed. We can then find the f_es for all other cells by subtraction.

Step 5. Making a Decision and Interpreting the Results of the Test. Comparing the test statistic with the critical region,

$$\chi^2(\text{obtained}) = 10.78$$
$$\chi^2(\text{critical}) = 3.841$$

we see that the test statistic falls into the critical region and, therefore, we reject the null hypothesis of independence. The pattern of cell frequencies observed in Table 10.3 is unlikely to have occurred by chance alone. The variables are dependent. Specifically, based on these sample data, the probability of securing employment in the field of social work depends on the accreditation status of the program. *(For practice in conducting and interpreting the chi square test for independence, see problems 10.2 to 10.15.)*

Let me stress exactly what the chi square test does and does not tell us. A significant chi square means that the variables are (very likely) dependent on each other in the population: Accreditation status makes a difference in whether or not a person is working as a social worker. What chi square does not tell us is the exact nature of the relationship. In other words, chi square, by itself, does not tell us if the graduates of the accredited programs or those of the unaccredited programs are more likely to be working as social workers. To make this determination, we must perform some additional calculations. We can figure out how the independent variable (accreditation status) is affecting the dependent variable (employment as a social worker) by computing **column percents** or by calculating percentages within each column of the bivariate table. This procedure is analogous to calculating percentages for frequency distributions (see Chapter 2).

To calculate column percentages, divide each cell frequency by the total number of cases in the column (the column marginal) and multiply the result by 100. For Table 10.3, starting in the upper-left-hand cell, we see that there are 30 cases in this cell and 55 cases in the column. In other words, 30 of the 55 graduates of accredited programs are working as social workers. The column percentage for this cell is therefore: $(30/55) \times 100 = 54.55\%$. For the lower-left-hand cell, the column percent is $(25/55) \times 100 = 45.45\%$. For the two cells in the right-hand column (graduates of unaccredited programs), the column percents are $(10/45) \times 100 = 22.22\%$ and $(35/45) \times 100 = 77.78\%$. Table 10.6 displays all column percents.

Column percents help to make the relationship between the two variables more obvious. Using Table 10.6, we can easily see that nearly 55% of the students from accredited programs are working as social workers versus 22.22% of the students from unaccredited programs. We already knew that this

TABLE 10.6   COLUMN PERCENTS FOR TABLE 10.3

Employment Status	Accreditation Status		Totals
	Accredited	Not Accredited	
Working as a social worker	54.55%	22.22%	40.00%
Not working as a social worker	45.45%	77.78%	60.00%
Totals	100.00%	100.00%	100.00%
	(55)	(45)	

STEP BY STEP	Computing Column Percents

Step 1: Start with the far left-hand column of the bivariate table. Find the marginal (subtotal) for this column. That is, add up all of the cell frequencies in this column to find the total number of cases in the column.

Step 2: Start with the top cell in the column and divide the number of cases in the cell by the column marginal (total number of cases in the column). Multiply the result by 100 and round to two places of accuracy beyond the decimal point. This value is the percentage of cases in the column that are in the top cell.

Step 3: Go down one cell (staying in the far left-hand column) and repeat Step 2 for this cell. This is the percentage of all cases in the column that are in this cell.

Step 4: Repeat step 2 for all remaining cells (if any) in the far left-hand column.

Step 5: Move one column to the right and repeat steps 1 and 2 for all cells in that column. Continue until you have computed all column percentages for all columns

relationship is significant (unlikely to be caused by random chance) and now, with the aid of column percents, we know how the two variables are related. According to these results, graduating from an accredited program would be a decided advantage for people seeking to become social workers.

Let's highlight two points in summary:

1. Chi square is a test of statistical significance. It tests the null hypothesis that the variables are independent in the population. If we reject the null hypothesis, we are concluding, with a known probability of error (equal to the alpha level), that the variables are dependent on each other in the population. In the terms of our example, this means that accreditation status makes a difference in the likelihood of finding employment as a social worker. By itself, however, chi square does not tell us the exact nature of the relationship.
2. Computing column percents allows us to examine the bivariate relationship in more detail. By comparing the column percents for the various scores of the independent variable, we can see exactly how the independent variable affects the dependent variable. In this case, the column percents reveal that graduates of accredited programs are more likely to find jobs as social workers. We will explore column percentages more extensively when we discuss bivariate association in Chapter 11.

10.6 THE CHI SQUARE TEST: AN ADDITIONAL EXAMPLE

To this point, we have confined our attention to 2 × 2 tables. For purposes of illustration, we will work through the computational routines and decision-making process for a larger table. As you will see, larger tables require more computations (because they have more cells), but, in all other essentials, we handle them in the same way as the 2 × 2 table.

A researcher is concerned with the possible effects of marital status on the academic progress of college students. Do married students, with their extra burden of family responsibilities, suffer academically as compared to unmarried

Application 10.1

Do members of different groups have different levels of "narrow-mindedness"? A random sample of 47 white and black Americans have been rated as high or low on a scale that measures intolerance of viewpoints or belief systems different from their own. Here are the results:

Intolerance	Group White	Black	Totals
High	15	5	20
Low	10	17	27
Totals	25	22	47

The frequencies we would expect to find if the null hypothesis (H_0: the variables are independent) were true are:

Intolerance	Group White	Black	Totals
High	10.64	9.36	20.00
Low	14.36	12.64	27.00
Totals	25.00	22.00	47.00

We find expected frequencies on a cell-by-cell basis by using the formula

$$f_e = (\text{Row marginal} \times \text{Column marginal})/N$$

and the calculation of chi square will be organized into a computational table.

(1) f_o	(2) f_e	(3) $f_o - f_e$	(4) $(f_o - f_e)^2$	(5) $(f_o - f_e)^2/f_e$
15	10.64	4.36	19.01	1.79
5	9.36	−4.36	19.01	2.03
10	14.36	−4.36	19.01	1.32
17	12.64	4.36	19.01	1.50
47	47.00	0.00		6.64

$$\chi^2(\text{obtained}) = 6.64$$

Step 1. Making Assumptions and Meeting Test Requirements.

Model: Independent random samples
Level of measurement is nominal

Step 2. Stating the Null Hypothesis.

H_0: The two variables are independent
(H_1: The two variables are dependent)

Step 3. Selecting the Sampling Distribution and Establishing the Critical Region.

Sampling distribution = χ^2 distribution
Alpha = .05
Degrees of freedom = 1
$\chi^2(\text{critical}) = 3.841$

Step 4. Computing the Test Statistic.

$$\chi^2(\text{obtained}) = \sum \frac{(f_o - f_e)^2}{f_e}$$

$$\chi^2(\text{obtained}) = 6.64$$

Step 5. Making a Decision and Interpreting the Results of the Test.
With an obtained χ^2 of 6.64, we would reject the null hypothesis of independence. For this sample, there is a statistically significant relationship between group membership and intolerance.

To complete the analysis, it would be useful to know exactly how the two variables are related. We can determine this by computing and analyzing column percents:

Intolerance	Group White	Black
High	60.00%	22.73%
Low	40.00%	77.27%
Totals	100.00%	100.00%

The column percents show that 60% of whites in this sample are high on intolerance versus only 22.73% of blacks. We have already concluded that the relationship is significant, and now we know the pattern of the relationship: The white respondents were more likely to be high on intolerance.

TABLE 10.7 GRADE-POINT AVERAGE (GPA) BY MARITAL STATUS FOR 453 COLLEGE STUDENTS

GPA	Marital Status		Totals
	Married	Not Married	
Good	70	90	160
Average	60	110	170
Poor	45	78	123
Totals	175	278	453

TABLE 10.8 EXPECTED FREQUENCIES FOR TABLE 10.7

GPA	Marital Status		Totals
	Married	Not Married	
Good	61.8	98.2	160
Average	65.7	104.3	170
Poor	47.5	75.5	123
Totals	175.0	278.0	453

TABLE 10.9 COMPUTATIONAL TABLE FOR TABLE 10.7

(1)	(2)	(3)	(4)	(5)
f_o	f_e	$f_o - f_e$	$(f_o - f_e)^2$	$(f_o - f_e)^2/f_e$
70	61.8	8.2	67.24	1.09
90	98.2	−8.2	67.24	0.69
60	65.7	−5.7	32.49	0.49
110	104.3	5.7	32.49	0.31
45	47.5	−2.5	6.25	0.13
78	75.5	2.5	6.25	0.08
$N = 453$	$N = 453.0$	0.0	χ^2(obtained) = 2.79	

students? Is academic performance dependent on marital status? A random sample of 453 students is gathered, and each student is classified as either married or unmarried, and—using grade-point average (GPA) as a measure—as a good, average, or poor student. Table 10.7 presents the results.

For the top-left-hand cell (married students with good GPAs) the expected frequency would be $(160 \times 175)/453$, or 61.8. For the other cell in this row, expected frequency is $(160 \times 278)/453$, or 98.2. In similar fashion, all expected frequencies are computed (being very careful to use the correct row and column marginals). Table 10.8 displays the frequencies that would be expected (f_e) if the variables are not related and only random chance is operating.

The next step is to solve the formula for χ^2(obtained), being very careful to use the proper f_os and f_es for each cell. Once again, we will use a computational table (Table 10.9) to organize the calculations and then test the obtained chi square for its statistical significance. Remember that obtained chi square is equal to the total of column 5.

Now we can test the value of the obtained chi square (2.79) for its significance.

Step 1. Making Assumptions and Meeting Test Requirements.

Model: Independent random samples
Level of measurement is nominal

Step 2. Stating the Null Hypothesis.

H_0: The two variables are independent
(H_1: The two variables are dependent)

Step 3. Selecting the Sampling Distribution and Establishing the Critical Region.
Sampling distribution = χ^2 distribution
Alpha = .05
Degrees of freedom = $(r - 1)(c - 1) = (3 - 1)(2 - 1) = 2$
χ^2(critical) = 5.991

Step 4. Computing the Test Statistic.
$$\chi^2(\text{obtained}) = \sum \frac{(f_o - f_e)^2}{f_e}$$
$$\chi^2(\text{obtained}) = 2.79$$

Step 5. Making a Decision and Interpreting the Results of the Test. The test statistic, χ^2(obtained) = 2.79, does not fall into the critical region, which, for alpha = 0.05, df = 2, begins at the χ^2(critical) of 5.991. Therefore, we fail to reject the null. The observed frequencies are not significantly different from the frequencies we would expect to find if the variables were independent and only random chance were operating. Based on these sample results, we can conclude that the academic performance of college students is not dependent on their marital status.

Even though we failed to reject the null hypothesis, we still might compute column percentages to see if there is any pattern to the relationship. To do this, divide each cell frequency by its column total and then multiply by 100. Starting with the upper-left-hand cell, we see that 70 of the 175 married students have "good" GPAs. The column percent for this cell is $(70/175) \times 100 = 40.00\%$. For the middle cell in this column (married students with "average" GPAs), the column percent is $(60/175) \times 100 = 34.29\%$. Table 10.10 presents the column percents for Table 10.7.

TABLE 10.10 COLUMN PERCENTS FOR TABLE 10.7

GPA	Marital Status		Totals
	Married	Not Married	
Good	40.00%	32.37%	160
Average	34.29%	39.57%	170
Poor	25.71%	28.06%	123
Totals	100.00%	100.00%	453
	(175)	(278)	

The differences in percents from column to column are relatively small (and not significant), but we can see that married students are slightly more likely to have good GPAs (40% versus about 32% for unmarried students) and that unmarried students have a very small tendency to have more poor GPAs.

10.7 THE LIMITATIONS OF THE CHI SQUARE TEST

As with any other test, chi square has limits, and you should be aware of several potential difficulties. First, even though chi square is very flexible and handles many different types of variables, it becomes difficult to interpret when the variables have many categories. For example, two variables with five categories each would generate a 5 × 5 table with 25 cells—far too many combinations of scores to be easily absorbed or understood. As a very rough rule of thumb, the chi square test is easiest to interpret and understand when both variables have 4 or fewer scores.

Two further limitations of the test are related to sample size. When sample size is small, we can no longer assume that the chi square distribution accurately describes the sampling distribution of all possible sample outcomes. For chi square, a small sample is defined as one where a high percentage of the cells have expected frequencies (f_e) of 5 or less. Various rules of thumb have been developed to help the researcher decide what constitutes a "high percentage of cells." Probably the safest course is to take corrective action whenever *any* cell has expected frequencies of 5 or less.

In the case of 2 × 2 tables, we can adjust the value of χ^2(obtained) by applying Yates's correction for continuity, given in Formula 10.4:

FORMULA 10.4

$$\chi_c^2 = \sum \frac{(|f_o - f_e| - 0.5)^2}{f_e}$$

where
χ_c^2 = corrected chi square
$|f_o - f_e|$ = the absolute values of the difference between the observed and expected frequency for each cell

The correction factor is applied by reducing the absolute value[3] of the term ($f_o - f_e$) by .5 before squaring the difference and dividing by the expected frequency for the cell.

For tables larger than 2 × 2, there is no correction formula for computing χ^2(obtained) for small samples. You may be able to combine some of the categories of the variables and thereby increase cell sizes. Obviously, however, you should take this course of action only when it is sensible to do so. In other words, do not erase distinctions that have clear theoretical justifications merely to conform to the requirements of a statistical test. When you feel that categories cannot be combined to build up cell frequencies, and the percentage of cells with expected frequencies of 5 or less is small, it is probably justifiable to continue with the uncorrected chi square test as long as you regard the results with a suitable amount of caution.

A second potential problem related to sample size occurs with large samples. I pointed out in Chapter 8 that all tests of hypothesis are sensitive to sample size. That is, the probability of rejecting the null hypothesis increases as the

---

[3] Absolute values ignore plus and minus signs.

number of cases increases, regardless of any other factor. Chi square is especially sensitive to sample size, and larger samples may lead to the decision to reject the null when the actual relationship is trivial. In fact, chi square is more responsive to changes in sample size than other test statistics because the value of χ^2(obtained) will increase at the same rate as sample size. That is, if sample size is doubled, the value of χ^2(obtained) will be doubled. *(For an illustration of this principle, see problem 10.14.)*

You should be aware of this relationship between sample size and the value of chi square because it, once again, raises the distinction between statistical significance and theoretical importance. On one hand, tests of significance play a crucial role in research. When we are working with random samples, we must know if our research results could have been produced by mere random chance.

On the other hand, as with any other statistical technique, tests of hypothesis are limited in the range of questions they can answer. Specifically, these tests will tell us whether our results are statistically significant or not. They will not necessarily tell us if the results are important in any other sense. To deal more directly with questions of importance, we must use an additional set of statistical techniques called measures of association. We previewed these techniques, called measures of association, in this chapter when we used column percentages, and they will be the subject of Part III of this text.

SUMMARY

1. The chi square test for independence is appropriate for situations in which the variables of interest have been organized into table format. The null hypothesis is that the variables are independent or that the classification of a case into a particular category on one variable has no effect on the probability that the case will be classified into any particular category of the second variable.

2. Because chi square is nonparametric and requires only nominally measured variables, its model assumptions are easily satisfied. Furthermore, because it is computed from bivariate tables, in which the number of rows and columns can be easily expanded, we can use the chi square test in many situations in which other tests are inapplicable.

3. In the chi square test, we first find the frequencies that would appear in the cells if the variables were independent (f_e) and then compare those frequencies, cell by cell, with the frequencies actually

observed in the cells (f_o). If the null is true, expected and observed frequencies should be close in value. The greater the difference between the observed and expected frequencies, the more likely we will reject the null hypothesis.

4. The chi square test has several important limitations. It is often difficult to interpret when tables have many (more than 4 or 5) dimensions. Also, as sample size (N) decreases and many cells have fewer than five cases, the chi square test becomes less trustworthy, and corrective action may be required. Finally, with very large samples, we may declare relatively trivial relationships to be statistically significant. As is the case with all tests of hypothesis, statistical significance is not the same thing as "importance" in any other sense. As a general rule, statistical significance is a necessary but not sufficient condition for theoretical or practical importance.

SUMMARY OF FORMULAS

Chi square (obtained)	10.1	$\chi^2(\text{obtained}) = \sum \dfrac{(f_o - f_e)^2}{f_e}$
Expected frequencies	10.2	$f_e = (\text{Row marginal} \times \text{Column marginal})/N$

Degrees of freedom, bivariate tables	10.3	$df = (r - 1)(c - 1)$
Yates's correction for continuity	10.4	$\chi_c^2 = \sum \dfrac{(|f_o - f_e| - 0.5)^2}{f_e}$

GLOSSARY

Bivariate table. A table that displays the joint frequency distributions of two variables.

Cell. The cross-classification category of the variables in a bivariate table.

χ^2(critical). The score on the sampling distribution of all possible sample chi squares that marks the beginning of the critical region.

χ^2(obtained). The test statistic as computed from sample results.

Chi square test. A nonparametric test of hypothesis for variables that have been organized into a bivariate table.

Column. The vertical dimension of a bivariate table. By convention, each column represents a score on the independent variable.

Column percent. Percent computed within each column of a bivariate table.

Expected frequency (f_e). The cell frequencies that would be expected in a bivariate table if the variables were independent.

Independence. The null hypothesis in the chi square test. Two variables are independent if, for all cases, the classification of a case on one variable has no effect on the probability that the case will be classified in any particular category of the second variable.

Marginal. The row and column subtotal in a bivariate table.

Nonparametric. A "distribution-free" test. These tests do not assume a normal sampling distribution.

Observed frequency (f_o). The cell frequencies actually observed in a bivariate table.

Row. The horizontal dimension of a bivariate table, conventionally representing a score on the dependent variable.

MULTIMEDIA RESOURCES

The Wadsworth Sociology Resource Center: Virtual Society at
http://www.thomsonedu.com/sociology

Visit the companion website for the seventh edition of *Statistics: A Tool for Social Research* to access a wide range of student resources. Begin by clicking on the Student Resources section of the book's website to access the following study tools:

- Basic math review
- Flash cards

- Additional chapter problems
- Internet links
- Table of random numbers
- MicroCase and SPSS examples and exercises
- Hypothesis testing for variables measured at the ordinal level

PROBLEMS

10.1 For each table below, calculate the obtained chi square. (*HINT: Calculate the expected frequencies for each cell with Formula 10.2. Double-check to make sure you are using the correct row and column marginals for each cell. It may be helpful to record the expected frequencies in* table format as well—see Tables 10.2, 10.4, and 10.8. Next, use a computational table to organize the calculation for Formula 10.1—see Tables 10.5 and 10.9. For each cell, subtract expected frequency from observed frequency and record the result in column 3. Square the value *in column 3 and record the result in column 4, and then divide the value in column 4 by the expected frequency for that cell and record the result in column 5. Remember that the sum of column 5 in the computational table is obtained chi square. As you proceed, double-check to make sure that you are using the correct values for each cell.)*

a.

20	25	45
25	20	45
45	45	90

b.

10	15	25
20	30	50
30	45	75

c.

25	15	40
30	30	60
55	45	100

d.

20	45	65
15	20	35
35	65	100

10.2 SOC A sample of 25 cities have been classified as high or low on their homicide rates and on the number of handguns sold within the city limits. Is there a relationship between these two variables? Explain your results in a sentence or two.

	Homicide Rate		
Volume of Gun Sales	Low	High	Totals
High	8	5	13
Low	4	8	12
Totals	12	13	25

10.3 SW A local politician is concerned that a program for the homeless in her city is discriminating against blacks and other minorities. The data below were taken from a random sample of black and white homeless people.

	Race		
Received Services?	Black	White	Totals
Yes	6	7	13
No	4	9	13
Totals	10	16	26

a. Is there a statistically significant relationship between race and whether or not a person has received services from the program?
b. Compute column percents for the table to determine the pattern of the relationship. Which group was more likely to get services?

10.4 PS Many analysts have noted a "gender gap" in elections for the U.S. presidency with women more likely to vote for the Democratic candidate. A sample of university faculty has been asked about their political party preference. Do the responses indicate a significant relationship between gender and party preference?

	Gender		
Party Preference	Male	Female	Totals
Democrats	10	15	25
Republicans	15	10	25
Totals	25	25	50

a. Is there a statistically significant relationship between gender and party preference?
b. Compute column percents for the table to determine the pattern of the relationship. Which gender is more likely to prefer the Democrats?

10.5 PA Is there a relationship between salary levels and unionization for public employees? The data below represent this relationship for fire departments in a random sample of 100 cities of roughly the same size. Salary data have been dichotomized at the median. Summarize your findings.

	Status		
Salary	Union	Nonunion	Totals
High	21	29	50
Low	14	36	50
Totals	35	65	100

a. Is there a statistically significant relationship between these variables?
b. Compute column percents for the table to determine the pattern of the relationship. Which group was more likely to get high salaries?

10.6 SOC A program of pet therapy has been running at a local nursing home. Are the participants in the program more alert and responsive than nonparticipants? The results, drawn from a random sample of residents, are reported below.

	Status		
Alertness	Participants	Nonparticipants	Totals
High	23	15	38
Low	11	18	29
Totals	34	33	67

a. Is there a statistically significant relationship between participation and alertness?

b. Compute column percents for the table to determine the pattern of the relationship. Which group was more likely to be alert?

10.7 SOC The state Department of Education has rated a sample of local school systems for compliance with state-mandated guidelines for quality. Is the quality of a school system significantly related to the affluence of the community as measured by per capita income?

	Per Capita Income		
Quality	Low	High	Totals
Low	16	8	24
High	9	17	26
Totals	25	25	50

a. Is there a statistically significant relationship between these variables?

b. Compute column percents for the table to determine the pattern of the relationship. Are high- or low-income communities more likely to have high-quality schools?

10.8 CJ A local judge has been allowing some individuals convicted of "driving under the influence" to work in a hospital emergency room as an alternative to fines, suspensions, and other penalties. A random sample of offenders has been drawn. Do participants in this program have lower rates of recidivism for this offense?

	Status		
Recidivist?	Participants	Nonparticipants	Totals
Yes	60	123	183
No	55	108	163
Totals	115	231	346

a. Is there a statistically significant relationship between these variables?

b. Compute column percents for the table to determine the pattern of the relationship. Which group is more likely to be rearrested for driving under the influence?

10.9 SOC Is there a relationship between length of marriage and satisfaction with marriage? The necessary information has been collected from a random sample of 100 respondents drawn from a local community. Write a sentence or two explaining your decision.

	Length of Marriage (in years)			
	Less than 5	5–10	More than 10	Totals
Satisfaction				
Low	10	20	20	50
High	20	20	10	50
Totals	30	40	30	100

a. Is there a statistically significant relationship between these variables?

b. Compute column percents for the table to determine the pattern of the relationship. Which group is more likely to be highly satisfied?

10.10 PS Is there a relationship between political ideology and class standing? Are upper-class students significantly different from underclass students on this variable? The table below reports the relationship between these two variables for a random sample of 267 college students.

	Class Standing		
Ideology	Underclass	Upper-Class	Totals
Liberal	43	40	83
Moderate	50	50	100
Conservative	40	44	84
Totals	133	134	267

a. Is there a statistically significant relationship between these variables?

b. Compute column percents for the table to determine the pattern of the relationship. Which group is more likely to be conservative?

10.11 SOC At a large urban college, about half of the students live off campus in various arrangements, and the other half live in dormitories on campus. Does academic performance depend on living arrangements? The results based on a random sample of 300 students are presented here.

Residential Status

GPA	Off Campus with Roommates	Off Campus with Parents	On Campus	Totals
Low	22	20	48	90
Moderate	36	40	54	130
High	32	10	38	80
Totals	90	70	140	300

a. Is there a statistically significant relationship between these variables?

b. Compute column percents for the table to determine the pattern of the relationship. Which group is more likely to have a high GPA?

10.12 SOC An urban sociologist has built up a database describing a sample of the neighborhoods in her city and has developed a scale by which each area can be rated for its "quality of life" (this includes measures of pollution, noise, open space, services available, and so on). She has also asked samples of residents of these areas about their level of satisfaction with their neighborhoods. Is there significant agreement between the sociologist's objective ratings of quality and the respondents' self-reports of satisfaction?

Quality of Life

Satisfaction	Low	Moderate	High	Totals
Low	21	15	6	42
Moderate	12	25	21	58
High	8	17	32	57
Totals	41	57	59	157

a. Is there a statistically significant relationship between these variables?

b. Compute column percents for the table to determine the pattern of the relationship? Which group is most likely to say that their satisfaction is high?

10.13 SOC Does support for the legalization of marijuana vary by region of the country? The table displays the relationship between the two variables for a random sample of 1,020 adult citizens. Is the relationship significant?

Region

Legalize?	North	Midwest	South	West	Totals
Yes	60	65	42	78	245
No	245	200	180	150	775
Totals	305	265	222	228	1,020

a. Is there a statistically significant relationship between these variables?

b. Compute column percents for the table to determine the pattern of the relationship. Which region most likely to favor the legalization of marijuana?

10.14 SOC A researcher is concerned with the relationship between attitudes toward violence and violent behavior. If attitudes "cause" behavior (a very debatable proposition), then people who have positive attitudes toward violence should have high rates of violent behavior. A pretest was conducted on 70 respondents and, among other things, they were asked, "Have you been involved in a violent incident of any kind over the past six months?" The researcher established the following relationship:

Attitude Toward Violence

Involvement	Favorable	Unfavorable	Totals
Yes	16	19	35
No	14	21	35
Totals	30	40	70

The chi square calculated on these data is .23, which is not significant at the .05 level (confirm this conclusion with your own calculations). Undeterred by this result, the researcher proceeded with the project and gathered a random sample of 7,000. In terms of percentage distributions, the results for the full sample were exactly the same as for the pretest:

Attitude Toward Violence

Involvement	Favorable	Unfavorable	Totals
Yes	1,600	1,900	3,500
No	1,400	2,100	3,500
Totals	3,000	4,000	7,000

However, the chi square obtained is a very healthy 23.4 (confirm with your own calculations), a value that is statistically significant at the 0.05 level. Why is the full-sample chi square significant when the pretest was not? What happened? Do you think that the second result is important?

10.15 SOC Some results from the annual General Social Survey are presented below. See Appendix G for the complete questions and other details about the survey. For each table, conduct the chi square test of significance and compute column percents. Write a sentence or two of interpretation for each test.

a. Support for the legal right to an abortion "for any reason" by age:

	Age			
Support?	Younger Than 30	30–49	50 and Older	Totals
Yes	154	360	213	727
No	179	441	429	1,049
Totals	333	801	642	1,776

b. Support for the death penalty by age:

	Age			
Support?	Younger Than 30	30–49	50 and Older	Totals
Favor	361	867	675	1,903
Oppose	144	297	252	693
Totals	505	1,164	927	2,596

c. Fear of walking alone at night by age:

	Age			
Fear?	Younger Than 30	30–49	50 and Older	Totals
Yes	147	325	300	772
No	202	507	368	1,077
Totals	349	832	668	1,849

d. Support for legalizing marijuana by age:

	Age			
Legalize?	Younger Than 30	30–49	50 and Older	Totals
Should	128	254	142	524
Should Not	224	534	504	1,262
Totals	352	788	646	1,786

e. Support for suicide when a person has an incurable disease by age:

	Age			
Support?	Younger Than 30	30–49	50 and Older	Totals
Yes	225	537	367	1,129
No	107	270	266	643
Totals	332	807	633	1,772

SPSS for Windows

Using *SPSS for Windows* to Conduct the Chi Square Test

SPSS DEMONSTRATION 10.1 Do Child-Care Practices Vary by Social Class?

The **Crosstabs** procedure in SPSS produces bivariate tables and a wide variety of statistics. This procedure is very commonly used in social science research at all levels, and you will see many references to **Crosstabs** in chapters to come. I will introduce the command here, and we will return to it often in later sessions.

Do people in different social classes raise their children differently? Which social class would be most likely to approve of spanking as a disciplinary technique? We can answer these questions, at least for the 2004 GSS sample, by constructing a bivariate table to display the relationship between *class* and *spanking*. We will also request a chi square test for the table.

Begin by clicking **Analyze**, then click **Descriptive Statistics** and **Crosstabs**. The **Crosstabs** dialog box will appear with the variables listed in a box on the left. Highlight *spanking* (the label for this variable is FAVOR SPANKING TO DISCIPLINE CHILD) and click the arrow to move the variable name into the Rows box, and then highlight *class* (the label for this variable is SUBJECTIVE CLASS IDEN-TIFICATION) and move it into the **Columns** box. Click the **Statistics** button at the bottom of the window and click the box next to **Chi-square.** Then, click the Cells button and select "column" in the **Percentages** box. This will generate column percents for the table. Click **Continue** and **OK,** and the output below will be produced. *(NOTE: This output has been slightly edited for clarity. It will not exactly match the output on your screen.)*

FAVOR SPANKING TO DISCIPLINE CHILD * SUBJECTIVE CLASS IDENTIFICATION Cross-tabulation

FAVOR SPANKING TO DISCIPLINE CHILD	SUBJECTIVE CLASS IDENTIFICATION				Total
	LOWER CLASS	WORKING CLASS	MIDDLE CLASS	UPPER CLASS	
STRONGLY AGREE	11	38	44	7	100
	33.3%	20.9%	23.5%	38.9%	23.8%
AGREE	13	100	87	7	207
	39.4%	54.9%	46.5%	38.9%	49.3%
DISAGREE	8	36	40	1	85
	24.2%	19.8%	21.4%	5.6%	20.2%
STRONGLY DISAGREE	1	8	16	3	28
	3.0%	4.4%	8.6%	16.7%	6.7%
Total	33	182	187	18	420
	100.0%	100.0%	100.0%	100.0%	100.0%

Chi Square Tests

	Value	df	Asymp. Sig. (2-sided)
Pearson Chi Square	14.245(a)	9	.114
Likelihood Ratio	14.140	9	.117
Linear-by-Linear Association	.758	1	.384
N of Valid Cases	420		

aFour cells (25.0%) have expected count less than 5. The minimum expected count is 1.20.

The crosstab table is large (4 × 4) and it will probably take a bit of effort to absorb the information. Begin with the cells. Each cell displays the number of cases in the cell and the column percent for that cell. For example, 11 lower-class respondents "strongly agreed" with spanking; these were 33.3% of all lower-class respondents. If you focus on the bottom row, you can see that the percentage of each class that strongly disagrees with spanking is lowest for the lower class and increases as class increases. That is, the column percentages in this row increase steadily from left to right. Only 3.0% of lower-class respondents strongly opposed spanking, but this percentage increased to 16.7% for the upper-class respondents.

The results of the chi square test are reported in the output block that follows the table. The value of chi square (obtained) is 14.245, the degrees of freedom are 9, and the exact significance of the chi square is .114. This is well above the standard indicator of a significant result (alpha = .05), so we may conclude that there is no statistically significant relationship between *class* and *spanking*. Support for spanking does not depend on social class.

SSPS DEMONSTRATION 10.2 Do Sexual Attitudes Vary by Social Class?

Is there a relationship between social class and attitudes about extramarital sex? Which class is most opposed? Run the **Crosstabs** procedure again with *xmarsex* as the row variable (in place of *spanking*) and class as the column variable. Don't forget to request chi square and column percents. The output is reproduced below. (*NOTE: This output has been slightly edited for clarity. It will not exactly match the output on your screen.*)

SEX WITH PERSON OTHER THAN SPOUSE * SUBJECTIVE CLASS IDENTIFICATION Cross-tabulation

SEX WITH PERSON OTHER THAN SPOUSE	SUBJECTIVE CLASS IDENTIFICATION				Total
	LOWER CLASS	WORKING CLASS	MIDDLE CLASS	UPPER CLASS	
ALWAYS WRONG	23	167	166	7	363
	85.2%	80.7%	80.6%	70.0%	80.7%
ALMOST ALWAYS WRONG	2	22	30	1	55
	7.4%	10.6%	14.6%	10.0%	12.2%
SOMETIMES WRONG	1	14	7	2	24
	3.7%	6.8%	3.4%	20.0%	5.3%
NOT WRONG AT ALL	1	4	3	0	8
	3.7%	1.9%	1.5%	0%	1.8%
Total	26	223	218	13	480
	100.0%	100.0%	100.0%	100.0%	100.0%

Chi Square Tests

	Value	df	Asymp. Sig. (2-sided)
Pearson Chi Square	9.413(a)	9	.400
Likelihood Ratio	8.011	9	.533
Linear-by-Linear Association	.000	1	.985
N of Valid Cases	450		

[a]Eight cells (50.0%) have expected count less than 5. The minimum expected count is .18.

Chi square is 9.413, degrees of freedom are 9, and the exact probability of getting this pattern of cell frequencies by random chance alone is .400. There is no significant relationship between the variables, but the column percents show that opposition to extramarital sex generally decreases as class increases. Focusing on the top row, we see that 85.2% of the lower-class respondents thought extramarital sex was "always wrong" versus only 70.0% of the upper class respondents.

What other variables might be significantly related to *xmarsex*? See exercise 10.2 for an opportunity to investigate further.

SPSS DEMONSTRATION 10.3 Is Ignorance Bliss?

We can test the accuracy of the ancient folk wisdom that "ignorance is bliss" with the 2004 GSS. Are more-educated people less happy? To test the relationship, we will use *degree* as our measure of education. To measure the dependent variable, we will use an item from the GSS (*happy*), which asked respondents to rate their overall level of happiness. As originally coded, *degree* has five categories (see Appendix G), but we will simplify the analysis and distinguish only between people with a high school education or less and people with at least some education beyond high school. When you recode *degree,* remember to choose "Recode into Different Variable" and give the recoded variable a new name (say, *rdeg*). The recode instruction in the Old → New box of the **Recode into Different Variable** window should look like this:

$$0 \text{ thru } 1 \rightarrow 1$$
$$2 \text{ thru } 4 \rightarrow 2$$

Follow the instructions in Demonstration 10.1 for the **Crosstabs** command. Specify *happy* as the row variable and recoded *degree* as the column variable. Don't forget to request chi square and column percents. The output should resemble the following table. (*NOTE: This output has been slightly edited for clarity. It will not exactly match the output on your screen.*)

GENERAL HAPPINESS * RDEG Cross-tabulation

GENERAL HAPPINESS	RDEG		Total
	1.00 (HS or less)	2.00 (At least some college)	
VERY HAPPY	114	92	206
	27.5%	39.3%	31.7%
PRETTY HAPPY	230	117	347
	55.4%	50.0%	53.5%
NOT TOO HAPPY	71	25	96
	17.1%	10.7%	14.84%
Total	415	234	649
	100.0%	100.0%	100.0%

Chi Square Tests

	Value	df	Asymp. Sig. (2-sided)
Pearson Chi Square	11.614	2	.003
Likelihood Ratio	11.651	2	.003
Linear-by-Linear Association	11.423	1	.001
N of Valid Cases	649		

The table displays the observed frequencies for the cells along with column percents. The value of chi square (11.614), the degrees of freedom (2), and the exact probability that this pattern of cell frequencies occurred by chance (.003) are reported in the output block below the table. The reported significance is less than .05, so we would reject the null hypothesis of independence. There is a relationship between level of education and degree of happiness.

Remember that chi square tells us only that the overall relationship between the variables is significant. To assess the specific idea that "ignorance is bliss," we need to analyze the column percents. If the idea is true, a higher percentage of the less educated (category 1 of *rdeg*) should be happy. Unfortunately for the old adage, the bivariate table shows the reverse pattern. While 27.5% of the less-educated respondents are "very happy," 39.3% of the more educated place themselves in this category.

Exercises

10.1 For a follow-up on Demonstration 10.1, pick two variables that measure social class. If necessary, recode these variables so that they have only two or three categories. Run the **Crosstabs** procedure with *spanking* as the row variable and your new measures of social class as the column variables. How do these relationships compare with the table generated in Demonstration 10.1? Are these tables consistent with the idea that approval of spanking does not depend on social class?

10.2 Find three more variables that might be related to *xmarsex* (*sex* and *relig* seem like possibilities). Use the **Crosstabs** procedure to see if any of your variables has a significant relationship with *xmarsex*. Which of the variables had the most significant relationship?

10.3 Since we already recoded *degree* for Demonstration 10.3, let's see if education has any significant relationship with *grass* and *cappun*. Run the **Crosstabs** procedure with the dependent variables in the **Rows** box and recoded *degree* in the **Columns** box. Write a paragraph summarizing the results of these three tests. Which relationships are significant? At what levels?

Part III

Bivariate Measures of Association

The chapters in this part cover the computation and analysis of a class of statistics known as measures of association. These statistics are extremely useful in scientific research and commonly reported in the professional literature. They provide, in a single number, an indication of the strength and—if applicable—direction of a bivariate relationship.

It is important to remember the difference between statistical significance, covered in Part II, and association, the topic of Part III. Tests for statistical significance provide answers to certain questions: Were the differences or relationships observed in the sample caused by mere random chance? What is the probability that the sample results reflect patterns in the populations from which the samples were selected? Measures of association address different questions: How strong is the relationship between the variables? What is the direction or pattern of the relationship? Thus, the information that measures of association provide complements tests of significance. Association and significance are two different things and, while the most satisfying results are those that are *both* statistically significant and strong, it is common to find mixed results: relationships that are statistically significant but weak, not statistically significant but strong, and so forth.

Chapter 11 introduces the basic ideas behind analysis of association in terms of bivariate tables and column percentages and presents measures of association appropriate for nominal-level variables. Chapter 12 presents measures of association for variables measured at the ordinal level, and Chapter 13 presents Pearson's *r*, the most important measure of association and the only one designed for interval-ratio-level variables.

11

Introduction to Bivariate Association and Measures of Association for Variables Measured at the Nominal Level

LEARNING OBJECTIVES

By the end of this chapter, you will be able to:

1. Explain how we can use measures of association to describe and analyze the importance of relationships (versus their statistical significance).
2. Define "association" in the context of bivariate tables and in terms of changing conditional distributions.
3. List and explain the three characteristics of a bivariate relationship: existence, strength, and pattern or direction.
4. Investigate a bivariate association by properly calculating percentages for a bivariate table and interpreting the results.
5. Compute and interpret measures of association for variables measured at the nominal level.

11.1 STATISTICAL SIGNIFICANCE AND THEORETICAL IMPORTANCE

As we have seen over the past several chapters, tests of statistical significance are extremely important in social science research. As long as social scientists must work with random samples rather than populations, these tests are indispensable for dealing with the possibility that our research results are the products of mere random chance. However, tests of significance are, typically, only the first step in the analysis of research results. These tests do have limitations, and statistical significance is not necessarily the same thing as relevance or importance. Furthermore, sample size affects all tests of significance: Tests performed on large samples may result in decisions to reject the null hypothesis when, in fact, the observed differences are slight.

Now we will turn our attention to **measures of association.** Whereas tests of significance detect nonrandom relationships, measures of association provide information about the strength and direction of relationships. This information allows us to assess the importance of relationships and test the power and validity of our theories. The theories that guide scientific research are almost always stated in cause-and-effect terms (for example: "variable X causes variable Y"). As an example, recall our discussion of the contact hypothesis in Chapter 1. In that theory, the causal (or independent) variable was equal-status contacts between groups, and the effect (or dependent) variable was level of individual prejudice. The theory asserts that equal-status contact between members of different groups *causes* prejudice to decline. If the theory is true, we should expect to find a strong relationship between variables that measure equal-status contacts and variables that measure prejudice. Furthermore, we should find that prejudice declines as involvement increases. Measures of association help us trace causal relationships between variables and they are our most important and powerful statistical tools for documenting, measuring, and analyzing cause-and-effect relationships.

As useful as they are, measures of association, like any class of statistics, do have their limitations. Most importantly, these statistics cannot *prove* that two variables are causally related. Even if there is a strong (and statistically significant) association between two variables, we cannot necessarily conclude that one variable is a cause of the other. A common adage in the social sciences is, "Correlation (or association) is not the same thing as causation," and you would do well to keep this caution in mind. We can use a statistical association between variables as evidence for a causal relationship, but association by itself is not proof that a causal relationship exists.

Another important use for measures of association is prediction. If two variables are associated, we can predict the score of a case on one variable from the score of that case on the other variable. For example, if equal-status contacts and prejudice are associated, we can predict that people who have experienced many such contacts will be less prejudiced than those who have had few or no contacts. Note that prediction and causation can be two separate things. If variables are associated, we can predict from one to the other even if the variables are not causally related.

This chapter introduces the concept of **association** between variables in the context of bivariate tables and then demonstrates how to use percentages to analyze associations between variables. We will then proceed to the logic, calculation, and interpretation of several widely used measures of association. By the end of this chapter, you will have an array of statistical tools you can use to analyze the strength and direction of associations between variables.

11.2 ASSOCIATION BETWEEN VARIABLES AND BIVARIATE TABLES

Most generally, two variables are said to be associated if the distribution of one of them changes under the various categories or scores of the other. For example, suppose that an industrial sociologist was concerned with the relationship between job satisfaction and productivity for assembly-line workers. If these two variables are associated, then scores on productivity will change under the different conditions of satisfaction. Highly satisfied workers will have scores on productivity different from workers who are low on satisfaction, and levels of productivity will vary by levels of satisfaction.

This relationship will become clearer with the use of bivariate tables. As you recall (see Chapter 10), bivariate tables display the scores of cases on two different variables. By convention, the **independent** or **X variable** (that is, the variable taken as causal) is arrayed in the columns and the **dependent** or **Y variable** in the rows.[1] That is, each column of the table (the vertical dimension) represents a score or category of the independent variable (X), and each row (the horizontal dimension) represents a score or category of the dependent variable (Y).

Table 11.1 displays a relationship between productivity and job satisfaction for a fictitious sample of 173 factory workers. We focus on the columns to detect the presence of an association between variables displayed in table format. Each column shows the pattern of scores on the dependent variable for each score on the independent variable. For example, the left-hand column indicates that 30 of the 60 workers who were low on job satisfaction were low

---

[1] In the material that follows, we will often, for the sake of brevity, refer to the independent variable as X and the dependent variable as Y.

TABLE 11.1 PRODUCTIVITY BY JOB SATISFACTION (frequencies)

Productivity (Y)	Job Satisfaction (X)			Totals
	Low	Moderate	High	
Low	30	21	7	58
Moderate	20	25	18	63
High	10	15	27	52
Totals	60	61	52	173

on productivity, 20 were moderately productive, and 10 were highly productive. The middle column shows that 21 of the 61 moderately satisfied workers were low on productivity, 25 were moderately productive, and 15 were high on productivity. Of the 52 workers who are highly satisfied (the right-hand column), 7 were low on productivity, 18 were moderate, and 27 were high on productivity.

By inspecting the table from column to column, we can observe the effects of the independent variable on the dependent variable (provided, of course, that the independent variable is in the columns). These "within-column" frequency distributions are called the **conditional distributions of Y** because they display the distribution of scores on the dependent variable (Y) for each condition (or score) of the independent variable (X).

Table 11.1 indicates that productivity and satisfaction are associated: The distribution of scores on Y (productivity) changes across the various conditions of X (satisfaction). For example, half of the workers who were low on satisfaction were also low on productivity (30 out of 60). On the other hand, over half of the workers who were high on satisfaction were high on productivity (27 out of 52).

Although it is intended to be a test of significance, the chi square statistic provides another way to detect the existence of an association between two variables that have been organized into table format. Any nonzero value for obtained chi square indicates that the variables are associated. For example, the obtained chi square for Table 11.1 is 24.2, a value that affirms our previous conclusion, based on the conditional distributions of Y, that an association of some sort exists between job satisfaction and productivity.

Often, the researcher will have already conducted a chi square test before considering matters of association. In such cases, it will not be necessary to inspect the conditional distributions of Y to ascertain whether or not the two variables are associated. If the obtained chi square is zero, the two variables are independent and not associated. Any value other than zero indicates some association between the variables. Remember, however, that statistical significance and association are two different things. It is perfectly possible for two variables to be associated (as indicated by a nonzero chi square) but still independent (if we fail to reject the null hypothesis).

In this section, we have defined, in a general way, the concept of association between two variables. We have also looked at two different ways to detect the presence of an association. In the next section, we will extend the analysis beyond questions of the mere presence or absence of an association and, in a systematic way, see how we can develop additional, very useful information about the relationship between two variables.

11.3 THREE CHARACTERISTICS OF BIVARIATE ASSOCIATIONS

Bivariate associations possess three different characteristics, each of which must be analyzed for a full investigation of the relationship. Think of investigating these characteristics as answering three questions:

1. Does an association exist?
2. If an association does exist, how strong is it?
3. What is the pattern and/or the direction of the association?

We will consider each of these questions separately.

Does an Association Exist?   We have already discussed the general definition of association, and we have seen that we can detect an association by observing the conditional distributions of Y in a table or by using chi square. In Table 11.1, we know that the two variables are associated to some extent because the conditional distributions of productivity (Y) are different across the various categories of satisfaction (X) and because the chi square statistic is a nonzero value.

Comparisons from column to column in Table 11.1 are relatively easy to make because the column totals are roughly equal. This will not usually be the case, and it is helpful to compute percentages to control for varying column totals. These column percentages, introduced in Chapter 10, are computed within each column separately and make the pattern of association more visible.

The general procedure for detecting association with bivariate tables is to compute percentages within the columns (vertically or down each column) and then compare column to column across the table (horizontally or across the rows). See the step-by-step box in Chapter 10 (p. 236) to review computing column percentages. Table 11.2 presents column percentages calculated from the data in Table 11.1. Note that this table reports the row and column marginals in parentheses. Besides controlling for any differences in column totals, tables in percentage form are usually easier to read because changes in the conditional distributions of Y are easier to detect.

In Table 11.2, we can see that the largest cell changes position from column to column. For workers who are low on satisfaction, the single largest cell is in the top row (low on productivity). For the middle column (moderate on satisfaction), the largest cell is in the middle row (moderate on productivity), and, for the right-hand column (high on satisfaction), it is in the bottom row (high on productivity). Even a cursory glance at the conditional distributions of Y in Table 11.2 reinforces our conclusion that an association does exist between these two variables.

TABLE 11.2   PRODUCTIVITY BY JOB SATISFACTION (percentages)

| Productivity (Y) | Job Satisfaction (X) | | | Totals |
	Low	Moderate	High	
Low	50.00%	34.43%	13.46%	33.52% (58)
Moderate	33.33%	40.98%	34.62%	36.42% (63)
High	16.67%	24.59%	51.92%	30.06% (52)
Totals	100.00% (60)	100.00% (61)	100.00% (52)	100.00% (173)

TABLE 11.3 PRODUCTIVITY BY HEIGHT: NO ASSOCIATION

	Height (X)		
Productivity (Y)	Short	Medium	Tall
Low	33.33%	33.33%	33.33%
Moderate	33.33%	33.33%	33.33%
High	33.33%	33.33%	33.33%
Totals	100.00%	100.00%	100.00%

TABLE 11.4 PRODUCTIVITY BY HEIGHT: PERFECT ASSOCIATION

	Height (X)		
Productivity (Y)	Short	Medium	Tall
Low	0%	0%	100%
Moderate	0%	100%	0%
High	100%	0%	0%
Totals	100%	100%	100%

If two variables are not associated, then the conditional distributions of Y will not change across the columns. The distribution of Y would be the same for each condition of X. Table 11.3 illustrates a "perfect nonassociation" between the height of the workers and their productivity. Table 11.3 is only one of many patterns that indicate "no association." The important point is that the conditional distributions of Y are the same. Levels of productivity do not change at all for the various heights and, therefore, no association exists between these variables. Also, the obtained chi square computed from this table would have a value of zero, again indicating no association.

How Strong Is the Association?   Once we establish the existence of the association, we need to develop some idea of how strong the association is. This is essentially a matter of determining the amount of change in the conditional distributions of Y. At one extreme, of course, there is the case of "no association" where the conditional distributions of Y do not change at all (see Table 11.3). At the other extreme is a perfect association, the strongest possible relationship. In general, a perfect association exists between two variables if each value of the dependent variable is associated with one and only one value of the independent variable.[2] In a bivariate table, all cases in each column would be located in a single cell and there would be no variation in Y for a given value of X (see Table 11.4).

A perfect relationship would be taken as very strong evidence of a causal relationship between the variables, at least for the sample at hand. In fact, the results presented in Table 11.4 would indicate that, for this sample, height is the

---

[2] Each measure of association introduced in the following chapters incorporates its own definition of a "perfect association," and these definitions vary somewhat, depending on the specific logic and mathematics of the statistic. That is, for different measures computed from the same table, some measures will possibly indicate perfect relationships when others will not. I will note these variations in the mathematical definitions of a perfect association at the appropriate times.

sole cause of productivity. Also, in the case of a perfect relationship, predictions from one variable to the other could be made without error. If we knew that a particular worker was short, for example, we could be sure that he or she is highly productive.

Of course, the huge majority of relationships will fall somewhere between the two extremes of no association and perfect association. We need to develop some way of describing these intermediate relationships consistently and meaningfully. For example, Tables 11.1 and 11.2 show that there is an association between productivity and job satisfaction. How could this relationship be described in terms of strength? How close is the relationship to perfect? How far away from no association?

To answer these questions, researchers rely on statistics called measures of association, which provide precise, objective indicators of the strength of a relationship. Virtually all of these statistics are designed so that they have a lower limit of 0.00 and an upper limit of 1.00 (±1.00 for ordinal and interval-ratio measures of association). A measure that equals 0.00 indicates no association between the variables (the conditional distributions of Y do not vary), and a measure of 1.00 (±1.00 in the case of ordinal and interval-ratio measures) indicates a perfect relationship. The exact meaning of values between 0.00 and 1.00 varies from measure to measure, but, for all measures, the closer the value is to 1.00, the stronger the relationship (the greater the change in the conditional distributions of Y).

We will begin to consider the many different measures of association used in social research later in this chapter. At this point, let's consider the **maximum difference,** a less formal way of assessing the strength of a relationship based on comparing column percentages across the rows. This technique is "quick and dirty": easy to apply (at least for small tables) but limited in its usefulness. To compute the maximum difference, compute the column percentages as usual and then skim the table across each of the rows to find the largest difference—in any row—between column percentages. For example, the largest difference in column percentages in Table 11.2 is in the top row between the "Low" column and the "High" column: 50.00% – 13.46% = 36.54%. The maximum difference in the middle row is between "moderates" and "highs" (40.98% – 33.33% = 7.65%) and, in the bottom row, it is between "highs" and "lows" (51.92% – 16.67% = 35.25%). Both of these values are less than the difference in the top row.

Once you have found the maximum difference, use the scale in Table 11.5 to describe the strength of the relationship:

TABLE 11.5 THE RELATIONSHIP BETWEEN THE MAXIMUM DIFFERENCE AND THE STRENGTH OF THE RELATIONSHIP

Maximum Difference	Strength
If the maximum difference is:	The strength of the relationship is:
Less then 10 percentage points	Weak
Between 10 and 30 percentage points	Moderate
More than 30 percentage points	Strong

Using this scale, we can describe the relationship between productivity and job satisfaction in Table 11.2 as strong.

Be aware that the relationships between the size of the maximum difference and the descriptive terms (weak, moderate, and strong) in Table 11.5 are arbitrary and approximate. We will get more precise and useful information when we compute and analyze the measures of association that I will start to present later in this chapter. Also, maximum differences are easiest to find and most useful for smaller tables. In large tables, with many (say, more than three) columns and rows, it will probably be too cumbersome to bother with this statistic; we would use measures of association only as indicators of the strength for these tables. Finally, note that the maximum difference is based on just two values (the high and low column percentages within any row). As with the range (see Chapter 4), this statistic can give a misleading impression of the overall strength of the relationship. Within these limits, however, the maximum difference can provide a useful, quick, and easy way of characterizing the strength of relationships (at least for smaller tables).

As a final caution, do not mistake chi square as an indicator of the strength of a relationship. Even very large values for chi square do not necessarily mean that the relationship is strong. Remember that significance and association are two separate matters and chi square, by itself, is not a measure of association. While a nonzero value indicates that there is some association between the variables, the magnitude of chi square bears no particular relationship to the strength of the association. *(For practice in computing percentages and judging the existence and strength of an association, see any of the problems at the end of this chapter.)*

What Is the Pattern and/or the Direction of the Association? Investigating the pattern of the association requires that we ascertain which values or categories of one variable are associated with which values or categories of the other. We have already remarked on the pattern of the relationship between productivity and satisfaction. Table 11.2 indicates that low scores on satisfaction are associated with low scores on productivity, moderate satisfaction with moderate productivity, and high satisfaction with high productivity.

When one or both variables in a bivariate table are nominal in level of measurement, our analysis will end with a consideration of the *pattern* of the relationship. However, when both variables are ordinal in level of measurement, we can go on to describe the association in terms of *direction*.[3] The direction of the association can be either positive or negative. An association is positive if the variables vary in the same direction. That is, in a **positive association,** high scores on one variable are associated with high scores on the other variable, and low scores on one variable are associated with low scores on the other. In a positive association, as one variable increases in value, the other also increases; and as one variable decreases, the other also decreases. Table 11.6 displays, with fictitious data, a positive relationship between education and use of public libraries. As education increases (as you move from left to right across the table), library use also increases (the percentage of "high" users increases). The association between job satisfaction and productivity, as displayed in Tables 11.1 and 11.2, is also a positive association.

In a **negative association,** the variables vary in opposite directions. High scores on one variable are associated with low scores on the other, and increases in one variable are accompanied by decreases in the other. Table 11.7 displays a negative relationship, again with fictitious data, between education and

---

[3] Variables measured at the nominal level have no numerical order to them (by definition). Therefore, associations including nominal-level variables, while they may have a pattern, cannot be said to have a direction.

TABLE 11.6   LIBRARY USE BY EDUCATION: A POSITIVE RELATIONSHIP

	Education		
Library Use	Low	Moderate	High
Low	60%	20%	10%
Moderate	30%	60%	30%
High	10%	20%	60%
Total	100%	100%	100%

TABLE 11.7   AMOUNT OF TELEVISION VIEWING BY EDUCATION: A NEGATIVE RELATIONSHIP

	Education		
Television Viewing	Low	Moderate	High
Low	10%	20%	60%
Moderate	30%	60%	30%
High	60%	20%	10%
Totals	100%	100%	100%

television viewing. The amount of television viewing decreases as education increases. In other words, as you move from left to right across the top of the table (as education increases), the percentage of heavy viewers decreases.

Measures of association for ordinal and interval-ratio variables are designed so that they will take on positive values for positive associations and negative values for negative associations. Thus, a measure of association preceded by a plus sign indicates a positive relationship between the two variables, with the value +1.00 indicating a perfect positive relationship. A negative sign indicates a negative relationship, with −1.00 indicating a perfect negative relationship. We will consider the direction of relationships in more detail in Chapter 12.

11.4 INTRODUCTION TO MEASURES OF ASSOCIATION

The column percentages provide very useful information about the bivariate association and should always be computed and analyzed. However, they can be awkward and cumbersome to use, especially for larger tables. Measures of association, on the other hand, characterize the strength (and, for ordinal-level variables, the direction) of bivariate relationships in a single number, a more compact and convenient format for interpretation and discussion.

There are many measures of association but we will confine our attention to a few of the most widely used. We will cover these statistics by the level of measurement for which they are most appropriate. In this chapter, we will consider measures appropriate for nominal variables and, in the next chapter, we will cover measures of association for ordinal-level variables. Finally, in Chapter 13, we will consider Pearson's r, a measure of association or correlation for interval-ratio-level variables. For relationships with variables at different levels of measurement (for example, one nominal-level variable and one ordinal-level variable), we generally use the measure of association appropriate for the lower level of measurement.

Application 11.1

Why are many Americans attracted to movies that emphasize graphic displays of violence? One idea is that "slash" movie fans feel threatened by violence in their daily lives and use these movies as a means of coping with their fears. In the safety of the theater, violence can be vicariously experienced, and feelings and fears can be expressed privately. Also, highly violent movies almost always, as a necessary plot element, provide a role model of one character who does deal with violence successfully (usually, of course, with more violence).

Is frequency of attendance at high-violence movies associated with fear of violence? The following table reports the joint frequency distributions of "fear" and "attendance" in percentages for a fictitious sample of 600.

The conditional distributions of attendance (Y) do change across the values of fear (X), so these variables are associated. The clustering of cases in the diagonal from upper left to lower right suggests a substantial relationship in the predicted direction. People who are low on fear attend violent movies infrequently and people who are high on fear are frequent attendees. Because the maximum difference in column percentages in the table is 30 (in both the top and middle rows), we can characterize the relationship as moderate to strong.

These results do suggest an important relationship between fear and attendance. Notice, however, that these results pose an interesting causal problem. The table supports the idea that fearful and threatened people attend violent movies as a coping mechanism (X causes Y) but is also consistent with the reverse causal argument: Watching violent movies increases fears for one's personal safety (Y causes X). The results support *both* causal arguments and remind us that association is not the same thing as causation.

ATTENDANCE BY FEAR

Attendance	Fear		
	Low	Moderate	High
Rare	50%	20%	30%
Occasional	30%	60%	30%
Frequent	20%	20%	40%
Totals	100%	100%	100%
	(200)	(200)	(200)

11.5 MEASURES OF ASSOCIATION FOR VARIABLES MEASURED AT THE NOMINAL LEVEL: CHI SQUARE–BASED MEASURES

When working with nominal-level variables, social science researchers rely heavily on measures of association based on the value of chi square. When the value of chi square is already known, these measures are easy to calculate. To illustrate, let's reconsider Table 10.3, which displayed, with fictitious data, a relationship between accreditation and employment for social work majors. For the sake of convenience, Table 11.8 reproduces that table.

We saw in Chapter 10 that this relationship is statistically significant ($\chi^2 = 10.78$, which is significant at $\alpha = .05$) but the question now concerns the *strength* of the association. A brief glance at Table 11.8 shows that the conditional distributions of employment status do change, so the variables are associated.

TABLE 11.8 EMPLOYMENT OF 100 SOCIAL WORK MAJORS BY ACCREDITATION STATUS OF UNDERGRADUATE PROGRAM (fictitious data)

Employment Status	Accreditation Status		Totals
	Accredited	Not Accredited	
Working as a social worker	30	10	40
Not working as a social worker	25	35	60
Totals	55	45	100

TABLE 11.9   EMPLOYMENT BY ACCREDITATION STATUS (percentages)

| | Accreditation Status | | |
Employment Status	Accredited	Not Accredited	Totals
Working as a social worker	54.55%	22.22%	40.00%
Not working as a social worker	45.45%	77.78%	60.00%
Totals	100.00%	100.00%	100.00%

To emphasize this point, it is always helpful to calculate column percentages, as in Table 11.9.

So far, we know that the relationship between these two variables is statistically significant and that there is an association of some kind between accreditation and employment. To assess the strength of the association, we will compute a **phi (ϕ).** This statistic is a frequently used chi square–based measure of association appropriate for 2 × 2 tables (that is, tables with two rows and two columns).

Calculating Phi.   One of the attractions of phi is that it is easy to calculate. Simply divide the value of the obtained chi square by N and take the square root of the result. Expressed in symbols, the formula for phi is:

FORMULA 11.1
$$\phi = \sqrt{\frac{\chi^2}{N}}$$

For the data displayed in Table 11.8, the chi square was 10.78. Therefore, phi is:

$$\phi = \sqrt{\frac{\chi^2}{N}}$$

$$\phi = \sqrt{\frac{10.78}{100}}$$

$$\phi = 0.33$$

For a 2 × 2 table, phi ranges in value from 0 (no association) to 1.00 (perfect association). The closer to 1.00, the stronger the relationship and the closer to 0.00, the weaker the relationship. For Table 11.8, we already knew that the relationship was statistically significant at the 0.05 level. Phi, as a measure of association, adds information about the strength of the relationship. As for the pattern of the association, the column percentages in Table 11.9 show that graduates of accredited programs were more often employed as social workers.

Calculating Cramer's V.   For tables larger than 2 × 2 (specifically, for tables with more than two columns and more than two rows), the upper limit of phi can exceed 1.00. This makes phi difficult to interpret, and we must use a more general form of the statistic called **Cramer's V** for larger tables. The formula for Cramer's V is:

FORMULA 11.2
$$V = \sqrt{\frac{\chi^2}{(N)(\min r - 1, c - 1)}}$$

where (min $r - 1$, $c - 1$) = the minimum value of $r - 1$ (number of rows minus 1) or $c - 1$ (number of columns minus 1)

TABLE 11.10   ACADEMIC ACHIEVEMENT BY CLUB MEMBERSHIP (frequencies)

Academic Achievement	Membership			
	Fraternity or Sorority	Other Organization	No Memberships	Totals
Low	4	4	17	25
Moderate	15	6	4	25
High	4	16	5	25
Totals	23	26	26	75

In words: To calculate V, find the lesser of the number of rows minus 1 ($r-1$) or the number of columns minus 1 ($c-1$), multiply this value by N, divide the result into the value of chi square, and then find the square root. Cramer's V has an upper limit of 1.00 for any size table and will be the same value as phi if the table has either two rows or two columns. As with phi, we can interpret Cramer's V as an index that measures the strength of the association between two variables.

To illustrate the computation of V, suppose you had gathered the data in Table 11.10, which shows the relationship between membership in student organizations and academic achievement for a sample of college students. The obtained chi square for this table is 31.5, a value that is significant at the .05 level. Cramer's V is:

$$V = \sqrt{\frac{\chi^2}{(N)(\min r - 1, c - 1)}}$$

$$V = \sqrt{\frac{31.50}{(75)(2)}}$$

$$V = \sqrt{\frac{31.50}{150}}$$

$$V = \sqrt{0.21}$$

$$V = 0.46$$

Because Table 11.10 has the same number of rows and columns, we may use either ($r - 1$) or ($c - 1$) in the denominator. In either case, the value of the denominator is N multiplied by (3 − 1), or 2. Table 11.11 gives column percentages to help identify the pattern of this relationship. Fraternity and sorority members tend to be moderate, members of other organizations tend to be high, and nonmembers tend to be low in academic achievement.

Interpreting Phi and Cramer's V.   It will be helpful to have some general guidelines for interpreting the value of measures of association for nominal-level variables, similar to the guidelines we used for interpreting the maximum difference in column percentages. Table 11.12 presents the general relationship between the value of the statistic and the strength of the relationship for phi and Cramer's V. As was the case for Table 11.5, the relationships in Table 11.12 are arbitrary and meant as general guidelines only. Using these guidelines, we can characterize the relationships in Table 11.8 ($\phi = 0.33$) and Table 11.10 ($V = 0.46$) as strong.

TABLE 11.11   ACADEMIC ACHIEVEMENT BY CLUB MEMBERSHIP (percentages)

Academic Achievement	Membership			
	Fraternity or Sorority	Other Organization	No Memberships	Totals
Low	17.39	15.39	65.39	33.33%
Moderate	65.22	23.08	15.39	33.33%
High	17.39	61.54	19.23	33.33%
Totals	100.00	100.00	100.00	100.00%

TABLE 11.12   THE RELATIONSHIP BETWEEN THE VALUE OF NOMINAL-LEVEL MEASURES OF ASSOCIATION AND THE STRENGTH OF THE RELATIONSHIP

Value	Strength
If the value is:	The strength of the relationship is:
Less than 0.10	Weak
Between 0.11 and 0.30	Moderate
Greater than 0.30	Strong

The Limitations of Phi and V.   One limitation of phi and Cramer's V is that they are only general indicators of the strength of the relationship. Of course, the closer these measures are to 0.00, the weaker the relationship and the closer to 1.00, the stronger the relationship. Values between 0.00 and 1.00 can be described as weak, moderate, or strong according to the general convention introduced earlier but have no direct or meaningful interpretation. On the other hand, phi and V are easy to calculate (once you obtain the value of chi square) and are commonly used indicators of the importance of

---

STEP BY STEP | **Calculating and Interpreting Phi and Cramer's *V***

Calculating Phi

Step 1: To calculate phi, solve Formula 11.1

$$\phi = \sqrt{\frac{\chi^2}{N}}$$

a. Divide the value of chi square by N.
b. Take the square root of the quantity you found in step a.
c. Consult Table 11.12 to help interpret the value of phi.

Calculating Cramer's *V*

Step 1: To calculate Cramer's V, solve Formula 11.2

$$V = \sqrt{\frac{\chi^2}{(N)(\min r - 1, c - 1)}}$$

a. Determine the number of rows (r) and columns (c) in the table. Subtract 1 from the lesser of these two numbers to find (min $r - 1, c - 1$).
b. Multiply the value you found in step a by N.
c. Divide the value of chi square by the quantity you found in step b.
d. Take the square root of the quantity you found in step c.
e. Consult Table 11.12 to help interpret the value of V.

Application 11.2

A random sample of students at a large urban university has been classified as either "traditional" (18–23 years of age and unmarried) or "nontraditional" (24 or older or married). Subjects have also been classified as "vocational" if their primary motivation for college attendance is career or job oriented or "academic" if their motivation is to pursue knowledge for its own sake. Are these two variables associated?

MOTIVATION FOR COLLEGE ATTENDANCE
BY TYPE OF STUDENT

	Type		
Motivation	Traditional	Nontraditional	Totals
Vocational	25	60	85
Academic	75	15	90
Totals	100	75	175

Always begin your analysis of bivariate tables by computing column percentages (assuming that the independent variable is in the columns). This will allow you to detect the pattern of the association and, by finding the maximum difference, to assess the strength of the association. Finally, we will calculate an appropriate measure of association.

MOTIVATION FOR COLLEGE ATTENDANCE
BY TYPE OF STUDENT

	Type	
Motivation	Traditional	Nontraditional
Vocational	25.00%	80.00%
Academic	75.00%	20.00%
Totals	100.00%	100.00%

The maximum difference is 55, which indicates a strong relationship between these two variables. The pattern is clear: Traditional students are more likely to be academically motivated and nontraditional students are more vocational.

Because this is a 2 × 2 table, we can compute phi as a measure of association. The chi square for the table is 51.89, so phi is:

$$\phi = \sqrt{\frac{\chi^2}{N}}$$

$$\phi = \sqrt{\frac{51.89}{175}}$$

$$\phi = \sqrt{0.30}$$

$$\phi = 0.55$$

The value of phi, like the maximum difference, indicates a strong relationship between the two variables.

an association.[4] *(For practice, you can compute phi and Cramer's V for any of the problems at the end of this chapter. Problems with 2 × 2 tables will minimize computations. Remember that for tables with either two rows or two columns, phi and Cramer's V will have the same value.)*

11.6 LAMBDA: A PROPORTIONAL REDUCTION IN ERROR MEASURE OF ASSOCIATION FOR NOMINAL-LEVEL VARIABLES

The Logic of Proportional Reduction in Error (PRE). In recent years, measures based on a logic known as **proportional reduction in error (PRE)** have been developed to complement the older chi square–based measures of association. Most generally stated, the logic of these measures requires us to make two different predictions about the scores of cases. In the first prediction, we ignore information about the independent variable and, therefore, make many errors in predicting the score on the dependent variable. In the second

---

[4] Two other chi square–based measures of association, T^2 and C (the contingency coefficient), are sometimes reported in the literature. Both of these measures have serious limitations. T^2 has an upper limit of 1.00 only for tables with an equal number of rows and columns, and the upper limit of C varies, depending on the dimensions of the table. These characteristics make these measures more difficult to interpret and thus less useful than phi or Cramer's V.

prediction, we take account of the score of the case on the independent variable to help predict the score on the dependent variable. If there is an association between the variables, we will make fewer errors when taking the independent variable into account. PRE measures of association express the proportional reduction in errors between the two predictions. Applying these general thoughts to the case of nominal-level variables will make the logic clearer.

For nominal-level variables, we first predict the category into which each case will fall on the dependent variable (Y) while ignoring the independent variable (X). Because we would be predicting blindly in this case, we would make many errors (that is, we would often predict the value of a case on the dependent variable incorrectly).

The second prediction allows us to take the independent variable into account. If the two variables are associated, the additional information supplied by the independent variable will reduce our errors of prediction (that is, we should misclassify fewer cases). The stronger the association between the variables, the greater the reduction in errors. In the case of a perfect association, we would make no errors at all when predicting a score on Y from a score on X. When there is no association between the variables, however, knowledge of the independent variable will not improve the accuracy of our predictions. We would make just as many errors of prediction with knowledge of the independent variable as we did without knowledge of it.

An illustration should clarify these principles. Suppose you were placed in the rather unusual position of having to predict whether each of the next 100 people you meet will be shorter or taller than 5 feet 9 inches under the condition that you would have no knowledge of these people at all. With absolutely no information about these people, your predictions will be wrong often (you will frequently misclassify a tall person as short and vice versa).

Now assume that you must go through this ordeal twice, but, on the second round, you know the sex of the person whose height you must predict. Because height is associated with gender and females are, on the average, shorter than males, the optimal strategy would be to predict that all females are short and all males are tall. You will still make errors on this second round, but, if the variables are associated, the number of errors will be fewer. That is, using information about the independent variable will reduce the number of errors (if, of course, the two variables are related). How can these unusual thoughts be translated into a useful statistic?

Lambda.   One hundred individuals have been categorized by gender and height, and Table 11.13 displays the data. It is clear, even without percentages,

TABLE 11.13   HEIGHT BY GENDER

Height	Gender		Totals
	Male	Female	
Tall	44	8	52
Short	6	42	48
Totals	50	50	100

that the two variables are associated. To measure the strength of this association, a PRE measure called **lambda** (symbolized by the lowercase Greek letter λ) will be calculated. Following the logic introduced in the previous section, we must find two quantities. First, the number of prediction errors made while ignoring the independent variable (gender) must be found. Then, we will find the number of prediction errors made while taking gender into account. We will compare these two sums to derive the statistic.

First, the information given by the independent variable (gender) can be ignored by working only with the row marginals. Two different predictions can be made about height (the dependent variable) by using these marginals. We can predict either that all subjects are tall or that all subjects are short.[5] For the first prediction (all subjects are tall), 48 errors will be made. That is, for this prediction, all 100 cases would be placed in the top row. Because only 52 of the cases actually belong in this row, this prediction would result in $(100 - 52)$, or 48, errors. If we had predicted that all subjects were short, on the other hand, we would have made 52 errors $(100 - 48 = 52)$. We will take the *lesser* of these two numbers and refer to this quantity as E_1 for the number of errors made while ignoring the independent variable. So, $E_1 = 48$.

In the second step in the computation of lambda, we predict score on Y (height) again but this time we take X (gender) into account. To do this, follow the same procedure as in the first step but this time move from column to column. Because each column is a category of X, we thus take X into account in making our predictions. For the left-hand column (males), we predict that all 50 cases will be tall and make six errors $(50 - 44 = 6)$. For the second column (females), our prediction is that all females are short, and eight errors will be made. By moving from column to column, we have taken X into account and have made a total of 14 errors of prediction, a quantity we will label E_2 ($E_2 = 6 + 8 = 14$).

If the variables are associated, we will make fewer errors under the second procedure than under the first or, in other words, E_2 will be smaller than E_1. In this case, we made fewer errors of prediction while taking gender into account ($E_2 = 14$) than while ignoring gender ($E_1 = 48$), so gender and height are clearly associated. Our errors were reduced from 48 to only 14. To find the *proportional* reduction in error, use Formula 11.3:

FORMULA 11.3
$$\lambda = \frac{E_1 - E_2}{E_1}$$

For the sample problem, the value of lambda would be:

$$\lambda = \frac{E_1 - E_2}{E_1}$$

$$\lambda = \frac{48 - 14}{48}$$

---

[5] Other predictions are, of course, possible, but these are the only two permitted by lambda.

TABLE 11.14   ATTITUDE TOWARD CAPITAL PUNISHMENT BY RELIGIOUS DENOMINATION
(fictitious data)

	Religion				
Attitude	Catholic	Protestant	Other	None	Totals
Favors	10	9	5	14	38
Neutral	14	12	10	6	42
Opposed	11	4	25	10	50
Totals	35	25	40	30	130

$$\lambda = \frac{34}{48}$$

$$\lambda = 0.71$$

The value of lambda ranges from 0.00 to 1.00. Of course, a value of 0.00 means that the variables are not associated at all (E_1 is the same as E_2), and a value of 1.00 means that the association is perfect (E_2 is zero and scores on the dependent variable can be predicted without error from the independent variable). Unlike phi or V, however, the numerical value of lambda between the extremes of 0.00 and 1.00 has a precise meaning: It is an index of the extent to which the independent variable (X) helps us to predict (or, more loosely, understand) the dependent variable (Y). When multiplied by 100, the value of lambda indicates the strength of the association in terms of the percentage reduction in error. Thus, we would interpret the lambda above by concluding that knowledge of gender improves our ability to predict height by 71%.

An Additional Example of Calculating and Interpreting Lambda.   In this section, we will work through another example to state the computational routine for lambda in general terms. Suppose a researcher was concerned with the relationship between religious denomination and attitude toward capital punishment and had collected the data presented in Table 11.14.

To find E_1, the number of errors made while ignoring X (religion, in this case), subtract the largest row total from N. For Table 11.14, E_1 will be:

$$E_1 = N - (\text{Largest row total})$$

$$E_1 = 130 - 50$$

$$E_1 = 80$$

To find E_2, begin with the left-hand column (Catholics) and subtract the largest cell frequency from the column total. Repeat this procedure for each column in the table and then add the subtotals together:

For Catholics: $35 - 14$ = 21
For Protestants: $25 - 12$ = 13
For Other: $40 - 25$ = 15
For None: $30 - 14$ = $\underline{16}$
$E_2 = 65$

| STEP BY STEP | **Calculating and Interpreting Lambda** |

Step 1: To find E_1, subtract the largest row subtotal (marginal) from N.

Step 2: Starting with the far left-hand column, subtract the largest cell frequency in the column from the column total. Repeat this step for all columns in the table.

Step 3: Add up all the values you found in step 2. The result is E_2.

Step 4: To solve Formula 11.3: $\lambda = \dfrac{E_1 - E_2}{E_1}$

a. Subtract E_2 from E_1.
b. Divide the quantity you found in step a by E_1. The resultant quantity is lambda.

Step 5: To interpret lambda, multiply the value of lambda by 100. This percentage tells us the extent to which our predictions of the dependent variable are improved by taking the independent variable into account. In addition, we can interpret lambda using the descriptive terms in Table 11.12.

Substitute the values of E_1 and E_2 into Formula 11.3:

$$\lambda = \frac{80 - 65}{80}$$

$$\lambda = \frac{15}{80}$$

$$\lambda = 0.19$$

A lambda of 0.19 means that we are 19% better off using religion to predict attitude toward capital punishment (as opposed to predicting blindly). At best, this relationship is moderate in strength (see Table 11.12).

The Limitations of Lambda. Lambda has two characteristics of special note. First, lambda is asymmetric. This means that the value of the statistic will vary, depending on which variable is taken as independent. For example, in Table 11.14, the value of lambda would be 0.14 if attitude toward capital punishment had been taken as the independent variable (verify this with your own computation). Thus, you should exercise some caution in the designation of an independent variable. If you consistently follow the convention of arraying the independent variable in the columns and compute lambda as outlined above, the asymmetry of the statistic should not be confusing.

Second, when one of the row totals is much larger than the others, lambda can be misleading. It can be 0.00 even when other measures of association are greater than 0.00 and the conditional distributions for the table indicate that there is an association between the variables. This anomaly is a function of the way lambda is calculated and suggests that great caution should be exercised in the interpretation of lambda when the row marginals are very unequal. In fact, in the case of very unequal row marginals, a chi square–based measure of association would be the preferred measure of association. *(For practice in computing lambda, see any of the problems at the end of this chapter. As with phi and Cramer's V, it's probably a good idea to start with small samples and 2 × 2 tables.)*

STATISTICS IN EVERYDAY LIFE: The Truth About Astrology

According to recent polls,* about 30% of Americans believe in astrology and, no doubt, many more regularly consult their horoscope for fun or for curiosity. Is there any basis for this widespread belief? Do people born under different signs really vary significantly from each other? How strong are the relationships between astrological sign and attitudes and outlook?

The General Social Survey records the astrological sign of respondents and provides an opportunity to test the relationship between a person's sign and a number of dependent variables for a large, representative sample of Americans. Because there are 12 different astrological signs, the bivariate tables used in these tests are large and cannot be reproduced here. Instead, the table below lists the dependent variables and summarizes the results of four separate chi square tests. The dependent variables included measures of overall happiness and health, whether the respondent saw other people as generally helpful, and whether the respondent was generally sad or depressed. For each dependent variable, I constructed a bivariate table with astrological sign as the independent variable. The table below reports some results from these tests.

The four dependent variables are listed in the left-hand column followed by the results of the chi square test, including the value of χ^2(obtained) and the degrees of freedom (df) for the table. The next column, labeled "Sig.?," notes whether or not the relationship is significant and, in parentheses (α), also presents the exact probability of getting the frequencies observed in the table by random chance alone. Finally, I present Cramer's V for each relationship and provide some information on the pattern of the relationships.

The results of these tests are very consistent: None of the relationships approaches statistical significance. All, in fact, are considerably larger than 0.05, the standard indicator of a significant result. It is extremely likely that the pattern of cell frequencies in the table occurred by random chance alone and that there is no relationship with zodiac sign in the population. Cramer's V indicates that the relationships are weak (although two of the relationships might be considered "moderate to weak"), further reinforcing the conclusion that zodiac sign has no important relationship with any of these variables.

These tests are not good news for adherents of astrology but they, no doubt, will weather the storm. It could be argued that the tests reported above are crude and, at any rate, true believers are not easily deterred by mere scientific evidence.

Dependent Variable	Chi Square Test			V	Pattern
	χ^2	df	Sig.? (α)		
Happiness	24.24	22	No (0.34)	0.10	Happiest = Pisces (41%) Least Happy = Aquarius (20%)
Health	30.87	33	No (0.57)	0.09	Healthiest = Aries (37%) Unhealthiest = Sagittarius (22%)
Helpful	21.21	22	No (0.51)	0.11	Most likely to see people as helpful = Pisces (59%) Least likely to see people as helpful = Libra (40%)
Sad or depressed	17.59	22	No (0.66)	0.07	Most depressed[a] = Aries (22%) Least depressed[a] = Taurus (14%)

[a] Percent who say that "sad and blue" is a "very good, good, or fair" description of them.

* For example: The Harris poll reported that 30% of Americans are believers (http://www.harrisinteractive.com/harris_poll/index.asp?PID=359) and a Fox News poll (http://www.foxnews.com/ story/0,2933,99945,00.html) reported a very similar result (29%).

SUMMARY

1. Analyzing the association between variables provides information that is complementary to tests of significance. The latter are designed to detect nonrandom relationships, whereas measures of association are designed to quantify the importance or strength of a relationship.

2. Relationships between variables have three characteristics: the existence of an association, the strength of the association, and the direction or pattern of the association. We can investigate these three characteristics by calculating percentages for a bivariate table in the direction of the independent variable (vertically) and then comparing in the opposite direction (horizontally). It is often useful (as well as quick and easy) to assess the strength of a relationship by finding the maximum difference in column percentages in any row of the table.

3. Tables 11.1 and 11.2 can be analyzed in terms of these three characteristics. Clearly, a relationship does exist between job satisfaction and productivity because the conditional distributions of the dependent variable (productivity) are different for the three different conditions of the independent variable (job satisfaction). Even without a measure of association, we can see that the association is substantial in that the change in Y (productivity) across the three categories of X (satisfaction) is marked. The maximum difference of 36.54% confirms that the relationship is substantial (moderate to strong). Furthermore, the relationship is positive in direction.

Productivity increases as job satisfaction rises, and workers who report high job satisfaction tend also to be high on productivity. Workers with little job satisfaction tend to be low on productivity.

4. Given the nature and strength of the relationship, it could be predicted with fair accuracy that highly satisfied workers tend to be highly productive ("happy workers are busy workers"). These results might be taken as evidence of a causal relationship between these two variables, but they cannot, by themselves, prove that a causal relationship exists: Association is not the same thing as causation. In fact, although we have presumed that job satisfaction is the independent variable, we could have argued the reverse causal sequence ("busy workers are happy workers"). The results presented in Tables 11.1 and 11.2 are consistent with both causal arguments.

5. Phi, Cramer's V, and lambda are measures of association and each is appropriate for a specific situation. We use phi for nominal level variables in a 2 × 2 table and Cramer's V for tables larger than 2 × 2. While phi and Cramer's V are based on chi square, lambda is a PRE measure and can give us more exact information about values between 0.00 and 1.00. These statistics express information about the strength of the relationship *only*. In all cases, be sure to analyze the column percentages as well as the measure of association to maximize the information you have about the relationship.

SUMMARY OF FORMULAS

Phi 11.1 $$\phi = \sqrt{\frac{\chi^2}{N}}$$

Cramer's V 11.2 $$V = \sqrt{\frac{\chi^2}{(N)(\min r - 1, c - 1)}}$$

Lambda 11.3 $$\lambda = \frac{E_1 - E_2}{E_1}$$

GLOSSARY

Association. The relationship between two (or more) variables. Two variables are said to be associated if the distribution of one variable changes for the various categories or scores of the other variable.

Conditional distribution of Y. The distribution of scores on the dependent variable for a specific score or category of the independent variable when the variables have been organized into table format.

Cramer's V. A chi square–based measure of association. Appropriate for nominally measured variables that have been organized into a bivariate table of any number of rows and columns.

Dependent variable. In a bivariate relationship, the variable that is considered the effect.

Independent variable. In a bivariate relationship, the variable that is taken as the cause.

Lambda (λ). A proportional-reduction-in-error (PRE) measure of association for variables measured at the nominal level that have been organized into a bivariate table.

Maximum difference. A way to assess the strength of an association between variables that have been organized into a bivariate table. The maximum difference is the largest difference between column percentages for any row of the table.

Measure of association. Statistic that quantifies the strength of the association between variables.

Negative association. A bivariate relationship where the variables vary in opposite directions. As one

variable increases, the other decreases, and high scores on one variable are associated with low scores on the other.

Phi (φ). A chi square–based measure of association. Appropriate for nominally measured variables that have been organized into a 2 × 2 bivariate table.

Positive association. A bivariate relationship where the variables vary in the same direction. As one variable increases, the other also increases, and high scores on one variable are associated with high scores on the other.

Proportional reduction in error (PRE). The logic that underlies the definition and computation of lambda. The statistic compares the number of errors made when predicting the score of a case on the dependent variable while ignoring the independent variable with the number of errors made while taking the independent variable into account.

X. Symbol used for any independent variable.

Y. Symbol used for any dependent variable.

MULTIMEDIA RESOURCES

The Wadsworth Sociology Resource Center: Virtual Society at
http://www.thomsonedu.com/sociology

Visit the companion website for the seventh edition of *Statistics: A Tool for Social Research* to access a wide range of student resources. Begin by clicking on the Student Resources section of the book's website to access the following study tools:

- Basic math review
- Flash cards

- Additional chapter problems
- Internet links
- Table of random numbers
- MicroCase and SPSS examples and exercises
- Hypothesis testing for variables measured at the ordinal level

PROBLEMS

11.1 PA Various supervisors in the city government of Shinbone, Kansas, have been rated on the extent to which they practice authoritarian styles of leadership and decision making. The efficiency of each department has also been rated, and the results are summarized in table format. Use column percentages, the maximum difference, and measures of association to describe the strength and pattern of this association.

Efficiency	Authoritarianism		
	Low	High	Totals
Low	10	12	22
High	17	5	22
Totals	27	17	44

11.2 SOC The administration of a local college campus has proposed an increase in the mandatory student fee to finance an upgrading of the intercollegiate football program. A sample of the faculty has completed a survey on the issues. Is there any association between support for raising fees and the gender, discipline, or tenured status of the faculty? Use column percentages, the maximum difference, and measures of association to describe the strength and pattern of these associations.

a. Support for raising fees by gender:

| Support | Gender | | Totals |
	Males	Females	
For	12	8	20
Against	15	12	27
Totals	27	20	47

b. Support for raising fees by discipline:

| Support | Discipline | | Totals |
	Liberal Arts	Science & Business	
For	6	13	19
Against	14	14	28
Totals	20	27	47

c. Support for raising fees by tenured status:

| Support | Status | | Totals |
	Tenured	Nontenured	
For	15	4	19
Against	18	10	28
Totals	33	14	47

11.3 PS How consistent are people in their voting habits? Do people vote for the same party from election to election? Below are the results of a poll in which people were asked if they had voted Democrat or Republican in each of the last two presidential elections. Use column percentages, the maximum difference, and measures of association to describe the strength and pattern of the association.

| 2004 Election | 2000 Election | | Totals |
	Democrat	Republican	
Democrat	117	23	140
Republican	17	178	195
Totals	134	201	335

11.4 SOC A needs assessment survey has been distributed in a large retirement community. Residents were asked to indicate the services or programs they thought should be added. Is there an association between gender and the perception that more parties should be scheduled? Use column percentages, the maximum difference, and measures of association to describe the strength and direction of the association. Write a few sentences describing the relationship.

| More Parties? | Gender | | Totals |
	Males	Females	
Yes	321	426	747
No	175	251	426
Totals	496	677	1,173

11.5 SW As the state director of mental health programs, you note that some local mental health facilities have very high rates of staff turnover. You believe that part of this problem stems from the fact that some of the local directors have scant training in administration and poorly developed leadership skills. Before implementing a program to address this problem, you collect some data to make sure that your beliefs are supported by the facts. Is there a relationship between staff turnover and the administrative experience of the directors? Use column percentages, the maximum difference, and measures of association to describe the strength and direction of the association. Write a few sentences describing the relationship.

| Turnover | Director Experienced? | | Totals |
	No	Yes	
Low	4	9	13
Moderate	9	8	17
High	15	5	20
Totals	28	22	50

11.6 CJ About half the neighborhoods in a large city have instituted programs to increase citizen involvement in crime prevention. Do these areas experience less crime? Use column percentages, the maximum difference, and measures of association to describe the strength and direction of the association. Describe the relationship in a few sentences.

Crime Rate	Program		Totals
	No	Yes	
Low	29	15	44
Moderate	33	27	60
High	52	45	97
Totals	114	87	201

11.7 GER Researchers have surveyed senior citizens who live in either a housing development specifically designed for retirees or an age-integrated neighborhood. Is type of living arrangement related to sense of social isolation?

Sense of Isolation	Living Arrangement		Totals
	Housing Development	Integrated Neighborhood	
Low	80	30	110
High	20	120	140
Totals	100	150	250

11.8 SOC A researcher has conducted a survey on sexual attitudes for a sample of 317 teenagers. The respondents were asked whether they considered premarital sex to be "always wrong" or "OK under certain circumstances." The tables below summarize the relationship between responses to this item and several other variables. For each table, assess the strength and pattern of the relationship and write a paragraph interpreting these results.

a. Attitudes toward premarital sex by gender:

Premarital Sex	Gender		Totals
	Female	Male	
Always wrong	90	105	195
Not always wrong	65	57	122
Totals	155	162	317

b. Attitudes toward premarital sex by courtship status:

Premarital Sex	Ever "Gone Steady"?		Totals
	No	Yes	
Always wrong	148	47	195
Not always wrong	42	80	122
Totals	190	127	317

c. Attitudes toward premarital sex by social class:

Premarital Sex	Social Class		Totals
	Blue Collar	White Collar	
Always wrong	72	123	195
Not always wrong	47	75	122
Totals	119	198	317

11.9 SOC Here are five dependent variables cross-tabulated against gender as an independent variable. Use column percentages, the maximum difference, and an appropriate measure of association to conduct the analysis. Summarize the results of your analysis in a paragraph that describes the strength and pattern of each relationship.

a. Support for the legal right to an abortion by gender:

Right to Abortion?	Gender		Totals
	Male	Female	
Yes	310	418	728
No	432	618	1,050
Totals	742	1,036	1,778

b. Support for capital punishment by gender:

Capital Punishment?	Gender		Totals
	Male	Female	
Favor	908	998	1,906
Oppose	246	447	693
Totals	1,154	1,445	2,599

c. Approve of suicide for people with incurable disease by gender:

Right to Suicide?	Gender		Totals
	Male	Female	
Yes	524	608	1,132
No	246	398	644
Totals	770	1,006	1,776

d. Support for sex education in public schools by gender:

Sex Education?	Gender		Totals
	Male	Female	
Favor	685	900	1,585
Oppose	102	134	236
Totals	787	1,034	1,821

e. Support for traditional gender roles by gender:

Women Should Take Care of Running Their Homes and Leave Running the Country to Men	Gender		Totals
	Male	Female	
Agree	116	164	280
Disagree	669	865	1,534
Totals	785	1,029	1,814

SPSS for Windows

Using *SPSS for Windows* to Test Bivariate Association for Significance and Association

SPSS DEMONSTRATION 11.1 Does Support for Capital Punishment Vary by Gender?

The death penalty has often been a "hot button" item in American politics and, as with abortion, gay marriage, and many others, a central issue in the "culture wars" that animate everyday life in this society. In this demonstration, we'll use SPSS and the 2004 GSS to explore some of the correlates of opinion on this issue. In the GSS, the variable *cappun* measured attitude toward capital punishment. Respondents were asked whether they favored or opposed the death penalty for the crime of murder. These two dichotomous choices probably oversimplify the range of people's attitudes and feelings, but they will keep the bivariate table as simple as possible and simplify the statistical analysis.

Let's begin by looking at the relationship with gender. Because males are commonly seen as more violent and aggressive, it makes sense to hypothesize that they would be more supportive of the death penalty. Is this true? In this test, the dependent variable will be the respondent's support for capital punishment and the independent variable will be gender.

To test this relationship, click **Analyze, Descriptive Statistics,** and **Crosstabs** and name *cappun* as the row (dependent) variable and *sex* as the column (independent) variable. Click the **Cells** button and request column percentages by clicking the box next to **Column** in the **Percentages** box. Also, click the **Statistics** button and request chi square, phi, Cramer's *V,* and lambda by clicking the appropriate boxes. Click **Continue** and **OK,** and the following output will be produced. (NOTE: *The output has been modified to improve readability.*)

FAVOR OR OPPOSE DEATH PENALTY FOR MURDER * RESPONDENTS' SEX Cross-tabulation

			RESPONDENTS' SEX		
			Male	Female	Total
FAVOR OR OPPOSE DEATH PENALTY FOR MURDER	Favor	Count	213	197	410
		%	75.0%	61.2%	67.7%
	Oppose	Count	71	125	196
		%	25.0%	38.8%	32.3%
	Total		284	322	606
			100.0%	100.0%	100.0%

Chi Square Tests

	Value	df	Asymp. Sig. (2-sided)	Exact Sig. (2-sided)	Exact Sig. (1-sided)
Pearson Chi Square	13.171(b)	1	.000		
Continuity Correction(a)	12.547	1	.000		
Likelihood Ratio	13.312	1	.000		
Fisher's Exact Test				.000	.000
Linear-by-Linear Association	13.149	1	.000		
N of Valid Cases	606				

[a]Computed only for a 2 × 2 table.

[b]Zero cells (.0%) have expected count of less than 5. The minimum expected count is 91.85.

Directional Measures

			Value	Asymp. Std. Error(a)	Approx. T(b)	Approx. Sig.
Nominal by Nominal	Lambda	Symmetric	.033	.041	.791	.429
		FAVOR OR OPPOSE DEATH PENALTY FOR MURDER Dependent	.000	.000	.(c)	.(c)
		RESPONDENTS' SEX Dependent	.056	.069	.791	.429
	Goodman and Kruskal tau	FAVOR OR OPPOSE DEATH PENALTY FOR MURDER Dependent	.022	.012		.000(d)
		RESPONDENTS' SEX Dependent	.022	.012		.000(d)

[a]Not assuming the null hypothesis.

[b]Using the asymptotic standard error assuming the null hypothesis.

[c]Cannot be computed because the asymptotic standard error equals zero.

[d]Based on chi square approximation.

Symmetric Measures

		Value	Approx. Sig.
Nominal by Nominal	Phi	.147	.000
	Cramer's V	.147	.000
N of Valid Cases		606	

The **Crosstabs** command produces a lot of output so we need to proceed carefully. The three questions about bivariate association introduced in this chapter can provide a useful framework for analysis:

1. *Is there an association?* The column percentages in the bivariate table change so there is an association between support for capital punishment and gender. This conclusion is verified by the fact that chi square is a nonzero value.

2. *How strong is the association?* We can assess strength in several different ways. First, the maximum difference is 75.0 − 61.2 or 13.8 percentage points. (This calculation is based on the top row, but because this is a 2 × 2 table, using the bottom row would result in exactly the same value: 38.8 − 25.0 = 13.8.) According to Table 11.5, this means that the relationship between the variables is moderate in strength.

Second, we can (and should) use a measure of association to assess the strength of the relationship. We have a choice of two different measures: lambda and (because this is a 2 × 2 table) phi. Looking in the "Directional Measures" output box, we see several different values for lambda. Recall that lambda is an asymmetric measure and that it changes value depending on which variable is seen as dependent. In this case, the dependent variable is "FAVOR OR OPPOSE DEATH PENALTY FOR MURDER," and the associated lambda is reported as .000. This indicates that there is no association between the variables, but we have already seen that the variables *are* associated. Remember that lambda can be zero when the variables are associated, but the row totals in the table are very unequal. That's the problem here ("Favor" is four times as popular as "Oppose"). We should disregard lambda in this case.

Turning to phi (under "Symmetric Measures"), we see a value of .147. This indicates that the relationship is moderate in strength (see Table 11.12), a conclusion that is consistent with the value of the maximum difference.

3. *What is the pattern of the relationship?* Because sex is a nominal-level variable, the relationship cannot have a direction (that is, it cannot be positive or negative). We can, however, discuss the pattern of the relationship: How do values of the variables seem to go together? Recall that we hypothesized that males would be more supportive of the death penalty than females would be. A glance at the column percentages in the bivariate table shows that this hypothesis is supported. That is, although the majority of both sexes support capital punishment, support is greater for males (75%) than for females (61.2%)

In summary, using chi square, the column percentages, and phi, we can say that there is a statistically significant, moderately strong relationship between gender and support for the death penalty. Furthermore, males are more supportive than females.

SPSS DEMONSTRATION 11.2 What Are the Correlates of Support for Capital Punishment?

It is common in social science research to analyze a particular variable (support for capital punishment, for example) by observing its relationship with an array of possible independent variables. We will follow that procedure in this demonstration using some of the nominal-level variables included in the General Social Survey: marital status (*marital*), race (*racecen1*), region (*region*), and religious denomination (*relig*).

To generate the output showing the relationships between each of these independent variables and *cappun,* repeat the instructions in Demonstration 11.1 and name *marital, racecen1, region,* and *relig* as independent or column variables. Make sure that you request chi square and phi. We will disregard lambda for this output because, as we saw in the previous demonstration, *cappun* has very unequal row totals and lambda will be zero. The output from this procedure is voluminous, so we will present our results in summary form. As you examine this output, note that the number of cases in some of the categories (for example, Native American for *racecen1* and "Other" for *relig*) is very small; we should exercise extra caution in interpreting these results.

	% in Favor	Significance of Chi Square	Phi or *V*
MARITAL			
Married	71.7%		
Widowed	62.3%		
Divorced	69.0%		
Separated	64.7%		
Never Married	59.8%	.146	.106
RACE			
White	72.2%		
Black	37.5%		
Native Am.	80.0%		
Asian Am.	69.2%		
Hispanic	57.9%	.000	.243
REGION			
New England	70.6%		
Mid Atlantic	59.5%		
E. N. Central	66.4%		
W. N. Central	70.5%		
S. Atlantic	71.9%		
E. S. Central	55.6%		
W. S. Central	66.7%		
Mountain	76.6%		
Pacific	69.3%	.429	.115
RELIGION			
Protestant	73.2%		
Catholic	67.8%		
Jew	71.4%		
None	51.6%		
Other	100.0%	.001	.177

Note, first of all, that only two of the relationships are statistically significant. If a relationship based on a random sample is not significant, it probably is not important in any other sense and should be disregarded. Turning to the two significant relationships, we see that they are both moderate in strength and that race has the stronger relationship with support of capital punishment. Black respondents are much less supportive of the death penalty than any other group, perhaps because it is perceived as being unjustly applied to blacks. Note also that Hispanics are much less supportive than the other groups. For religion, the pattern is that people with no affiliation ("None") are the least supportive. Protestants, Catholics, and Jews have roughly similar levels of support (although the sample includes only 7 Jews so we need to be cautious in making conclusions). Respondents in the "Other" category are unanimous in their support, but again, the very small size of this category prevents us from making any firm conclusions.

Overall, we may conclude that support for the death penalty is significantly and moderately related to gender, race, and religion but not to marital status or region of the country.

Exercises

11.1 What other variables might be associated with *cappun*? If necessary, use the **Recode** command to reduce the number of categories in the column variable. Try *degree* and *income98* as independent variables.

11.2 Select a different dependent variable (e.g., *grass*, *gunlaw*) and repeat the analysis in Demonstrations 11.1 and 11.2. That is, for each dependent variable, run the **Crosstabs** command with *sex*, *marital*, *racecen1*, *region*, and *relig* as the column (independent) variables. Make sure that you request chi square, lambda, phi, *V*, and column percents. Analyze each table in terms of significance and the strength and pattern of the association. Why do you suppose these patterns exist?

12

Association Between Variables Measured at the Ordinal Level

LEARNING OBJECTIVES

By the end of this chapter, you will be able to:

1. Calculate and interpret gamma and Spearman's rho.
2. Explain the logic of proportional reduction in error in terms of gamma.
3. Use gamma and Spearman's rho to analyze and describe a bivariate relationship in terms of the three questions introduced in Chapter 11.

12.1 INTRODUCTION

There are two common types of ordinal-level variables. Some have many possible scores and look, at least at first glance, like interval-ratio level variables. We will call these "continuous ordinal variables." An attitude scale that incorporated many different items and, therefore, had many possible values would produce this type of variable.

The second type, which we will call a "collapsed ordinal variable," has only a few (no more than five or six) values or scores and can be created either by collecting data in collapsed form or by collapsing a continuous ordinal scale. For example, we would produce collapsed ordinal variables by measuring social class as upper, middle, or lower or by reducing the scores on an attitude scale into just a few categories (such as high, moderate, and low).

A number of measures of association are available for use with collapsed ordinal-level variables. Rather than attempt a comprehensive coverage of all of these statistics, we will concentrate on **gamma (G).** For continuous ordinal variables, a statistic called **Spearman's rho (r_s)** is commonly used, and we will cover this measure of association later in this chapter.

Chapter 12 will expand your understanding of how we can describe and analyze bivariate associations, but remember that we are still trying to answer the three questions raised in Chapter 11: Are the variables associated? How strong is the association? What is the direction of the association?

12.2 PROPORTIONAL REDUCTION IN ERROR (PRE)

For nominal-level variables, the logic of PRE was based on two different "predictions" of the scores of cases on the dependent variable (Y): one that ignored the independent variable (X) and a second that took the independent variable into account. The value of lambda showed the extent to which taking the independent variable into account improved accuracy when predicting scores of the dependent variable. The PRE logic for variables measured at the ordinal level is similar, and gamma, like lambda, measures the proportional reduction in error gained by predicting one variable while taking the other into account. The major difference lies in how we make predictions.

In the case of gamma, we predict the order of pairs of cases rather than a score on the dependent variable. That is, we predict whether one case will have a higher or lower score than the other. First, we predict the order of a pair of cases on the dependent variable while ignoring their order on the independent variable. Second, we predict the order on the dependent variable while taking order on the independent variable into account.

As an illustration, assume that a researcher is concerned about the causes of "burnout" (that is, demoralization and loss of commitment) among elementary school teachers and wonders about the relationship between levels of burnout and years of service. One way to state the research question would be to ask if teachers with more years of service have higher levels of burnout. Another way to ask the same question is: Do teachers who *rank higher* on years of service also *rank higher* on burnout? If we knew that teacher A had more years of service than teacher B, would we be able to predict that teacher A was also more "burned out" than teacher B? That is, would knowledge of the order of this pair of cases on one variable help us predict their order on the other?

If the two variables are associated, we will reduce our errors when our predictions about one of the variables are based on knowledge of the other. Furthermore, the stronger the association, the fewer the errors we will make. When there is no association between the variables, gamma will be 0.00, and knowledge of the order of a pair of cases on one variable will not improve our ability to predict their order on the other. A gamma of ±1.00 denotes a perfect relationship: The order of all pairs of cases on one variable would be predictable without error from their order on the other variable.

With nominal-level variables, we analyzed the pattern of the relationship between the variables. That it, we looked to see which value on one variable (e.g., "male" on the variable gender) was associated with which value on the other variable (e.g., "tall" on the variable height). Recall that a defining characteristic of variables measured at the ordinal level is that the scores or values can be rank ordered from high to low or from more to less (see Chapter 1). This means that relationships between ordinal-level variables can have a direction as well as a pattern. In terms of the logic of gamma, the overall relationship between the variables is *positive* if cases tend to be ranked in the same order on both variables. For example, if Case A is ranked above Case B on one variable, it would also be ranked above Case B on the second variable. The relationship suggested above between years of service and burnout would be a positive relationship. In a *negative* relationship, the order of the cases would be reversed between the two variables. If Case A ranked above Case B on one variable, it would tend to rank below Case B on the second variable. If there is a negative relationship between prejudice and education and Case A was more educated than Case B (or ranked *above* Case B on education), then Case A would be less prejudiced (or would rank *below* Case B on prejudice).

12.3 GAMMA

Computation. Table 12.1 summarizes the relationship between "length of service" and "burnout" for a fictitious sample of 100 teachers. To compute gamma, two sums are needed. First, we must find the number of pairs of cases that are ranked the same on both variables (we will label this N_s) and then the

TABLE 12.1   BURNOUT BY LENGTH OF SERVICE (fictitious data)

Burnout	Length of Service			Totals
	Low	Moderate	High	
Low	20	6	4	30
Moderate	10	15	5	30
High	8	11	21	40
Totals	38	32	30	100

number of pairs of cases ranked differently on the variables (N_d). We find these sums by working with the cell frequencies.

To find the number of pairs of cases ranked the same (N_s), begin with the cell containing the cases that were ranked the lowest on both variables. In Table 12.1, this would be the upper-left-hand cell. *(NOTE: Not all tables are constructed with values increasing from left to right across the columns and from top to bottom across the rows. When using other tables, always be certain that you have located the proper cell.)* The 20 cases in the upper-left-hand cell all rank low on both burnout and length of service, and we will refer to these cases as "low-lows," or "LLs."

Now, form a pair of cases by selecting one case from this cell and one from any other cell—for example, the middle cell in the table. All 15 cases in this cell are moderate on both variables and, following our practice above, can be labeled moderate-moderates, or MMs. Any pair of cases formed between these two cells will be ranked the same on both variables. That is, all LLs are lower than all MMs on both variables (on X, low is less than moderate, and on Y, low is less than moderate). We obtain the total number of pairs of cases by multiplying the cell frequencies. So, the contribution of these two cells to the total N_s is (20)(15), or 300.

Gamma ignores all pairs of cases that are tied on either variable. For example, any pair of cases formed between the LLs and any other cell in the top row (low on burnout) or the left-hand column (low on length of service) will be tied on one variable. Also, any pair of cases formed within any cell will be tied on both X and Y. Gamma ignores all pairs of cases formed within the same row, column, or cell. This means that, in computing N_s, we will work with only the pairs of cases that can be formed between each cell and the cells below and to the right of it.

In summary: To find the total number of pairs of cases ranked the same on both variables (N_s), multiply the frequency in each cell by the total of all frequencies below and to the right of that cell. Repeat this procedure for each cell and add the resultant products. The total of these products is N_s. This procedure is displayed below for each cell in Table 12.1. Note that none of the cells in the bottom row or the right-hand column can contribute to N_s because they have no cells below and to the right of them. Figure 12.1 shows the direction of multiplication for each of the four cells that, in a 3 × 3 table, can contribute to N_s. Computing N_s for Table 12.1, we find that a total of 1,831 pairs of cases are ranked the same on both variables.

FIGURE 12.1 COMPUTING N_s IN A 3 × 3 TABLE

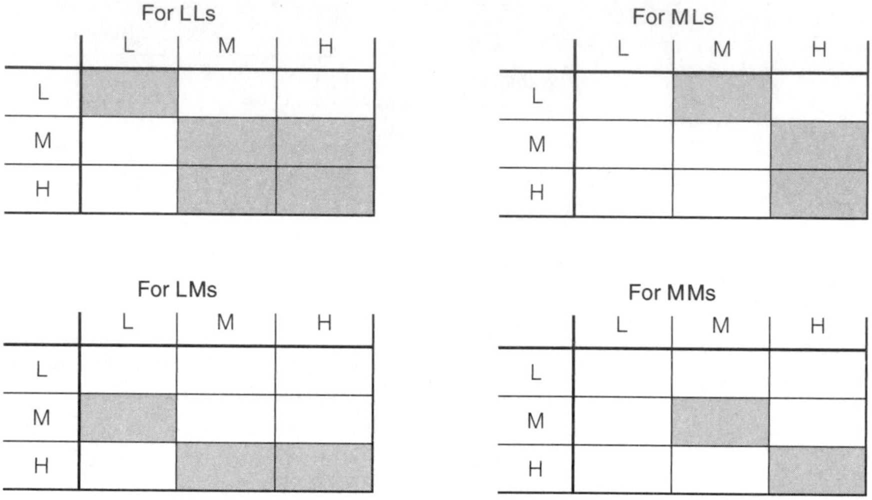

Contribution to N_s	
For LLs, 20(15 + 5 + 11 + 21)	= 1,040
For MLs, 6(21 + 5)	= 156
For HLs, 4(0)	= 0
For LMs, 10(11 + 21)	= 320
For MMs, 15(21)	= 315
For HMs, 5(0)	= 0
For LHs, 8(0)	= 0
For MHs, 11(0)	= 0
For HHs, 21(0)	= 0
	N_s = 1,831

Our next step is to find the number of pairs of cases ranked differently (N_d) on both variables. To find the total number of pairs of cases ranked in different order on the variables, multiply the frequency in each cell by the total of all frequencies below and to the left of that cell. Note that the pattern for computing N_d is the reverse of the pattern for N_s. This time, we begin with the upper-right-hand cell (high-lows, or HLs) and multiply the number of cases in the cell by the total frequency of cases below and to the left. The four cases in the upper-right-hand cell are low on Y and high on X and, if a pair is formed with any case from this cell and any cell below and to the left, the cases will be ranked differently on the two variables. For example, if a pair is formed between any HL case and any case from the middle cell (moderate-moderates, or MMs), the HL case would be less than the MM case on Y ("low" is less than "moderate") but more than the MM case on X ("high" is greater than "moderate"). The computation of N_d appears below and Figure 12.2 shows it graphically. In the computations, I have omitted cells that cannot contribute to N_d because they have no cells below and to the left of them.

FIGURE 12.2 COMPUTING N_d IN A 3 × 3 TABLE

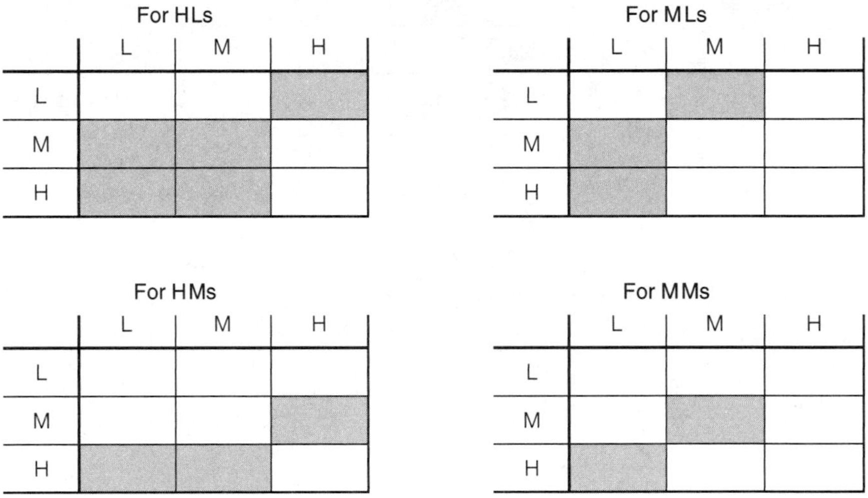

For HLs

	L	M	H
L			
M			
H			

For MLs

	L	M	H
L			
M			
H			

For HMs

	L	M	H
L			
M			
H			

For MMs

	L	M	H
L			
M			
H			

	Contribution to N_d
For HLs, 4(10 + 15 + 8 + 11)	= 176
For MLs, 6(10 + 8)	= 108
For HMs, 5(8 + 11)	= 95
For MMs, 15(8)	= 120
	N_d = 499

Table 12.1 has 499 pairs of cases ranked in different order and 1,831 pairs of cases ranked in the same order. The formula for computing gamma is:

FORMULA 12.1

$$G = \frac{N_s - N_d}{N_s + N_d}$$

where N_s = the number of pairs of cases ranked the same on both variables
N_d = the number of pairs of cases ranked differently on both variables

For Table 12.1, the value of gamma would be:

$$G = \frac{N_s - N_d}{N_s + N_d}$$

$$G = \frac{1{,}831 - 499}{1{,}831 + 499}$$

$$G = \frac{1{,}332}{2{,}330}$$

$$G = 0.57$$

Interpretation. A gamma of 0.57 indicates that we would make 57% fewer errors if we predicted the order of pairs of cases on one variable from the order of pairs of cases on the other (as opposed to predicting order while ignoring the other variable). Length of service is associated with degree of burnout, and the relationship is positive. Knowing the respective rankings of two teachers on

TABLE 12.2 THE RELATIONSHIP BETWEEN THE VALUE OF GAMMA AND THE STRENGTH OF THE RELATIONSHIP

Value	Strength
If the value is:	The strength of the relationship is:
Between 0.00 and 0.30	Weak
Between 0.31 and 0.60	Moderate
Greater than .61	Strong

length of service (Case A is higher on length of service than Case B) will help us predict their ranking on burnout (we would predict that Case A will also be higher than Case B on burnout).

Table 12.2 provides some additional assistance for interpreting gamma in a format similar to Tables 11.5 and 11.12. As before, the relationship between the values and the descriptive terms are arbitrary and intended as general guidelines only. Note, in particular, that the strength of the relationship is independent of its direction. That is, a gamma of -0.35 is exactly as strong as a gamma of $+0.35$ but opposite in direction.

To use the computational routine for gamma presented above, you must arrange the table in the manner of Table 12.1, with the column variable increasing in value as you move from left to right and the row variable increasing from top to bottom. Be careful to construct your tables according to this format and, if you are working with data already in table format, you may have to rearrange the table or rethink the direction of patterns. Gamma is a symmetrical measure of association—that is, the value of gamma will be the same regardless of which variable is taken as independent. *(To practice computing and interpreting gamma, see problems 12.1 to 12.10. Begin with some of the smaller, 2 × 2 tables until you are comfortable with these procedures.)*

12.4 DETERMINING THE DIRECTION OF RELATIONSHIPS

Nominal measures of association, such as phi and lambda, measure only the strength of a bivariate association. Ordinal measures of association, such as gamma, are more sophisticated and add information about the overall direction of the relationship (positive or negative). In one way, it is easy to determine direction: If the sign of the statistic is a plus, the direction is positive and a minus sign indicates a negative relationship. Often, however, direction is confusing when working with ordinal-level variables, and it will be helpful if we focus on the matter specifically. We'll discuss positive relationships first and then relationships in the negative direction.

With gamma, a positive relationship means that the scores of cases tend to be ranked in the same order on both variables. In more general terms, a positive relationship means that the variables change in the same direction. That is, as scores on one variable increase (or decrease), scores on the other variable also increase (or decrease). Cases tend to have scores in the same range on both variables (that is, low scores go with low scores, moderate with moderate, and so forth). Table 12.3 illustrates the general shape of a positive relationship: Cases tend to fall along a diagonal from upper left to lower right (assuming that tables have been constructed with the column variable increasing from left to right and the row variable from top to bottom).

TABLE 12.3   A GENERALIZED POSITIVE RELATIONSHIP

Variable Y	Variable X		
	Low	Moderate	High
Low	X		
Moderate		X	
High			X

TABLE 12.4   STATE STRUCTURE BY DEGREE OF STRATIFICATION (frequencies)*

Type of State	Degree of Stratification		
	Low	Medium	High
Stateless	77	5	0
Semistate	28	15	4
State	12	19	26
Totals	117	39	30

*Data are from the Human Relation Area File (HRAF), standard cross-cultural sample. These data are selected from the *Ethnographic Atlas* compiled by George Murdock and Douglas White.

TABLE 12.5   STATE STRUCTURE BY DEGREE OF STRATIFICATION (percentages)

Type of State	Degree of Stratification		
	Low	Medium	High
Stateless	66%	13%	0%
Semistate	24%	38%	13%
State	10%	49%	87%
Totals	100%	100%	100%

Tables 12.4 and 12.5 present an example of a positive relationship with actual data. The sample consists of 186 preindustrial societies from around the globe. Each has been rated in terms of its degree of stratification and the type of political institution it has. A society that is low on stratification is one in which people are essentially equal in terms of wealth and power; the degree of inequality increases from left to right across the columns of the table. In "stateless" societies, there is no formal political institution; the political institution becomes more elaborate and stronger as you move down the rows from top to bottom.

The gamma for this table is 0.86, so the relationship is strong and positive. Most cases fall in the diagonal from upper left to lower right. The percentages in Table 12.5 make it clear that societies with little inequality tend to be stateless and that the political institution becomes more elaborate as inequality increases. The great majority of the least stratified societies had no political institution and none of the highly stratified societies were stateless.

Negative relationships are the opposite of positive relationships. Low scores on one variable are associated with high scores on the other and high scores with low scores. This pattern means that the cases will tend to fall along a diagonal from lower left to upper right (at least for all tables in this text). Table 12.6 illustrates a generalized negative relationship. The cases with higher scores on Variable X tend to have lower scores on Variable Y and scores on Y decrease as scores on X increase.

Tables 12.7 and 12.8 present an example of a negative relationship with data taken from the 2002 General Social Survey, a survey administered to a representative sample of U.S. citizens. The independent variable is church attendance, and the dependent variable is approval of cohabitation ("Is it alright for a couple to live together without intending to get married?"). Note that rates of attendance increase from left to right, and approval of cohabitation increases from top to bottom of the table.

Once again, the percentages in Table 12.8 make the pattern obvious. The great majority of people who do not attend church ("never") were high on approval of cohabitation, and most people who were high on attendance were

TABLE 12.6   A GENERALIZED NEGATIVE RELATIONSHIP

	Variable X		
Variable Y	Low	Moderate	High
Low			X
Moderate		X	
High	X		

TABLE 12.7   APPROVAL OF COHABITATION* BY CHURCH ATTENDANCE (frequencies)

	Attendance			
Approval	Never	Monthly or Yearly	Weekly	Totals
Low	37	186	195	418
Moderate	25	126	46	197
High	156	324	52	532
Totals	218	636	293	1,147

*The labels for this variable have been changed to clarify the example. The original response categories were "approve," "neither approve nor disapprove," and "disapprove."

TABLE 12.8   APPROVAL OF COHABITATION* BY CHURCH ATTENDANCE (percentages)

	Attendance		
Approval	Never	Monthly or Yearly	Weekly
Low	17%	29%	67%
Moderate	11%	20%	16%
High	72%	51%	18%
Totals	100%	100%	101%

*The labels for this variable have been changed to clarify the example. The original response categories were "approve," "neither approve nor disapprove," and "disapprove."

STEP BY STEP	Computing and Interpreting Gamma

Double-check to make sure that the table is arranged with the column variable increasing from left to right and the row variable increasing from top to bottom.

Computation

Step 1: To compute N_s, start with the upper-left-hand cell. Multiply the number of cases in this cell by the total number of cases in all cells below and to the right. Repeat this process for each cell in the table: Multiply each cell frequency by the total number of cases in all cells below and to the right. Add up these subtotals to find N_s.

Step 2: To compute N_d, start with the upper-right-hand cell. Multiply the number of cases in this cell by the total number of cases in all cells below and to the left. Repeat this process for each cell in the table: Multiply each cell frequency by the total number of cases in all cells below and to the left. Add up these subtotals to find N_d.

Step 3: To find gamma, solve Formula 12.1:

$$G = \frac{N_s - N_d}{N_s + N_d}$$

a. Subtract N_d from N_s.
b. Add N_d and N_s.
c. Divide the quantity you found in step a by the quantity you found in step b.

Interpretation

Step 4: To interpret the *strength* of the relationship, always begin with the column percentages: the bigger the change in column percentages, the stronger the relationship. Next, you can use gamma to interpret the strength of the relationship in either or both of two ways:

a. Use Table 12.2 to describe strength in general terms.
b. To use the logic of proportional reduction in error, multiply gamma by 100. This value represents the percentage by which we improve our prediction of the dependent variable by taking the independent variable into account.

Step 5: To interpret the *direction* of the relationship, always begin by looking at the pattern of the column percentages. If the cases tend to fall in a diagonal from upper-left to lower-right, the relationship is positive. If the cases tend to fall in a diagonal from lower-left to upper-right, the relationship is negative. The sign of the gamma also tells the direction of the relationship. However, be very careful when interpreting direction with ordinal-level variables. Remember that coding schemes for these variables are arbitrary, and a positive gamma may mean that the actual relationship is negative and vice versa.

low on approval. As attendance increases, approval of cohabitation tends to decrease. The gamma for this table is −0.57, indicating a strong, negative relationship between attendance and approval of this living arrangement.

You should be aware of an additional complication. The coding for ordinal-level variables, such as approval of cohabitation, is arbitrary. A higher score may mean "more" or "less" of the variable being measured. For example, if we measured social class as upper, middle, and lower, we could assign scores to the categories in either of two ways:

A	B
(1) Upper	(3) Upper
(2) Middle	(2) Middle
(3) Lower	(1) Lower

While coding scheme B might seem preferable (because higher scores go with higher class position), *both* schemes are perfectly legitimate and the

---

Application 12.1

Forty nations have been rated as high or low on religiosity (based on the percentage of a random sample of citizens who described themselves as "a religious person") and as high or low in their support for single mothers (based on the percentage of a random sample of citizens who said they would approve of a woman choosing to be a single parent). Are more religious nations less approving of single mothers?

APPROVAL OF SINGLE MOTHERS
BY RELIGIOSITY OF NATION

| Approval | Religiosity | | Totals |
	Low	High	
Low	4 *(26.67%)*	9 *(36.00%)*	13
High	11 *(73.33%)*	16 *(64.00%)*	27
Totals	15 *(100.00%)*	25 *(100.00%)*	40

The column percentages show that nations that rank higher on religiosity are less approving of single mothers. The maximum difference of about 10 suggests a weak to moderate relationship.

Because both variables are ordinal in level of measurement, we can use gamma to measure the strength and direction of the relationship. The number of pairs of cases ranked in the same order on both variables (N_s) would be:

$$N_s = 4(16) = 64$$

The number of pairs of cases ranked in different order on both variables (N_d) would be:

$$N_d = 9(11) = 99$$

Gamma is:

$$G = \frac{N_s - N_d}{N_s + N_d} = \frac{64 - 99}{64 + 99} = \frac{-35}{163} = -0.21$$

A gamma of -0.21 means that, when predicting the order of pairs of cases on the dependent variable (approval of single mothers), we would make 21% fewer errors by taking the independent variable (religiosity) into account. There is a moderate to weak, negative association between these two variables. As religiosity increases, approval decreases (or more religious nations are less approving of single mothers).

---

direction of gamma will change, depending on which scheme we select. Using scheme B, we would find positive relationships between social class and education: as education increased, so would class. Using scheme A, however, the same relationship would appear to be negative because the numerical scores (1, 2, 3) are coded in reverse order: the highest social class is assigned the lowest score, and so forth. If you didn't check the coding scheme, you might conclude that the negative gamma means that class decreases as education increases when, actually, the opposite is true.

Unfortunately, we cannot avoid this source of confusion when working with ordinal-level variables. Coding schemes will always be arbitrary for these variables and you need to exercise additional caution when interpreting the direction of ordinal-level variables.

12.5 SPEARMAN'S RHO (r_s) **Introduction.** To this point, we have considered ordinal variables that have a limited number of categories (possible values) and are presented in tables. However, many ordinal-level variables have a broad range of scores and many distinct values. Such data may be collapsed into a few broad categories (such as high, moderate, and low), organized into a bivariate table, and analyzed with gamma. Collapsing scores in this manner may be beneficial and desirable in

TABLE 12.9   THE SCORES OF 10 SUBJECTS ON INVOLVEMENT IN JOGGING AND A MEASURE OF SELF-ESTEEM

Joggers	Involvement in Jogging (X)	Self-Esteem (Y)
Wendy	18	15
Debbie	17	18
Phyllis	15	12
Stacey	12	16
Evelyn	10	6
Tricia	9	10
Christy	8	8
Patsy	8	7
Marsha	5	5
Lynn	1	2

many instances, but some important distinctions between cases may be obscured or lost as a consequence.

For example, suppose a researcher wanted to test the claim that jogging is beneficial not only physically but also psychologically. Do joggers have an enhanced sense of self-esteem? To deal with this issue, 10 female joggers are measured on two scales, the first measuring involvement in jogging and the other measuring self-esteem. Table 12.9 reports the scores.

We could collapse these data and produce a bivariate table. For example, we could dichotomize both variables to create only two values (high and low) for both variables. Although collapsing scores in this way is certainly legitimate and often necessary,[1] two difficulties with this practice must be noted. First, the scores seem continuous, and there are no obvious or natural division points in the distribution that would allow us to distinguish, in a nonarbitrary fashion, between high scores and low ones. Second, and more important, grouping these cases into broader categories will lose information. That is, if both Wendy and Debbie are classified as "high" on involvement, the fact that they had different scores on the variable would be obscured. If these differences are important and meaningful, then we should opt for a measure of association that permits the retention of as much detail and precision in the scores as possible.

Computation.   Spearman's rho (r_s) is a measure of association for ordinal-level variables that have a broad range of many different scores and few ties between cases on either variable. Scores on ordinal-level variables cannot be manipulated mathematically except for judgments of "greater than" or "less than." To compute Spearman's rho, cases are first ranked from high to low on each variable and then the ranks (not the scores) are manipulated to produce the final measure. Table 12.10 displays the original scores and the rankings of the cases on both variables.

To rank the cases, first find the highest score on each variable and assign it rank 1. Wendy has the high score on X (18) and is thus ranked number 1. Debbie, on the other hand, is highest on Y and is ranked first on that variable. All

---

[1] For example, collapsing scores may be advisable when the researcher is not sure that fine distinctions between scores are meaningful.

TABLE 12.10   COMPUTING SPEARMAN'S RHO

	Involvement (X)	Rank	Self-Esteem (Y)	Rank	D	D^2
Wendy	18	1	15	3	−2.0	4
Debbie	17	2	18	1	1.0	1
Phyllis	15	3	12	4	−1.0	1
Stacey	12	4	16	2	2.0	4
Evelyn	10	5	6	8	−3.0	9
Tricia	9	6	10	5	1.0	1
Christy	8	7.5	8	6	1.5	2.25
Patsy	8	7.5	7	7	.5	.25
Marsha	5	9	5	9	0	0
Lynn	1	10	2	10	0	0
					$\Sigma D = 0$	$\Sigma D^2 = 22.50$

other cases are then ranked in descending order of scores. If any cases have the same score on a variable, assign them the average of the ranks they would have used had they not been tied. Christy and Patsy have identical scores of 8 on involvement. Had they not been tied, they would have used ranks 7 and 8. The average of these two ranks is 7.5, and this average of used ranks is assigned to all tied cases. (For example, if Marsha also had a score of 8, three ranks—7, 8, and 9—would have been used, and all three tied cases would have been ranked eighth.)

The formula for Spearman's rho is:

FORMULA 12.2
$$r_s = 1 - \frac{6\Sigma D^2}{N(N^2 - 1)}$$

where ΣD^2 = the sum of the differences in ranks, the quantity squared

To compute ΣD^2, the rank of each case on Y is subtracted from its rank on X (D is the difference between rank on Y and rank on X). Table 12.10 includes a column so that these differences may be recorded on a case-by-case basis. Note that the sum of this column (ΣD) is 0. That is, the negative differences in rank are equal to the positive differences, as will always be the case. You should find the total of this column as a check on your computations to this point. If the ΣD is not equal to 0, you have made a mistake either in ranking the cases or in subtracting the differences.

In the column headed D^2, each difference is squared to eliminate negative signs. The sum of this column is ΣD^2 and this quantity is entered directly into the formula. For our sample problem:

$$r_s = 1 - \frac{6\Sigma D^2}{N(N^2 - 1)}$$

$$r_s = 1 - \frac{6(22.5)}{10(100 - 1)}$$

$$r_s = 1 - \frac{135}{990}$$

$$r_s = 1 - 0.14$$

$$r_s = 0.86$$

STEP BY STEP	**Computing and Interpreting Spearman's Rho**

Computation

Step 1: Set up a computing table similar to Table 12.10 to help organize the computations. In the far left-hand column, list the cases in order with the case with the highest score on the independent variable (X) stated first.

Step 2: In the next column, list the scores on X.

Step 3: In the third column, list the rank of each case on X beginning with rank 1 for the highest score. If any cases have the same score, assign them the average of the ranks they would have used had they not been tied.

Step 4: In the fourth column, repeat steps 2 and 3 for the scores of the cases on the dependent variable (Y). List the scores on Y in the fourth column and then, in the fifth column, rank the cases on Y from high to low. Start by assigning the rank of 1 to the case with the highest score on Y and assign any tied cases the average of the ranks they would have used had they not been tied.

Step 5: For each case, subtract the rank on Y from the rank on X and write the difference (D) in the sixth column. Add the values in this column. If the sum is not zero, you have made a mistake and need to recompute.

Step 6: Square the value of each D and record the result in the seventh column. Add the values in this column and substitute the result into the numerator of Formula 12.2.

Step 7: Compute Spearman's rho by solving Formula 12.2:

$$r_s = 1 - \frac{6\Sigma D^2}{N(N^2 - 1)}$$

a. Multiply the ΣD^2 (the total of column 7 in the computing table) by 6.
b. Square N and subtract 1 from the result.
c. Multiply the quantity you found in step b by N.
d. Divide the quantity you found in step a by the quantity you found in step c.
e. Subtract the quantity you found in step d from 1. The result is r_s.

Interpretation

Step 8: We can use Spearman's rho to interpret the strength of a relationship in either or both of two ways:

a. Use Table 12.2 to characterize the strength of the relationship in general terms.
b. Square the value of r_s and multiply the result by 100. This value represents the percentage by which we improve our prediction of the dependent variable by taking the independent variable into account.

Step 9: To interpret the direction of the relationship, look at the sign of r_s. However, be careful when interpreting direction with ordinal-level variables. Remember that coding schemes for these variables are arbitrary; a positive r_s may mean that the actual relationship is negative and vice versa.

Interpretation.   Spearman's rho is an index of the strength of association between the variables; it ranges from 0 (no association) to ± 1.00 (perfect association). A perfect positive association ($r_s = +1.00$) would exist if there were no disagreements in ranks between the two variables (if cases were ranked in exactly the same order on both variables). A perfect negative relationship ($r_s = -1.00$) would exist if the ranks were in perfect disagreement (if the case ranked highest on one variable were lowest on the other, and so forth). A Spearman's rho of 0.86 indicates a strong, positive relationship between these two variables. The respondents who were highly involved in jogging also ranked high on self-esteem. These results support claims regarding the psychological benefits of jogging.

Spearman's rho is an index of the relative strength of a relationship, and values between 0 and ± 1.00 have no direct interpretation. However, if the value

Application 12.2

Five cities have been rated on an index that measures the quality of life. Also, the percentage of the population that has moved into each city over the past year has been determined. Have cities with higher quality-of-life scores attracted more new residents? The following table summarizes the scores, ranks, and differences in ranks for each of the five cities. Spearman's rho for these variables is:

$$r_s = 1 - \frac{6\Sigma D^2}{N(N^2 - 1)}$$

$$r_s = 1 - \frac{(6)(4)}{5(25 - 1)}$$

$$r_s = 1 - \left(\frac{24}{120}\right)$$

$$r_s = 1 - 0.20$$

$$r_s = 0.80$$

These variables have a strong, positive association. The higher the quality-of-life score, the greater the percentage of new residents. The value of r_s^2 is 0.64 ($0.80^2 = 0.64$), which indicates that we will make 64% fewer errors when predicting rank on one variable from rank on the other, as opposed to ignoring rank on the other variable.

City	Quality of Life	Rank	% New Residents	Rank	D	D²
A	30	1	17	1	0	0
B	25	2	14	3	−1	1
C	20	3	15	2	1	1
D	10	4	3	5	−1	1
E	2	5	5	4	1	1
					$\Sigma D = 0$	$\Sigma D^2 = 4$

of rho is squared, a PRE interpretation is possible. Rho squared (r_s^2) is the proportional reduction in errors of prediction when predicting rank on one variable from rank on the other variable, as compared to predicting rank while ignoring the other variable. In the example above, r_s was 0.86 and r_s^2 would be 0.74. Thus, our errors of prediction would be reduced by 74% if, when predicting the rank of a subject on self-esteem, we took into account the rank of the subject on involvement in jogging. *(For practice in computing and interpreting Spearman's rho, see problems 12.11 to 12.14. Because problem 12.11 has the fewest number of cases, it is a good choice for a first attempt at these procedures.)*

SUMMARY

1. Measures of association for variables with collapsed (gamma) and continuous (Spearman's rho) ordinal variables were covered. Both measures summarize the overall strength and direction of the association between the variables.

2. Gamma is a PRE-based measure that shows the improvement in our ability to predict the order of pairs of cases on one variable from the order of pairs of cases on the other variable, as opposed to ignoring the order of the pairs of cases on the other variable.

3. Spearman's rho is computed from the ranks of the scores of the cases on two continuous ordinal variables and, when squared, can be interpreted by the logic of PRE.

SUMMARY OF FORMULAS

Gamma   12.1 $G = \dfrac{N_s - N_d}{N_s + N_d}$

Spearman's rho   12.2 $r_s = 1 - \dfrac{6\Sigma D^2}{N(N^2 - 1)}$

GLOSSARY

Gamma (G). A measure of association appropriate for variables measured with "collapsed" ordinal scales that have been organized into table format; G is the symbol for gamma.

N_d. The number of pairs of cases ranked in different order on two variables.

N_s. The number of pairs of cases ranked in the same order on two variables.

Spearman's rho (r_s). A measure of association appropriate for ordinally measured variables that are "continuous" in form; r_s is the symbol for Spearman's rho.

MULTIMEDIA RESOURCES

The Wadsworth Sociology Resource Center: Virtual Society at
http://www.thomsonedu.com/sociology

Visit the companion website for the seventh edition of *Statistics: A Tool for Social Research* to access a wide range of student resources. Begin by clicking on the Student Resources section of the book's website to access the following study tools:

- Basic math review
- Flash cards

- Additional chapter problems
- Internet links
- Table of random numbers
- MicroCase and SPSS examples and exercises
- Hypothesis testing for variables measured at the ordinal level

PROBLEMS

For problems 12.1 to 12.10, calculate column percentages and use the percentages to help analyze the strength and direction of the association.

12.1 SOC A small sample of non-English-speaking immigrants to the United States has been interviewed about their level of assimilation. Is the pattern of adjustment affected by length of residence in the United States? For each table, compute gamma and summarize the relationship in terms of strength and direction. *(HINT: In 2 × 2 tables, only two cells can contribute to N_s or N_d. To compute N_s, multiply the number of cases in the upper-left-hand cell by the number of cases*

in the lower-right-hand cell. For N_d, multiply the number of cases in the upper-right-hand cell by the number of cases in the lower-left-hand cell.)

a. Facility in English:

| | Length of Residence | | |
English Facility	Less Than Five Years (Low)	More Than Five Years (High)	Totals
Low	20	10	30
High	5	15	20
Totals	25	25	50

b. Total family income:

	Length of Residence		
Income	Less Than Five Years (Low)	More Than Five Years (High)	Totals
Below National Average (Low)	18	8	26
Above National Average (High)	7	17	24
Totals	25	25	50

c. Extent of contact with country of origin:

	Length of Residence		
Contact	Less Than Five Years (Low)	More Than Five Years (High)	Totals
Rare (Low)	5	20	25
Frequent (High)	20	5	25
Totals	25	25	50

12.2 CJ In a random sample, 150 cities have been classified as small, medium, or large by population and as high or low on crime rate. Is there a relationship between city size and crime rate? Calculate gamma and describe the strength and direction of the relationship in a few sentences.

	City Size			
Crime Rate	Small	Medium	Large	Totals
Low	21	17	8	46
High	29	33	42	104
Totals	50	50	50	150

12.3 SOC Some research has shown that families vary by how they socialize their children to sports, games, and other leisure-time activities. In middle-class families, such activities are carefully monitored by parents and are, in general, dominated by adults (for example, Little League baseball). In working-class families, children more often organize and initiate such activities themselves, and parents are much less involved (for example, sandlot or playground baseball games). Are the data below consistent with these findings? Summarize your conclusions in a few sentences.

As a Child, Did You Play Mostly Organized or Sandlot Sports?	Social Class		
	White Collar	Blue Collar	Totals
Organized	155	123	278
Sandlot	101	138	239
Totals	256	261	517

12.4 Is support for sexual freedom related to age? Is the relationship between the variables different for different nations? The World Values Survey has been administered to random samples drawn from Canada, the United States, and Mexico. Respondents were asked if they agree or disagree that "individuals should have the chance to enjoy complete sexual freedom without being restricted." Compute gamma for each table. Is there a relationship? Describe the strength and direction of the relationship. Which age group is most supportive of sexual freedom? How does the relationship change from nation to nation?

a. Canada

"People should enjoy sexual freedom"	Age			
	18–34	35–54	55+	Totals
Agree	378	174	66	618
Neither agree nor disagree	626	710	586	1,922
Disagree	163	101	74	338
Totals	1,167	985	726	2,878

b. United States

"People should enjoy sexual freedom"	Age			
	18–34	35–54	55+	Totals
Agree	583	288	147	1,018
Neither agree nor disagree	877	982	1,061	2,920
Disagree	113	72	53	238
Totals	1,573	1,342	1,261	4,176

c. Mexico

"People should enjoy sexual freedom"	Age			
	18–34	35–54	55+	Totals
Agree	780	284	61	1,125
Neither agree nor disagree	1,300	847	275	2,422
Disagree	317	148	43	508
Totals	2,397	1,279	379	4,055

12.5 PA All applicants for municipal jobs in Shinbone, Kansas, are given an aptitude test, but the test has never been evaluated to see if test scores are in any way related to job performance. The following table reports aptitude test scores and job performance ratings for a random sample of 75 city employees.

Efficiency Ratings	Test Scores			
	Low	Moderate	High	Totals
Low	11	6	7	24
Moderate	9	10	9	28
High	5	9	9	23
Totals	25	25	25	75

a. Are these two variables associated? Describe the strength and direction of the relationship in a sentence or two.

b. Should the aptitude test continue to be administered? Why or why not?

12.6 SW A sample of children has been observed and rated for symptoms of depression. Their parents have been rated for authoritarianism. Is there any relationship between these variables? Write your conclusions in a few sentences.

Symptoms of Depression	Authoritarianism			
	Low	Moderate	High	Totals
Few	7	8	9	24
Some	15	10	18	43
Many	8	12	3	23
Totals	30	30	30	90

12.7 SOC Are prejudice and level of education related? State your conclusion in a few sentences.

Prejudice	Level of Education				
	Elementary School	High School	Some College	College Graduate	Totals
Low	48	50	61	42	201
High	45	43	33	27	148
Totals	93	93	94	69	349

12.8 SOC In a recent survey, a random sample of respondents was asked to indicate how happy they were with their situations in life. Are their responses related to income level? Describe the strength and direction of the relationship.

Happiness	Income			
	Low	Moderate	High	Totals
Not happy	101	82	36	219
Pretty happy	40	227	100	367
Very happy	216	198	203	617
Totals	357	507	339	1,203

12.9 The tables below test the relationship between income and a set of dependent variables. For each table, calculate percentages and gamma. Describe the strength and direction of each relationship in a few sentences. *Be careful in interpreting direction.*

a. Support by income for the legal right to an abortion:

Right to an Abortion?	Income			
	Low	Moderate	High	Totals
Yes	220	218	226	664
No	366	299	250	915
Totals	586	517	476	1,579

b. Support by income for capital punishment:

Capital Punishment?	Income			
	Low	Moderate	High	Totals
Favor	567	574	552	1,693
Oppose	270	183	160	613
Totals	837	757	712	2,306

c. Approval by income of suicide for people with an incurable disease:

Right to Suicide?	Income			Totals
	Low	Moderate	High	
Approve	343	341	338	1,022
Oppose	227	194	147	568
Totals	570	535	485	1,590

d. Support by income for sex education in public schools:

Sex Education?	Income			Totals
	Low	Moderate	High	
For	492	478	451	1,421
Against	85	68	53	206
Totals	577	546	504	1,627

e. Support by income for traditional gender roles:

Women Should Take Care of Running Their Homes and Leave Running the Country to Men	Income			Totals
	Low	Moderate	High	
Agree	130	71	39	240
Disagree	448	479	461	1,388
Totals	578	550	500	1,628

12.10 | SOC | What types of people are most concerned about the future of the environment? The World Values Survey includes an item asking people if they would agree to an increase in taxes if the extra money was used to prevent environmental damage. The tables below show the relationship between this variable and income for Canada, the United States, and Mexico. Income has been collapsed into three levels based on the relative income of each nation. Compute column percentages, the maximum difference, and an appropriate measure

of association for each table. Is there a relationship between income and concern for the environment? Describe the strength and direction of the relationship for each nation. Does the relationship between these variables change from nation to nation? If so, how?

a. Canada

Support higher tax to help the environment?	Income			Totals
	Low	Moderate	High	
Yes	157	380	281	818
No	90	222	117	429
	247	602	398	1,247

b. United States

Support higher tax to help the environment?	Income			Totals
	Low	Moderate	High	
Yes	330	655	631	1,616
No	263	405	360	1,028
	593	1,060	991	2,644

c. Mexico

Support higher tax to help the environment?	Income			Totals
	Low	Moderate	High	
Yes	695	627	236	1,558
No	531	400	96	1,027
	1,226	1,027	332	2,585

12.11 | SOC | In a random sample, 11 neighborhoods in Shinbone, Kansas, have been rated by an urban sociologist on a quality-of-life scale (which includes measures of affluence, availability of medical care, and recreational facilities) and a social cohesion scale. The results are presented below in scores. Higher scores indicate higher quality of life and greater social cohesion. Are the two variables associated? What is the strength and direction of the association? Summarize the relationship in a sentence or two. (*HINT: Don't forget to square the value of Spearman's rho for a PRE interpretation.*)

Neighborhood	Quality of Life	Social Cohesion
Queens Lake	17	8.8
North End	40	3.9
Brentwood	47	4.0
Denbigh Plantation	90	3.1
Phoebus	35	7.5
Kingswood	52	3.5
Chesapeake Shores	23	6.3
Windsor Forest	67	1.7
College Park	65	9.2
Beaconsdale	63	3.0
Riverview	100	5.3

12.12 SW Several years ago, a job-training program began, and a team of social workers screened the candidates for suitability for employment. Now the screening process is being evaluated, and the actual work performance of a sample of hired candidates has been rated. Did the screening process work? Is there a relationship between the original scores and performance evaluation on the job?

Case	Original Score	Performance Evaluation
A	17	78
B	17	85
C	15	82
D	13	92
E	13	75
F	13	72
G	11	70
H	10	75
I	10	92
J	10	70
K	9	32
L	8	55
M	7	21
N	5	45
O	2	25

12.13 SOC Below are the scores of a sample of 15 nations on a measure of ethnic diversity (the higher the number, the greater the diversity) and a measure of economic inequality (the higher the score, the greater the inequality). Are these variables related? Are ethnically diverse nations more economically unequal?

Nation	Diversity	Inequality
India	91	29.7
South Africa	87	58.4
Kenya	83	57.5
Canada	75	31.5
Malaysia	72	48.4
Kazakhstan	69	32.7
Egypt	65	32.0
United States	63	41.0
Sri Lanka	57	30.1
Mexico	50	50.3
Spain	44	32.5
Australia	31	33.7
Finland	16	25.6
Ireland	4	35.9
Poland	3	27.2

12.14 Twenty ethnic, racial, or national groups were rated by a random sample of white and black students on a social distance scale. Lower scores represent less social distance and less prejudice. How similar are these rankings?

	Group	Average Social Distance Scale Score	
		White Students	Black Students
1	White Americans	1.2	2.6
2	English	1.4	2.9
3	Canadians	1.5	3.6
4	Irish	1.6	3.6
5	Germans	1.8	3.9
6	Italians	1.9	3.3
7	Norwegians	2.0	3.8
8	American Indians	2.1	2.7
9	Spanish	2.2	3.0
10	Jews	2.3	3.3
11	Poles	2.4	4.2
12	Black Americans	2.4	1.3
13	Japanese	2.8	3.5
14	Mexicans	2.9	3.4
15	Koreans	3.4	3.7
16	Russians	3.7	5.1
17	Arabs	3.9	3.9
18	Vietnamese	3.9	4.1
19	Turks	4.2	4.4
20	Iranians	5.3	5.4

Using *SPSS for Windows* to Produce Ordinal-Level Measures of Association

SPSS DEMONSTRATION 12.1 Does Support for the Death Penalty Vary by Political Ideology?

In Demonstrations 11.1 and 11.2, we used column percentages and measures of association to look at the relationships between *cappun* (support for the death penalty) and several nominal-level variables. In this demonstration, we'll use political self-identification as liberal, moderate, or conservative (*polviews*) as an independent variable and continue the investigation. As coded in the GSS, *polviews* has seven values, ranging from extremely liberal to extremely conservative. We need to recode the variable into a more manageable number of values; we can do this by grouping all the liberals (scores 1–3) and conservatives (scores 5–7) together. We also need to keep the moderates (a score of 4) separate from the other two groups, so the recoding scheme will look like this:

1 − 3 = 1 (liberals)
4 = 2 (moderates)
5 − 7 = 3 (conservatives)

We will use the **Crosstabs** program with recoded *polviews* as the independent (column) variable and *cappun* as the dependent (row) variable. On the **Crosstabs** dialog window, click the **Statistics** button and request chi square and gamma. Don't forget to click the **Cells** button and request column percentages. Click **OK**, and the output will resemble the following table. (*NOTE: The table has been edited to improve readability.*)

FAVOR OR OPPOSE DEATH PENALTY FOR MURDER * recoded polviews
Cross-tabulation

		Recoded polviews			Totals
		1.00	2.00	3.00	
FAVOR OR OPPOSE DEATH PENALTY FOR MURDER	**FAVOR** Count	72	160	170	402
	%	51.1%	69.6%	76.6%	67.8%
	OPPOSE Count	69	70	52	191
	%	48.9%	30.4%	23.4%	32.2%
	Totals	141	230	222	593
		100.0%	100.0%	100.0%	100.0%

Chi Square Tests

	Value	df	Asymp. Sig. (2-sided)
Pearson Chi Square	26.247(a)	2	.000
Likelihood Ratio	25.561	2	.000
Linear-by-Linear Association	24.121	1	.000
N of Valid Cases	593		

[a]Zero cells (.0%) have expected count less than 5. The minimum expected count is 45.41.

Symmetric Measures

	Value	Asymp. Std. Error(a)	Approx. T(b)	Approx. Sig.
Ordinal by Ordinal Gamma	−.339	.066	−4.816	.000
N of Valid Cases	593			

[a]Not assuming the null hypothesis.
[b]Using the asymptotic standard error assuming the null hypothesis.

The chi square test indicates that this relationship is statistically significant. The column percentages are different, so there is a relationship between these variables. The gamma of –.339 indicates that the relationship is moderate in strength and negative in direction. As we noted earlier, interpreting direction with ordinal-level variables can be difficult. Remember that a negative gamma means that the cases tend to fall in a diagonal from lower left to upper right. In this case, that means that liberals (score of 1 on recoded *polviews*) tend to oppose the death penalty (a higher percentage of liberals oppose the death penalty than moderates or conservatives do), and conservatives tend to favor the death penalty (a higher percentage of conservatives favor capital punishment than moderates or liberals do). We can summarize the negative direction of the relationship by saying that support for capital punishment decreases as liberalism increases.

SPSS DEMONSTRATION 12.2 The Direction of Relationships: Do Attitudes Toward Sex Vary by Age?

Let's take a look at another relationship between two ordinal-level variables to review the interpretation of the direction of relationships. We'll start with *age* as an independent variable and **Recode** it into three categories as we did in Demonstration 9.1:

$$18 \text{ through } 36 \rightarrow 1$$
$$37 \text{ through } 52 \rightarrow 2$$
$$53 \text{ through } 89 \rightarrow 3$$

Recall that I used these particular cutting points to divide the sample into three categories with roughly equal numbers of cases.

Without looking at the data, I'm willing to bet that *age* will have a negative relationship with approval of premarital sex (*premarsx*). Run the **Crosstabs** procedure with *premarsx* as the row variable and *ager* (recoded *age*) as the column variable. Click the **Cells** button and make sure that the button next to **Columns** under **Percentages** is checked. Click the **Statistics** button and request gamma and, if you wish, chi square. The bivariate table for these two variables follows. (NOTE: The appearance of the table has been slightly changed to improve readability.)

Is there a relationship between these two variables? Do the conditional distributions change? Inspect the table column by column, and I think you will agree that there is a relationship.

Is this relationship positive or negative? Remember that in a positive association, high scores on one variable will be associated with high scores on the other and low scores will be associated with low. In a negative relationship, high scores on one variable are associated with low scores on the other.

Now go back and look at the table again. Find the single largest cell in each column and see if you can detect the pattern. The lowest score (youngest age group = 1) on *ager* is in the left-hand column. For this age group, the most common score (48.8% of this age group) on *premarsx* is 4 (not wrong at all). In other words, a little less than a majority of the youngest respondents supports the most

SEX BEFORE MARRIAGE* ager Cross-tabulation

			RECODED AGE			Totals
			1.00	2.00	3.00	
SEX BEFORE MARRIAGE	ALWAYS WRONG	Count	34	38	45	117
		%	24.1%	27.9%	33.8%	28.5%
	ALMOST ALWAYS WRONG	Count	10	11	18	39
		%	7.1%	8.1%	13.5%	9.5%
	SOMETIMES WRONG	Count	31	12	23	66
		%	22.0%	8.8%	17.3%	16.1%
	NOT WRONG AT ALL	Count	66	75	47	188
		%	46.8%	55.1%	35.3%	45.9%
		Totals	141	136	133	410
			100.0%	100.0%	100.0%	100.0%

Chi Square Tests

	Value	df	Asymp. Sig. (2-sided)
Pearson Chi Square	19.133(a)	6	.004
Likelihood Ratio	19.637	6	.003
Linear-by-Linear Association	5.658	1	.017
N of Valid Cases	410		

[a]Zero cells (.0%) have expected count less than 5. The minimum expected count is 12.65.

Symmetric Measures

	Value	Asymp. Std. Error(a)	Approx. T(b)	Approx. Sig.
Ordinal by Ordinal Gamma	−.141	.061	−2.287	.022
N of Valid Cases	410			

[a]Not assuming the null hypothesis.
[b]Using the asymptotic standard error assuming the null hypothesis.

permissive position on premarital sex. This score is also the most common response for the middle group on age, but the older respondents are much less likely to endorse this view. About one-third (35.3%) of the oldest age group felt than premarital sex was "not wrong at all." So, we could say that permissive attitudes tend to *decline* as age *increases*—the relationship is *negative* in direction.

Now look at the top row of the table. This is the least permissive position on premarital sex ("Always wrong"), and the percentage of cases in this row is about the same for the youngest and middle group but then increases for the oldest group. So, we could say that opposition to premarital sex *increases* as age *increases*—the variables change in the *same* direction, so the relationship is *positive*.

Which characterization of the relationship is correct? Both: The relationship is negative if you think of *premarsx* as measuring approval and positive if you think of the variable as measuring opposition to premarital sex. My point is to call your attention to the fact that the direction of a relationship can be ambiguous when we are dealing with ordinal-level variables like *premarsx*. In a positive association, the numerical *scores* always vary in the same direction. In a negative relationship, they always vary in opposite directions. However, because the coding for ordinal-level variables is arbitrary, higher scores don't necessarily mean that the quantity being measured is increasing, and lower scores don't always mean that the quantity is decreasing. Always pay careful attention to the coding scheme for the variable and exercise caution when analyzing the direction of relationships that involve ordinal-level variables.

Exercises

12.1 Follow up on Demonstration 12.1 with some new ordinal-level independent variables. What other factors might affect attitudes toward the death penalty? Among other possibilities, consider *degree* or *class* as independent variables. Other possibilities include *income98*, *prestg80*, or *attend* (these would have to be recoded).

12.2 Follow up on Demonstration 12.2 with other ordinal-level independent variables. What other factors might affect attitudes toward premarital sex? Consider *degree*, or *class*, or *attend* as independent variables, recoding as appropriate. The GSS includes several more measures of sexual attitudes (e.g., *marhomo* and *xmarsex*). Select one or two of the independent variables you used in your analysis of *premarsx* and see if they have similar relationships with these other measures of sexual attitudes.

13

Association Between Variables Measured at the Interval-Ratio Level

LEARNING OBJECTIVES   By the end of this chapter, you will be able to:

1. Interpret a scattergram.
2. Calculate and interpret slope (b), Y intercept (a), and Pearson's r and r^2.
3. Find and explain the least-squares regression line and use it to predict values of Y.
4. Explain the concepts of total, explained, and unexplained variance.
5. Use regression and correlation techniques to analyze and describe a bivariate relationship in terms of the three questions introduced in Chapter 11.

13.1 INTRODUCTION

This chapter presents a set of statistical techniques for analyzing the association or correlation between variables measured at the interval-ratio level.[1] As we shall see, these techniques are rather different in their logic and computation from those covered in Chapters 11 and 12. Let me stress at the outset, therefore, that we are still asking the same three questions: Is there a relationship between the variables? How strong is the relationship? What is the direction of the relationship? You might become preoccupied with some of the technical details and computational routines in this chapter, so remind yourself occasionally that our ultimate goals are unchanged: We are trying to understand bivariate relationships, explore possible causal ties between variables, and improve our ability to predict scores.

13.2 SCATTERGRAMS

As we have seen over the past several chapters, properly percentaged tables provide important information about bivariate associations between nominal- and ordinal-level variables. In addition to measures of association such as phi or gamma, the conditional distributions and patterns of cell frequency almost always provide useful information and a better understanding of the relationship between variables.

By the same token, the usual first step in analyzing a relationship between interval-ratio variables is to construct and examine a type of graph called a **scattergram.** As with bivariate tables, scattergrams allow us to quickly identify several important features of the relationship. An example will illustrate their construction and use. Suppose a researcher is interested in analyzing how dual-wage-earner families (that is, families where both husband and wife have jobs outside the home) cope with housework. Specifically, the researcher wonders if the number of children in the family is related to the amount of time the

---

[1]*The term *correlation* is commonly used instead of *association* when discussing the relationship between interval-ratio variables. I will use the two terms interchangeably.

husband contributes to housekeeping chores. Table 13.1 displays the relevant data for a sample of 12 dual-wage-earner families.

A scattergram, like a bivariate table, has two dimensions. The scores of the independent (X) variable are arrayed along the horizontal axis and the scores of the dependent (Y) variable along the vertical axis. Each dot on the scattergram represents a case in the sample and is located at a point determined by the scores of the case. The scattergram in Figure 13.1 shows the relationship between "number of children" and "husband's housework" for the sample of 12 families presented in Table 13.1. Family A has a score of 1 on the X variable (number of children) and 1 on the Y variable (husband's housework) and is represented by the dot above the score of 1 on the X axis and directly to the right of the score of 1 on the Y axis. All 12 cases are similarly represented by dots on Figure 13.1. Also note that, as with all tables, graphs, and charts, the scattergram is clearly titled and both axes are labeled.

TABLE 13.1   NUMBER OF CHILDREN AND HUSBAND'S CONTRIBUTION TO HOUSEWORK (fictitious data)

Family	Number of Children	Hours per Week Husband Spends on Housework
A	1	1
B	1	2
C	1	3
D	1	5
E	2	3
F	2	1
G	3	5
H	3	0
I	4	6
J	4	3
K	5	7
L	5	4

FIGURE 13.1   HUSBAND'S HOUSEWORK BY NUMBER OF CHILDREN

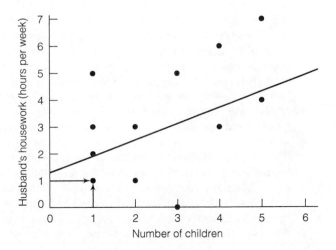

The overall pattern of the dots or cases summarizes the nature of the relationship between the two variables. We can clarify the pattern by drawing a straight line through the cluster of dots such that the line touches every dot or comes as close to doing so as possible. In Section 13.3, I will explain a precise technique for fitting this line to the pattern of the dots. For now, an "eyeball" approximation will suffice. This summarizing line is called the **regression line** and has already been added to the scattergram.

Scattergrams, even when they are crudely drawn, can serve a variety of purposes. They provide at least impressionistic information about the existence, strength, and direction of the relationship and can also be used to check the relationship for linearity (that is, how well the pattern of dots can be approximated with a straight line). Finally, we can use the scattergram to predict the score of a case on one variable from the score of that case on the other variable. We will briefly examine each of these uses.

To ascertain the existence of a relationship, we can return to the basic definition of an association stated in Chapter 11: Two variables are associated if the distributions of Y (the dependent variable) change for the various conditions of X (the independent variable). In Figure 13.1, scores on X (number of children) are arrayed along the horizontal axis. The dots above each score on X are the scores (or conditional distributions) of Y. That is, the dots represent scores on Y for each value of X. Figure 13.1 shows that there is a relationship between these variables because these conditional distributions of Y (the dots above each score on X) change as X changes. The existence of an association is further reinforced by the fact that the regression line lies at an angle to the X axis. If these two variables had not been associated, the conditional distributions of Y would not have changed, and the regression line would have been parallel to the horizontal axis.

We can judge the strength of the bivariate association by observing the spread of the dots around the regression line. In a perfect association, every single dot would be on the regression line. The more the dots are clustered around the regression line, the stronger the relationship.

The angle of the regression line tells us the direction of the relationship. Figure 13.1 shows a positive relationship: As X (number of children) increases, husband's housework (Y) also increases. Husbands in families with more children tend to do more housework. If the relationship had been negative, the regression line would have sloped in the opposite direction to indicate that high scores on one variable were associated with low scores on the other.

To summarize these points about the existence, strength, and direction of the relationship, Figure 13.2 shows a perfect positive and a perfect negative relationship and a "zero relationship" between two variables.

One key assumption underlying the statistical techniques to be introduced later in this chapter is that the two variables have an essentially **linear relationship.** In other words, the observation points or dots in the scattergram must form a pattern that can be approximated with a straight line. Significant departures from linearity would require the use of statistical techniques beyond the scope of this text. Figure 13.3 presents examples of some common curvilinear relationships. If the scattergram shows that the variables have a nonlinear relationship, you should use the techniques described in this chapter with great caution or not at all. Checking for the linearity of the relationship is perhaps the most important reason for

FIGURE 13.2   POSITIVE, NEGATIVE, AND ZERO RELATIONSHIPS

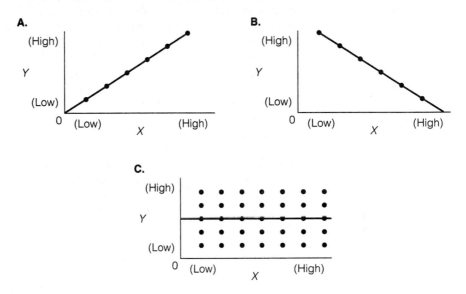

FIGURE 13.3   SOME NONLINEAR RELATIONSHIPS

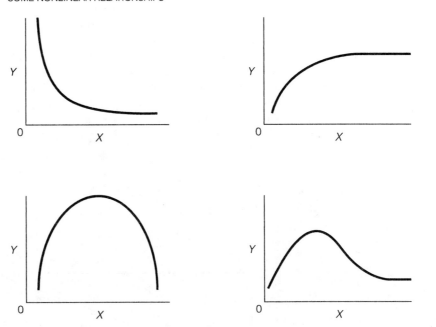

constructing at least a crude, hand-drawn scattergram before proceeding with the statistical analysis. If the relationship is nonlinear, you might need to treat the variables as if they were ordinal rather than interval-ratio in level of measurement. *(For practice in constructing and interpreting scatter-grams, see problems 13.1 to 13.4.)*

13.3 REGRESSION AND PREDICTION

A final use of the scattergram is to predict scores of cases on one variable from their score on the other. To illustrate, suppose that, based on the relationship between number of children and husband's housework displayed in Figure 13.1, we want to predict the number of hours of housework a husband with six children would do each week. The sample has no families with six children, but if we extend the axes and regression line in Figure 13.1 to incorporate this score, a prediction is possible. Figure 13.4 reproduces the scattergram and illustrates how the prediction would be made.

We find the predicted score on *Y*—which is symbolized as **Y'** to distinguish predictions of *Y* from actual *Y* scores—by first locating the relevant score on *X* (*X* = 6 in this case) and then drawing a straight line from that point to the regression line. From the regression line, we draw another straight line parallel to the *X* axis across to the *Y* axis. The predicted *Y* score (Y') is found at the point where this line crosses the *Y* axis. In our example, we would predict that, in a dual-wage-earner family with six children, the husband would devote about five hours per week to housework.

Of course, this prediction technique is crude, and the value of Y' can change, depending on how accurately the freehand regression line is drawn. One way to eliminate this source of error would be to find the straight line that most accurately summarizes the pattern of the observation points and so best describes the relationship between the two variables. How can we find the "best-fitting" straight line?

Recall that our criterion for the freehand regression line was that it touch all the dots or come as close to doing so as possible. Also, recall that the dots above each value of *X* can be thought of as conditional distributions of *Y*, the dependent variable. Within each conditional distribution of *Y*, the mean is the point around which the variation of the scores is at a minimum. In Chapter 3, we noted that the mean of any distribution of scores is the point around which the variation of the scores, as measured by squared deviations, is minimized:

$$\Sigma(X_i - \overline{X})^2 = \text{minimum}$$

FIGURE 13.4   PREDICTING HUSBAND'S HOUSEWORK

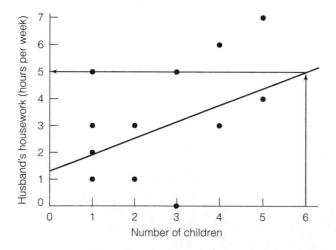

Thus, a regression line that is drawn so that it passes through each **conditional mean of Y** would be the straight line that comes as close as possible to all the scores.

We can find conditional means by summing all Y values for each value of X and then dividing by the number of cases. For example, four families had one child ($X = 1$), and the husbands of these four families devoted 1, 2, 3, and 5 hours per week to housework. Thus, for $X = 1$, $Y = 1, 2, 3$, and 5, and the conditional mean of Y for $X = 1$ is 2.75 (11/4 = 2.75). Husbands in families with one child worked an average of 2.75 hours per week doing housekeeping chores. Conditional means of Y can be computed in the same way for each value of X. They are displayed in Table 13.2 and plotted in Figure 13.5.

Let's quickly remind ourselves of the reason for these calculations. We are seeking the single best-fitting regression line for summarizing the relationship between X and Y, and we have seen that a line drawn through the conditional means of Y will minimize the spread of the observation points. It will come as close to all the scores as possible and will therefore be the single best-fitting regression line.

Now, a line drawn through the points in Figure 13.5 (the conditional means of Y) will be the best-fitting line we are seeking, but you can see from the scattergram that the line will not be straight. In fact, only rarely (when there is a perfect relationship between X and Y) will conditional means fall in a

TABLE 13.2 CONDITIONAL MEANS OF Y (HUSBAND'S HOUSEWORK) FOR VARIOUS VALUES OF X (NUMBER OF CHILDREN)

Number of Children (X)	Husband's Housework (Y)	Conditional Means of Y
1	1,2,3,5	2.75
2	3,1	2.00
3	5,0	2.50
4	6,3	4.50
5	7,4	5.50

FIGURE 13.5 CONDITIONAL MEANS OF Y

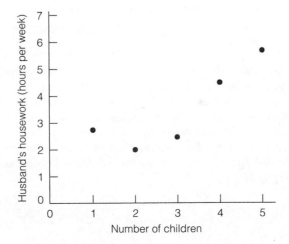

perfectly straight line. Because we still must meet the condition of linearity, we will revise our criterion and define the regression line as the unique straight line that touches all conditional means of Y or comes as close to doing so as possible. Formula 13.1 defines the **"least-squares"** regression line, or the single straight regression line that best fits the pattern of the data points.

FORMULA 13.1 $$Y = a + bX$$

where Y = score on the dependent variable
a = the Y intercept or the point where the regression
line crosses the Y axis
b = the slope of the regression line or the amount of
change produced in Y by a unit change in X
X = score on the independent variable

The formula introduces two new concepts. First, the **Y intercept (a)** is the point at which the regression line crosses the vertical, or Y, axis. Second, the **slope (b)** of the least-squares regression line is the amount of change produced in the dependent variable (Y) by a unit change in the independent variable (X). Think of the slope of the regression line as a measure of the effect of the X variable on the Y variable. If the variables have a strong association, then changes in the value of X will be accompanied by substantial changes in the value of Y, and the slope (b) will have a high value. The weaker the effect of X on Y (the weaker the association between the variables), the lower the value of the slope (b). If the two variables are unrelated, the least-squares regression line would be parallel to the X axis, and b would be 0.00 (the line would have no slope).

With the least-squares formula (Formula 13.1), we can predict values of Y in a much less arbitrary and impressionistic way than through mere eyeballing. This is because the least-squares regression line as defined by Formula 13.1 is the single straight line that best fits the data because it comes as close as possible to all of the conditional means of Y. Before seeing how predictions of Y can be made, however, we must first calculate a and b. *(For practice in using the regression line to predict scores on Y from scores on X, see problems 13.1 to 13.3 and 13.5.)*

13.4 THE COMPUTATION OF *a* AND *b*

Computing the Slope (b).   Because we need the value of b to solve for a, we will begin with the computation of the slope of the least-squares regression line. The definitional formula for the slope is:

FORMULA 13.2 $$b = \frac{\Sigma(X - \overline{X})(Y - \overline{Y})}{\Sigma(X - \overline{X})^2}$$

The numerator of this formula is called the *covariation* of X and Y. It is a measure of how X and Y vary together, and its value will reflect both the direction and strength of the relationship. These days, of course, computers are used to compute complex statistics such as b. However, for smaller samples (and for the end-of-chapter problems in this text), b is sometimes still calculated using handheld calculators. In these situations, Formula 13.2 is awkward to use. Instead, we can use the following computational formula derived from Formula 13.2:

FORMULA 13.3 $$b = \frac{N\Sigma XY - (\Sigma X)(\Sigma Y)}{N\Sigma X^2 - (\Sigma X)^2}$$

where b = the slope

N = the number of cases

ΣXY = the summation of the cross products of the scores

ΣX = the summation of the X scores

ΣY = the summation of the Y scores

ΣX^2 = the summation of the squared score on X

Admittedly, this formula appears formidable at first glance, but we can solve it without too much difficulty if we organize the computations into table format. The computing table displayed in Table 13.3 has a column for each of the four quantities needed to solve the formula. The data are from the sample of dual-wage-earner families (see Table 13.1).

In Table 13.3, the first two columns list the original X and Y scores for each case. The third column contains the squared scores on X, and the fourth lists the squared scores on Y. The fifth column lists the cross products of the scores for each case. In other words, we determine the entries in the last column by multiplying both scores for each case.

We can now replace the symbols in Formula 13.3 with the proper sums:

$$b = \frac{N \Sigma XY - (\Sigma X)(\Sigma Y)}{N \Sigma X^2 - (\Sigma X)^2}$$

$$b = \frac{(12)(125) - (32)(40)}{(12)(112) - (32)^2}$$

$$b = \frac{(1{,}500 - 1{,}280)}{(1{,}344 - 1{,}024)}$$

$$b = \frac{220}{320}$$

$$b = 0.69$$

TABLE 13.3   COMPUTATION OF THE SLOPE (b)

X	Y	X^2	Y^{2*}	XY
1	1	1	1	1
1	2	1	4	2
1	3	1	9	3
1	5	1	25	5
2	3	4	9	6
2	1	4	1	2
3	5	9	25	15
3	0	9	0	0
4	6	16	36	24
4	3	16	9	12
5	7	25	49	35
5	4	25	16	20
$\Sigma X = 32$	$\Sigma Y = 40$	$\Sigma X^2 = 112$	$\Sigma Y^2 = 184$	$\Sigma XY = 125$

$$\overline{X} = 32/12 = 2.67$$
$$\overline{Y} = 40/12 = 3.33$$

*The quantity ΣY^2 is not used in the computation of b. We will use it later, however, when we compute Pearson's r (see Section 13.5).

STEP BY STEP	Computing the Slope (*b*)

Step 1: Set up a computing table like Table 13.3. List the *X* and *Y* scores in the first two columns, the squared *X* and *Y* scores in columns 3 and 4, and the cross products in column 5. To calculate the cross products, multiply the *X* and *Y* scores for each case.

Step 2: Sum each of the columns in the computing table.

Step 3: To compute the slope, solve Formula 13.3:

$$b = \frac{N\sum XY - (\sum X)(\sum Y)}{N\sum X^2 - (\sum X)^2}$$. We will solve the denominator first and then the numerator:

  a. Find the sum of the squared *X* scores ($\sum X^2$) at the bottom of column 3 in the computing table. Multiply this sum by *N*.

b. Find the sum of the *X* scores ($\sum X$) at the bottom of column 1. Square this value.

c. Subtract the quantity you found in step b from the quantity you found in step a.

d. Find the sum of the cross products ($\sum XY$) at the bottom of column 5. Multiply this quantity by *N*.

e. Find the sums of the *X* ($\sum X$) and *Y* ($\sum Y$) scores at the bottom of columns 1 and 2. Multiply these quantities together.

f. Subtract the quantity you found in step e from the quantity you found in step d.

g. Divide the quantity you found in step f by the quantity you found in step c. The result is the value of the slope (*b*).

A slope of 0.69 indicates that, for each unit change in *X*, there is an increase of .69 units in *Y*. For our example, the addition of each child (an increase of 1 unit in *X*) results in an increase of .69 hour of housework being done by the husband (an increase of .69 units—or hours—in *Y*).

Computing the Y Intercept (*a*).   Once we've calculated the slope, finding the intercept (*a*) is relatively easy. To compute the mean of *X* and the mean of *Y*, divide the sums of columns 1 and 2 of Table 13.3 by *N* and enter these figures into Formula 13.4:

FORMULA 13.4

$$a = \overline{Y} - b\overline{X}$$

For our sample problem, the value of *a* would be

$$a = \overline{Y} - b\overline{X}$$
$$a = 3.33 - (0.69)(2.67)$$
$$a = 3.33 - 1.84$$
$$a = 1.49$$

Thus, the least-squares regression line will cross the *Y* axis at the point where *Y* equals 1.49.

Using the Least-Squares Regression Equation to Predict Scores on Y. The full least-squares regression line for our sample data can now be specified:

$$Y = a + bX$$
$$Y = (1.49) + (0.69)X$$

We can use this formula to estimate or predict scores on *Y* for any value of *X*. In Section 13.3, we used the freehand regression line to predict a score on *Y* (husband's housework) for a family with six children (*X* = 6). We predicted that, in families of six children, husbands would contribute about five hours per week to housekeeping chores. By using the least-squares regression line, we

STEP BY STEP	Computing the *Y* Intercept (*a*)

Step 1: To compute the *Y* intercept, solve Formula 13.4: $a = \bar{Y} - b\bar{X}$

   a. Find the mean of the *X* and *Y* scores. To find, $\bar{X}$ divide the sum of column 1 of the computing table by *N*. To find $\bar{Y}$, divide the sum of column 2 of the computing table by *N*.

   b. Multiply $\bar{X}$ by *b*.

   c. Subtract the value you computed in Step b from $\bar{Y}$. The result is the value of the *Y* intercept (*a*).

can see how close our impressionistic, eyeball prediction was:

$$Y = a + bX$$
$$Y = (1.49) + (0.69)(6)$$
$$Y = (1.49) + (4.14)$$
$$Y = 5.63$$

Based on the least-squares regression line, we would predict that in a dual-wage-earner family with six children, husbands would devote 5.63 hours a week to housework. What would be our prediction of husband's housework for a family of seven children ($X = 7$)?

Note that our predictions of *Y* scores are basically "educated guesses." We will be unlikely to predict values of *Y* exactly except in the (relatively rare) case where the bivariate relationship is perfect and perfectly linear. Note also, however, that the accuracy of our predictions will increase as relationships become stronger. This is because the dots are more closely clustered around the least-squares regression line in stronger relationships. *(The slope and Y intercept may be computed for any problem at the end of this chapter, but see problems 13.1 to 13.5 in particular. These problems have smaller data sets and will provide good practice until you are comfortable with these calculations.)*

13.5 THE CORRELATION COEFFICIENT (PEARSON'S *r*)

I pointed out in Section 13.4 that the slope of the least-squares regression line (*b*) is a measure of the effect of *X* on *Y*. Because the slope is the amount of change produced in *Y* by a unit change in *X*, *b* will increase in value as the relationship increases in strength. However, *b* does not vary between zero and one and is therefore awkward to use as a measure of association. Instead, researchers rely heavily (almost exclusively) on a statistic called **Pearson's *r*,** or the correlation coefficient, to measure association between interval-ratio variables. Like gamma and Spearman's *r*, the ordinal measures of association discussed in Chapter 12, Pearson's *r* varies from 0.00 to ±1.00, with 0.00 indicating no association and +1.00 and −1.00 indicating perfect positive and perfect negative relationships, respectively. The definitional formula for Pearson's *r* is:

FORMULA 13.5

$$r = \frac{\Sigma(X - \bar{X})(Y - \bar{Y})}{\sqrt{[\Sigma(X - \bar{X})^2][\Sigma(Y - \bar{Y})^2]}}$$

Note that the numerator of this formula is the covariation of *X* and *Y*, as was the case with Formula 13.2. When computing with a hand calculator, computational Formula 13.6 is preferred over the definitional formula.

FORMULA 13.6

$$r = \frac{N\sum XY - (\sum X)(\sum Y)}{\sqrt{[N\sum X^2 - (\sum X)^2][N\sum Y^2 - (\sum Y)^2]}}$$

A computing table such as Table 13.3 is strongly recommended as a way of organizing the quantities needed to solve this equation. For our sample problem involving dual-wage-earner families, we can directly substitute the quantities displayed in Table 13.3 into Formula 13.6:

$$r = \frac{N\sum XY - (\sum X)(\sum Y)}{\sqrt{[N\sum X^2 - (\sum X)^2][N\sum Y^2 - (\sum Y)^2]}}$$

$$r = \frac{(12)(125) - (32)(40)}{\sqrt{[(12)(112) - (32)^2][(12)(184) - (40)^2]}}$$

$$r = \frac{1{,}500 - 1{,}280}{\sqrt{(1{,}344 - 1{,}024)(2{,}208 - 1{,}600)}}$$

$$r = \frac{220}{\sqrt{(320)(608)}}$$

$$r = \frac{220}{\sqrt{194{,}560}}$$

$$r = \frac{220}{441.09}$$

$$r = 0.50$$

STEP BY STEP	**Computing Pearson's *r***

Step 1: Set up a computing table like Table 13.3.

Step 2: Sum each of the columns in the computing table.

Step 3: To compute *r*, solve Formula 13.6:

$$r = \frac{N\sum XY - (\sum X)(\sum Y)}{\sqrt{[N\sum X^2 - (\sum X)^2][N\sum Y^2 - (\sum Y)^2]}}$$

We will solve the denominator first and then the numerator:

a. Find the sum of the squared *X* scores ($\sum X^2$) at the bottom of column 3 in the computing table. Multiply this sum by *N*.

b. Find the sum of the *X* scores ($\sum X$) at the bottom of column 1. Square this value.

c. Subtract the quantity you found in step b from the quantity you found in step a.

d. Find the sum of the squared *Y* scores ($\sum Y^2$) at the bottom of column 4 in the computing table. Multiply this sum by *N*.

e. Find the sum of the *Y* scores ($\sum Y$) at the bottom of column 2. Square this value.

f. Subtract the quantity you found in step e from the quantity you found in step d.

g. Multiply the quantity you found in step f by the quantity you found in step c.

h. Take the square root of the quantity you found in step g.

i. Find the sum of the cross products ($\sum XY$) at the bottom of column 5. Multiply this quantity by *N*.

j. Find the sums of the *X* ($\sum X$) and *Y* ($\sum Y$) scores at the bottom of columns 1 and 2. Multiply these quantities together.

k. Subtract the quantity you found in step j from the quantity you found in step i.

l. Divide the quantity you found in step k by the quantity you found in step h. The result is the value of Pearson's *r*.

An *r* value of .50 indicates a moderately strong, positive linear relationship between the variables. As the number of children in the family increases, the hourly contribution of husbands to housekeeping duties also increases. *(Every problem at the end of this chapter requires the computation of Pearson's r. It is probably a good idea to practice with smaller data sets and easier computations first—see problem 13.1 in particular.)*

13.6 INTERPRETING THE CORRELATION COEFFICIENT: r^2

Pearson's *r* is an index of the strength of the linear relationship between two variables. While a value of 0.00 indicates no linear relationship and a value of ±1.00 indicates a perfect linear relationship, values between these extremes have no direct interpretation. We can, of course, describe relationships in terms of how closely they approach the extremes (for example, coefficients approaching 0.00 can be described as "weak" and those approaching ±1.00 as "strong"), but this description is somewhat subjective. Also, we can use the guidelines stated in Table 12.2 for gamma to attach descriptive terms to the specific values of Pearson's *r*. In other words, values between 0.00 and 0.30 would be described as weak, values between 0.30 and 0.60 would be moderate, and values greater than 0.60 would be strong. Remember that these labels are arbitrary guidelines and will not be appropriate or useful in all possible research situations.

The Coefficient of Determination.   Fortunately, we can develop a less arbitrary, more direct interpretation of *r* by calculating an additional statistic called the **coefficient of determination.** This statistic, which is simply the square of Pearson's *r* (r^2), can be interpreted with logic akin to proportional reduction in error (PRE). As you recall, the logic of PRE measures of association is to predict the value of the dependent variable under two different conditions. First, *Y* is predicted while ignoring the information supplied by *X* and, second, the independent variable is taken into account. With r^2, both the method of prediction and the construction of the final statistic are somewhat different and require the introduction of some new concepts.

Predicting *Y* Without *X*.   When working with variables measured at the interval-ratio level, the predictions of the *Y* scores under the first condition (while ignoring *X*) will be the mean of the *Y*. Given no information on *X*, this prediction strategy will be optimal because we know that the mean of any distribution is closer to all the scores than any other point in the distribution. I remind you of the principle of minimized variation introduced in Chapter 3 and expressed as:

$$\Sigma(Y - \overline{Y})^2 = \text{minimum}$$

The scores of any variable vary less around the mean than around any other point. If we predict the mean of *Y* for every case, we will make fewer errors of prediction than if we predict any other value for *Y*.

Of course, we will still make many errors in predicting *Y* even if we faithfully follow this strategy. The amount of error is evident in Figure 13.6, which displays the relationship between number of children and husband's housework with the mean of *Y* noted. The vertical lines from the actual scores to the predicted score represent the amount of error we would make when predicting *Y* while ignoring *X*.

FIGURE 13.6   PREDICTING Y WITHOUT X (dual-career families)

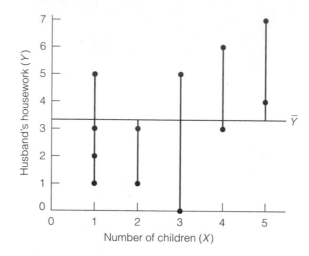

We can define the extent of our prediction error under the first condition (while ignoring X) by subtracting the mean of Y from each actual Y score and squaring and summing these deviations. The resultant figure, which can be noted as $\Sigma(Y - \overline{Y})^2$, is called the **total variation** in Y. We now have a visual representation (Figure 13.6) and a method for calculating the error we incur by predicting Y without knowledge of X. As we shall see below, we do not need to actually calculate the total variation to find the value of the coefficient of determination, r^2.

Predicting Y with X.   Our next step will be to determine the extent to which knowledge of X improves our ability to predict Y. If the two variables have a linear relationship, then predicting scores on Y from the least-squares regression equation will incorporate knowledge of X and reduce our errors of prediction. So, under the second condition, our predicted Y score for each value of X will be:

$$Y' = a + bX$$

Figure 13.7 displays the data from the dual-career families with the regression line, as determined by the above formula, drawn in. The vertical lines from each data point to the regression line represent the amount of error in predicting Y that remains even after X has been taken into account.

Explained, Unexplained, and Total Variation.   As was the case under the first condition, we can precisely define the reduction in error that results from taking X into account. Specifically, we can find two different sums and then compare them with the total variation of Y to construct a statistic that will indicate the improvement in prediction.

The first sum, called the **explained variation,** represents the improvement in our ability to predict Y when taking X into account. This sum is found by subtracting $\overline{Y}$ (our predicted Y score without X) from the score predicted by the regression equation (Y', or the Y score predicted with knowledge of X) for each case and then squaring and summing these differences. These operations can be summarized as $\Sigma(Y' - \overline{Y})^2$. We could then compare the resultant figure with the

FIGURE 13.7   PREDICTING Y WITH X (dual-career families)

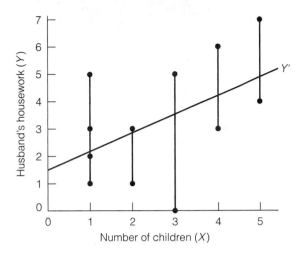

total variation in Y to ascertain the extent to which our knowledge of X improves our ability to predict Y. Specifically, it can be shown mathematically that:

FORMULA 13.7

$$r^2 = \frac{\Sigma(Y' - \overline{Y})^2}{\Sigma(Y - \overline{Y})^2} = \frac{\text{Explained variation}}{\text{Total variation}}$$

Thus, the coefficient of determination, or r^2, is the proportion of the total variation in Y attributable to or explained by X. Like other PRE measures, r^2 indicates precisely the extent to which X helps us predict, understand, or explain Y.

Above, we refer to the improvement in predicting Y with X as the explained variation. This term suggests that some of the variation in Y will be "unexplained" or not attributable to the influence of X. In fact, the vertical lines in Figure 13.7 represent the **unexplained variation,** or the difference between our best prediction of Y with X and the actual scores. The unexplained variation is thus the scattering of the actual scores around the regression line. We can find it by subtracting the predicted Y scores from the actual Y scores for each case and then squaring and summing the differences. These operations can be summarized as $\Sigma(Y - Y')^2$, and the resultant sum would measure the amount of error in predicting Y that remains even after X has been taken into account. We can find the proportion of the total variation in Y unexplained by X by subtracting the value of r^2 from 1.00. Unexplained variation is usually attributed to the influence of some combination of other variables, measurement error, and random chance.

As you may have recognized by this time, the explained and unexplained variations bear a reciprocal relationship with each other. As one of these sums increases in value, the other decreases. Furthermore, the stronger the linear relationship between X and Y, the greater the value of the explained variation and the lower the unexplained variation. In the case of a perfect relationship ($r = \pm1.00$), the unexplained variation would be 0 and r^2 would be 1.00. This would indicate that X explains or accounts for all the variation in Y and that we could predict Y from X without error. On the other hand, when X and Y are not linearly related ($r = 0.00$), the explained variation would be 0 and r^2 would be 0.00. In such a case, we would conclude that X explains none of the variation in Y and does not improve our ability to predict Y.

Application 13.1

Are nations that have more educated populations more tolerant? Are more educated nations therefore less likely to see homosexuality as wrong? Random samples from 10 nations have been asked if they agree that homosexuality is "never acceptable."

Information has also been gathered on the average years of school completed for people over 25 in each nation. How are these variables related? The data appear in the table below. Columns have been added for all necessary sums.

Nation	Average Years of Schooling (X)	Percent Agreeing That Homosexuality Is "Always Wrong" (Y)	X^2	Y^2	XY
China	6	88	36	7,744	528
Brazil	5	56	25	3,136	280
United States	12	45	144	2,025	540
Japan	10	42	100	1,764	420
Mexico	7	55	49	3,025	385
India	5	77	25	5,929	385
South Africa	6	61	36	3,721	366
Finland	10	37	100	1,369	370
United Kingdom	9	21	81	441	189
Germany	10	15	100	225	150
Totals	80	497	696	29,379	3,613

Data are from the World Values Survey and the Human Development Report published by the United Nations.

The slope (b) is:

$$b = \frac{N \sum XY - (\sum X)(\sum Y)}{N \sum X^2 - (\sum X)^2}$$

$$b = \frac{(10)(3,613) - (80)(497)}{(10)(696) - (80)^2}$$

$$b = \frac{36,130 - 39,760}{6,960 - 6,400}$$

$$b = \frac{-3,630}{560}$$

$$b = -6.48$$

A slope of -6.48 means that for every increase in years of education (a unit change in X), there is a decrease of 6.48 points in the percentage of people who feel that homosexuality is never acceptable.

The Y intercept (a) is:

$$a = \overline{Y} - b\overline{X}$$

$$a = \frac{497}{10} - (-6.48)\left(\frac{80}{10}\right)$$

$$a = 49.7 - (-6.48)(8)$$

$$a = 49.7 + 51.48$$

$$a = 101.18$$

The least-squares regression equation is:

$$Y = a + bX = 101.18 + (-6.48)X$$

The correlation coefficient is:

$$r = \frac{N \sum XY - (\sum X)(\sum Y)}{\sqrt{[N \sum X^2 - (\sum X)^2][N \sum Y^2 - (\sum Y)^2]}}$$

$$r = \frac{(10)(3,613) - (80)(497)}{\sqrt{[(10)(696) - (80)^2][(10)(29,379) - (497)^2]}}$$

$$r = \frac{36,130 - 39,760}{\sqrt{(6,960 - 6,400)(293,790 - 247,009)}}$$

$$r = \frac{-3,630}{\sqrt{(560)(46,781)}}$$

$$r = \frac{-3,630}{\sqrt{26,197,360}}$$

$$r = \frac{-3,630}{5,118.34}$$

$$r = -0.71$$

For these 10 nations, education and disapproval of homosexuality have a strong, negative relationship. Disapproval of homosexuality decreases as education increases. The coefficient of determination, r^2, is $(0.71)^2$, or 0.50. This indicates that 50% of the variance in attitude toward homosexuality is explained by education for this sample of 10 nations.

Relationships intermediate between these two extremes can be interpreted in terms of how much X increases our ability to predict or explain Y. For the dual-career families, we calculated an r of 0.50. Squaring this value yields a coefficient of determination of 0.25 ($r^2 = 0.25$), which indicates that number of children (X) explains 25% of the total variation in husband's housework (Y). When predicting the number of hours per week that husbands in such families would devote to housework, we will make 25% fewer errors by basing the predictions on number of children and predicting from the regression line, as opposed to ignoring this variable and predicting the mean of Y for every case. Also, 75% of the variation in Y is unexplained by X and presumably due to some combination of the influence of other variables, measurement error, and random chance. *(For practice in the interpretation of r^2, see any of the problems at the end of this chapter.)*

13.7 THE CORRELATION MATRIX

Social science research projects usually include many variables, and the data analysis phase of a project often begins with the examination of a **correlation matrix:** a table that shows the relationships between all possible pairs of variables. The correlation matrix gives a quick, easy-to-read overview of the interrelationships in the data set and may suggest strategies or "leads" for further analysis. These tables are commonly included in the professional research literature, and you'll find it useful to have some experience reading them.

Table 13.4 presents an example of a correlation matrix, using cross-national data. The matrix uses variable names as rows and columns, and the cells in the table show the bivariate correlation (usually a Pearson's r) for each

TABLE 13.4   A CORRELATION MATRIX SHOWING INTERRELATIONSHIPS FOR FIVE VARIABLES ACROSS 161 NATIONS

	(1) GDP per Capita	(2) Inequality	(3) Unemployment Rate	(4) Literacy	(5) Voter Turnout
(1) GDP per Capita	1.00	−0.43	−0.34	0.46	0.28
(2) Inequality	−0.43	1.00	0.33	−0.15	−0.36
(3) Unemployment Rate	−0.34	0.33	1.00	−0.48	−0.28
(4) Literacy	0.46	−0.15	−0.48	1.00	0.40
(5) Voter Turnout	0.28	−0.36	−0.28	0.40	1.00

VARIABLES:

(1) *GDP per Capita:* Gross domestic product (the total value of all goods and services) divided by population size. This variable is an indicator of the level of affluence and prosperity in the society. Higher scores mean greater prosperity.

(2) *Inequality:* An index of income inequality. Higher scores mean greater inequality.

(3) *Unemployment Rate:* The annual rate of joblessness.

(4) *Literacy Rate:* Number of people over 15 able to read and write per 1,000 population.

(5) *Voter Turnout:* Percentage of eligible voters who participated in the most recent election.

combination of variables. Note that the row headings duplicate the column headings. To read the table, begin with GDP per capita, the variable in the far left-hand column (column 1) and top row (row 1). Read down column 1 or across row 1 to see the correlations of this variable with all other variables, including the correlation of GDP per capita with itself (1.00) in the top left-hand cell. To see the relationships between other variables, move from column to column or row to row.

Note that the diagonal from upper left to lower right of the matrix presents the correlation of each variable with itself. Values along this diagonal will always be exactly 1.00 and, since this information is not useful, it could easily be deleted from the table.

Also note that the cells below and to the left of the diagonal are redundant with the cells above and to the right of the diagonal. For example, look at the second cell down (row 2) in column 1. This cell displays the correlation between GDP per capita and inequality, as does the cell in the top row (row 1) of column 2. In other words, the cells below and to the left of the diagonal are mirror images of the cells above and to the right of the diagonal. Commonly, research articles in the professional literature will delete the redundant cells to make the table more readable.

What does this matrix tell us? Starting with the top row of the table and reading left to right, we can see that GDP per capita has a moderate negative relationship with inequality and unemployment rate, which means that more affluent nations tend to have less inequality and lower rates of joblessness. GDP per capita also has a moderate positive relationship with literacy (more affluent nations have higher levels of literacy) and a weak to moderate positive relationship with voter turnout (more affluent nations tend to have higher levels of participation in the electoral process).

To assess the other relationships in the data set, move from column to column and row to row, one variable at a time. For each subsequent variable, there will be one less cell of new information. For example, consider inequality, the variable in column 2 and row 2. We have already noted its moderate negative relationship with GDP per capita and, of course, we can ignore the correlation of the variable with itself. This leaves only three new relationships, which we can read by moving down column 2 or across row 2. Inequality has a positive moderate relationship with unemployment (the greater the inequality, the greater the unemployment), a weak negative relationship with literacy (nations with more inequality tend to have lower literacy rates), and a moderate negative relationship with voter turnout (the greater the inequality, the lower the turnout).

For unemployment, the variable in column 3, there are only two new relationships: a moderate negative correlation with literacy (the higher the unemployment, the lower the literacy) and a weak to moderate negative relationship with voter turnout (the higher the unemployment rate, the lower the turnout). For voter turnout, the variable in column 5, there is only one new relationship. Voter turnout has a moderate positive relationship with literacy (turnout increases as literacy goes up).

In closing, note that the cells in a correlation matrix will often include other information in addition to the bivariate correlations. It is common, for example, to include the number of cases on which the correlation is based and, if relevant, an indication of the statistical significance of the relationship.

STATISTICS IN EVERYDAY LIFE: Correlation, Causation, and Cancer

Today, it is common knowledge in the United States that smoking tobacco causes lung cancer. However, this information was not part of the common wisdom just a few generations ago. In the 1950s, about half of all American men smoked, and smoking was equated with sophistication and adulthood. Since that time, a strong body of medical research has established links among smoking, cancer, and a number of other health risks and, just as importantly, these connections have been widely broadcast by both public and private agencies. The effect has been dramatic: Now, fewer than 25% of U.S. adults smoke.

Statistics played an important role in establishing the links that led to the public campaign against smoking in the United States and many other nations. More specifically, the link between smoking and cancer was established by a series of studies that featured measures of association, especially Pearson's r. The most convincing of these studies featured large samples of individuals who were followed over long periods of time. The studies collected a variety of medical information for each respondent, including their smoking habits and the

incidence of cancer and other health problems. For example, one study conducted by the office of the U.S. Surgeon General studied women from 1976 to 1988. The graph below clarifies the connection between smoking and cancer. The graph plots number of cigarettes per day for the smokers against the relative risk of contracting cancer. (The relative risk is the actual cancer death rate for the smokers compared to the cancer death rate for nonsmokers, controlling for age and a number of medical conditions.)

Even though this relationship is not linear, it is clear that there is a strong correlation between number of cigarettes smoked and cancer risk. Women who smoked more than 35 cigarettes a day had approximately twice the relative risk of women who smoked less than 14 cigarettes a day. Graphs like this—along with tons of other medical evidence, of course—persuaded the medical profession and the government that smoking was a very significant public health risk.

Some might try to downplay the risk of smoking by arguing that correlation is not the same thing as causation: Just because two variables change

RELATIVE RISK OF DEATH BY NUMBER OF CIGARETTES PER DAY (female smokers only)

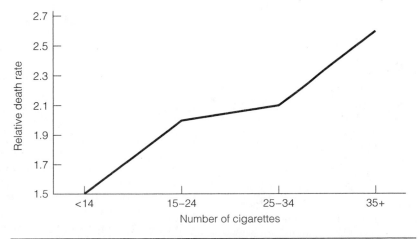

Source: http://www.cdc.gov/tobacco/sgr/sgr_forwomen/pdfs/chp3.pdf.

(continued)

together does not by itself prove that one variable causes the other. This is a perfectly valid point, but the body of research that documents the harmful effects of smoking has developed a convincing response to this argument. First, the best studies are longitudinal (that is, they follow people across time) and have been able to demonstrate a time order between the variables. These studies start with a large number of smokers in good health. If the smokers develop cancer later in the study, cancer must be the dependent variable. That is, in this design, smoking can be the cause of cancer but the reverse cannot be true: Given the time order, getting cancer could not have caused people to start smoking.

By itself, time order doesn't prove that the relationship is causal. It might be that people who smoke have some other characteristic that makes them susceptible to cancer. That is, some other variable might explain both the fact that they smoke and their cancer. For example, it may be that very

nervous or anxious people smoke to "calm their nerves" and are more likely to contract cancer because of their nervousness, not because of their smoking.

The best, most convincing studies use various means to control for the effect of third variables, but it is not possible to eliminate every single possible third variable. Instead, the case against smoking rests on several other arguments, including the fact that people who smoke more are more at risk (e.g., see the graph above) and the fact that people who quit smoking get a health dividend: They are less likely to contract cancer. Finally, tests show that the chemicals in tobacco smoke cause cancer in rats and other animals.

The case against tobacco rests on a strong correlation between smoking and cancer and a number of other relationships, which, taken together, provide overwhelming evidence for a causal relationship.

SUMMARY

This summary is based on the example used throughout the chapter.

1. We began with a question: Is the number of children in dual-wage-earner families related to the number of hours per week husbands devote to housework? We presented the observations in a scattergram (Figure 13.1), and our visual impression was that the variables were associated in a positive direction. The pattern formed by the observation points in the scattergram could be approximated with a straight line; thus, the relationship was roughly linear.

2. Values of Y can be predicted with the freehand regression line, but predictions are more accurate if we use the least-squares regression line, which is the line that best fits the data by minimizing the variation in Y. Using the formula that defines the least-squares regression line ($Y = a + bX$), we found a slope (b) of .69, which indicates that each

additional child (a unit change in X) is accompanied by an increase of .69 hour of housework per week for the husbands. We also predicted, based on this formula, that in a dual-wage-earner family with six children ($X = 6$), husbands would contribute 5.63 hours of housework a week ($Y' = 5.63$ for $X = 6$).

3. Pearson's r is a statistic that measures the overall linear association between X and Y. Our impression from the scattergram of a substantial positive relationship was confirmed by the computed r of .50. We also saw that this relationship yields an r^2 of .25, which indicates that X (number of children) accounts for 25% of the total variation in Y (husband's housework).

4. We acquired a great deal of information about this bivariate relationship. We know the strength and direction of the relationship and have also identified the regression line that best summarizes the

effect of X on Y. We know the amount of change we can expect in Y for a unit change in X. In short, we have a greater volume of more precise information about this association between interval-ratio variables than we ever did about associations between ordinal or nominal variables. This is possible because the data generated by interval-ratio measurement are more precise and flexible than those produced by ordinal or nominal measurement techniques.

SUMMARY OF FORMULAS

Least-squares regression line	13.1	$Y = a + bX$
Definitional formula for the slope	13.2	$b = \dfrac{\Sigma(X - \overline{X})(Y - \overline{Y})}{\Sigma(X - \overline{X})^2}$
Computational formula for the slope	13.3	$b = \dfrac{N\Sigma XY - (\Sigma X)(\Sigma Y)}{N\Sigma X^2 - (\Sigma X)^2}$
Y intercept	13.4	$a = \overline{Y} - b\overline{X}$
Definitional formula for Pearson's r	13.5	$r = \dfrac{\Sigma(X - \overline{X})(Y - \overline{Y})}{\sqrt{[\Sigma(X - \overline{X})^2][\Sigma(Y - \overline{Y})^2]}}$
Computational formula for Pearson's r	13.6	$r = \dfrac{N\Sigma XY - (\Sigma X)(\Sigma Y)}{\sqrt{[N\Sigma X^2 - (\Sigma X)^2][N\Sigma Y^2 - (\Sigma Y)^2]}}$
Coefficient of determination	13.7	$r^2 = \dfrac{\Sigma(Y' - \overline{Y})^2}{\Sigma(Y - \overline{Y})^2}$

GLOSSARY

Coefficient of determination (r^2). The proportion of all variation in Y that is explained by X. Found by squaring the value of Pearson's r.

Conditional means of Y. The mean of all scores on Y for each value of X.

Correlation matrix. A table that shows the correlations between all variables in a data set, including the correlation of a variable with itself.

Explained variation. The proportion of all variation in Y that is attributed to the effect of X.

Least squares. A principle for fitting the regression line. The least-squares regression line touches all conditional means of Y or comes as close to doing so as possible.

Linear relationship. A relationship between two variables in which the observation points (dots) in the scattergram can be approximated with a straight line.

Pearson's r. A measure of association for variables that have been measured at the interval-ratio level.

Regression line. The single, best-fitting straight line that summarizes the relationship between two variables. Regression lines are fitted to the data points by the least-squares criterion, whereby the line touches all conditional means of Y or comes as close to doing so as possible.

Scattergram. Graph that shows the relationship between two variables.

Slope (b). The amount of change in one variable per unit change in the other; b is the symbol for the slope of a regression line.

Total variation. The spread of the Y scores around the mean of Y.

Unexplained variation. The proportion of the total variation in Y that is not accounted for by X.

Y intercept (a). The point where the regression line crosses the Y axis.

Y'. Symbol for predicted score on Y.

MULTIMEDIA RESOURCES

The Wadsworth Sociology Resource Center: Virtual Society at
http://www.thomsonedu.com/sociology

Visit the companion website for the seventh edition of *Statistics: A Tool for Social Research* to access a wide range of student resources. Begin by clicking on the Student Resources section of the book's website to access the following study tools:

■ Basic math review
■ Flash cards

■ Additional chapter problems
■ Internet links
■ Table of random numbers
■ MicroCase and SPSS examples and exercises
■ Hypothesis testing for variables measured at the ordinal level

PROBLEMS

13.1 PS Why does voter turnout vary from election to election? For municipal elections in five different cities, information has been gathered on the percent of registered voters who actually voted, unemployment rate, average years of education for residents, and the percentage of all political ads that used "negative campaigning" (personal attacks and negative portrayals of the opponent's record). For each relationship:

a. Draw a scattergram and a freehand regression line.

b. Compute the slope (*b*) and find the *Y* intercept (*a*). (*HINT: Remember to compute* b *before computing* a. *A computing table such as Table 13.3 is highly recommended.*)

c. State the least-squares regression line and predict the voter turnout for a city in which the unemployment rate was 12%, a city in which the average years of schooling was 11, and an election in which 90% of the ads were negative. (*HINT: You will need to do three separate predictions. All will have voter turnout as the* Y *variable. The first will use unemployment rate as* X, *and* b *will be the slope of the relationship between unemployment rate and turnout. Next, use schooling and then percent negative ads as the* X *variable. Make sure you use the correct values for* b *for the different independent variables.*)

d. Compute *r* and *r²*. (*HINT: A computing table such as Table 13.3 is highly recommended. If you constructed one for computing* b, *you already have most of the quantities you will need to solve for* r.)

e. Describe the strength and direction of the relationships in a sentence or two. Which factor had the strongest effect on turnout?

Turnout and unemployment:

City	Turnout	Unemployment Rate
A	55	5
B	60	8
C	65	9
D	68	9
E	70	10

Turnout and level of education:

City	Turnout	Average Years of School
A	55	11.9
B	60	12.1
C	65	12.7
D	68	12.8
E	70	13.0

Turnout and negative campaigning:

City	Turnout	% of Negative Ads
A	55	60
B	60	63
C	65	55
D	68	53
E	70	48

13.2 SOC This table presents occupational prestige scores for a sample of fathers and their oldest son and oldest daughter.

Family	Father's Prestige	Son's Prestige	Daughter's Prestige
A	80	85	82
B	78	80	77
C	75	70	68
D	70	75	77
E	69	72	60
F	66	60	52
G	64	48	48
H	52	55	57

Analyze the relationship between father's and son's prestige and the relationship between father's and daughter's prestige. For each relationship:

a. Draw a scattergram and a freehand regression line.

b. Compute the slope (b) and find the Y intercept (a).

c. State the least-squares regression line. What prestige score would you predict for a son whose father had a prestige score of 72? What prestige score would you predict for a daughter whose father had a prestige score of 72?

d. Compute r and r^2.

e. Describe the strength and direction of the relationships in a sentence or two. Does the occupational prestige of the father have an impact on his children? Does it have the same impact for daughters as it does for sons?

13.3 GER   The residents of a housing development for senior citizens have completed a survey about how physically active they are and how many visitors they receive each week. Are these two variables related for the 10 cases reported here? Draw a scattergram and compute r and r^2. Find the least-squares regression line. What would be the predicted number of visitors for a person whose level of activity was a 5? How about a person who scored 18 on level of activity?

Case	Level of Activity	Number of Visitors
A	10	14
B	11	12
C	12	10
D	10	9
E	15	8
F	9	7
G	7	10
H	3	15
I	10	12
J	9	2

13.4 PS   The variables below were collected for a random sample of 10 precincts during the last national election. Draw scattergrams and compute r and r^2 for each combination of variables. Write a paragraph interpreting the relationship among these variables.

Precinct	Percent Democrat	Percent Minority	Voter Turnout
A	50	10	56
B	45	12	55
C	56	8	52
D	78	15	60
E	13	5	89
F	85	20	25
G	62	18	64
H	33	9	88
I	25	0	42
J	49	9	36

13.5 SOC/CJ   The table below presents the scores of 10 states on each of six variables: three

State	Crime Rates			Population		
	Homicide	Robbery	Car Theft	Growth	Density	Urban
Maine	1	19	104	3.8	41.7	52.6
New York	5	214	286	5.5	402.7	92.1
Ohio	4	138	344	4.7	277.8	81.2
Iowa	2	37	184	5.4	52.3	45.3
Virginia	6	89	252	14.4	181.5	78.1
Kentucky	5	81	230	9.6	102.3	48.8
Texas	6	145	447	22.8	81.5	84.8
Arizona	7	146	842	40.0	46.7	88.2
Washington	3	99	594	21.1	90.0	83.1
California	6	178	537	13.6	221.2	96.7

Growth: percentage change in population from 1990 to 2000.
Density: population per square mile of land area, 2001.
Urban: percent of population living in metropolitan areas, 2000.
Source: U.S. Bureau of the Census, *Statistical Abstracts of the United States: 2001*. Washington, DC, 2002.

measures of criminal activity and three measures of population structure. Crime rates are number of incidents per 100,000 population as of 2000.

For each combination of crime rate and population characteristic:

a. Draw a scattergram and a freehand regression line.

b. Compute the slope (b) and find the Y intercept (a).

c. State the least-squares regression line. What homicide rate would you predict for a state with a growth rate of -1? What robbery rate would you predict for a state with a population density of 250? What auto theft rate would you predict for a state in which 50% of the population lived in urban areas?

d. Compute r and r^2.

e. Describe the strength and direction of each of these relationships in a sentence or two.

13.6 Data on three variables have been collected for 15 nations. The variables are: fertility rate (average number of children born to each woman), average education for females (expressed as a percentage of average education for men), and maternal mortality (death rate for mothers per 100,000 live births). On education of females, a value of 100 means that men and women have equal levels of education, values over 100 mean that women have more education, and values less than 100 mean that women have less education than men.

Nation	Fertility	Education of Females	Maternal Mortality
Niger	7.2	50	590
Cambodia	4.8	74	470
Guatemala	4.7	86	190
Ghana	4.0	46	210
Bolivia	3.7	60	390
Egypt	3.2	41	170
Dominican Republic	3.0	87	230
Mexico	2.7	96	55
Vietnam	2.5	59	130
Turkey	2.2	49	130
United States	2.1	102	8
China	1.8	60	55
United Kingdom	1.7	102	7
Japan	1.4	98	8
Italy	1.2	99	7

a. Compute r and r^2 for each combination of variables.

b. Summarize these relationships in terms of strength and direction.

13.7 The basketball coach at a small local college believes that his team plays better and scores more points in front of larger crowds. The number of points scored and attendance for all home games last season are reported below. Do these data support the coach's argument?

Game	Points Scored	Attendance
1	54	378
2	57	350
3	59	320
4	80	478
5	82	451
6	75	250
7	73	489
8	53	451
9	67	410
10	78	215
11	67	113
12	56	250
13	85	450
14	101	489
15	99	472

13.8 The table below presents the scores of 15 states on three variables. Compute r and r^2 for each combination of variables. Write a paragraph interpreting the relationship among these three variables.

State	Per Capita Expenditures on Education (in dollars)	Percent High School Graduates*	Rank in Per Capita Income, 2001
Arkansas	1,102	82	48
Colorado	1,339	90	7
Connecticut	1,907	88	1
Florida	1,171	84	25
Illinois	1,621	86	9
Kansas	1,276	88	24
Louisiana	1,159	81	45
Maryland	1,412	86	5
Michigan	1,487	86	18
Mississippi	1,041	80	50
Nebraska	1,194	90	22
New Hampshire	1,262	88	6
North Carolina	1,163	79	32
Pennsylvania	1,223	86	15
Wyoming	1,549	90	20

*Based on percentage of population age 25 and older.

Source: U.S. Bureau of the Census, *Statistical Abstracts of the United States: 2001*. Washington, DC, 2002.

13.9 Fifteen individuals were selected from the 2004 General Social Survey data set, and their scores on five variables are reproduced here. Is there a relationship between occupational prestige and age? Between church attendance and number of children? Between number of children and hours of TV watching? Between age and hours of TV watching? Between age and number of children? Between hours of TV watching and occupational prestige? Appendix G explains the variables and codes in detail.

Occupational Prestige	Number of Children	Age	Church Attendance	Hours of TV per Day
32	3	34	3	1
50	0	41	0	3
17	0	52	7	2
69	3	67	0	5
17	0	40	0	5
52	0	22	2	3
32	3	31	0	4
50	0	23	8	4
19	9	64	1	6
37	4	55	0	2
14	3	66	5	5
51	0	22	6	0
45	0	19	3	7
44	0	21	4	1
46	4	58	2	0

Using *SPSS for Windows* to Produce Pearson's *r*

SPSS DEMONSTRATION 13.1 What Are the Correlates of Occupational Prestige?

How does the social class of our parents shape our ability to rise in the class structure? How much difference does our own effort and preparation for the world of work make? We can investigate these issues by analyzing the relationships among three variables: *prestg80,* or the prestige of the respondent's occupation; *educ,* the respondent's years of schooling; and *paeduc,* the level of education of the respondent's father.

If a person's social class position reflects her or his own preparation for the job market, there should be a strong positive correlation between *educ* and *prestg80.* On the other hand, if it's *who* you know and not *what* you know—if class position is greatly affected by the social class of one's family of origin—then there should be a strong positive relationship between *paeduc* and *prestg80.* By comparing the strength of these bivariate correlations, we may be able to make some judgment about the relative importance of these factors in determining occupational prestige.

To request Pearson's *r,* click **Analyze, Correlate,** and **Bivariate.** The Bivariate **Correlations** window appears with the variable list on the left. Find *educ, paeduc,* and *prestg80* in the list and click the arrow to move them into the **Variables** box. If you wish, you can get descriptive statistics for each variable by clicking the **Options** button and requesting means and standard deviations. Unless you request otherwise, the program will conduct two-tailed tests of significance on the correlations. Note that Spearman's rho (see Chapter 12) is also an option. Click **OK,** and your output should look like this:

Correlations

		HIGHEST YEAR OF SCHOOL COMPLETED	HIGHEST YEAR SCHOOL COMPLETED, FATHER	RS OCCUPATIONAL PRESTIGE SCORE (1980)
HIGHEST YEAR OF SCHOOL COMPLETED	Pearson Correlation	1.000	.418(**)	.561(**)
	Sig. (2-tailed)		.000	.000
	N	1,415	1,059	1,328
HIGHEST YEAR SCHOOL COMPLETED, FATHER	Pearson Correlation	.418(**)	1.000	.239(**)
	Sig. (2-tailed)	.000		.000
	N	1,059	1,059	995
RS OCCUPATIONAL PRESTIGE SCORE (1980)	Pearson Correlation	.561(**)	.239(**)	1.000
	Sig. (2-tailed)	.000	.000	
	N	1,328	995	1,328

*Correlation is significant at the 0.05 level (2-tailed).
**Correlation is significant at the 0.01 level (2-tailed).

The output is in the form of a correlation matrix showing the relationships among all variables, including the correlation of a variable with itself. For each possible relationship, Pearson's *r* is reported in the top row, the results of a test of significance in the second row, and sample size in the third row (*N* varies because not everyone answered every question).

For our questions, we need to focus on the correlations among *prestg80, educ,* and *paeduc.* Each of these relationships is statistically significant ($p < .000$) and positive. The relationship between *prestg80* and *educ* is stronger (0.561) than the relationship between *prestg80* and *paeduc* (0.239). This suggests that preparation for the job market (*educ*) is more important than the relative advantage or disadvantage conferred by one's family of origin (*paeduc*). Success may come more easily to people from an advantaged family background, but preparation (education) matters more. To further disentangle these relationships would require more sophisticated statistics and, fortunately, some will become available in Chapter 14.

SPSS DEMONSTRATION 13.2 Does Gender Affect the Correlates of Occupational Prestige?

Are the pathways to success and higher prestige open to all equally? We can begin to address this question by analyzing the relationships reported in Demonstration 13.1 for men and women separately. If gender has no effect on a person's opportunities for success, *prestg80, paeduc,* and *educ* should be related in the same way for both men and women.

To observe the effect of gender, we will split the GSS sample into two subfiles. Men and women will then be processed separately, and SPSS will produce one correlation matrix for men and another for women.

Click **Data** from the main menu and then click **Split File**. On the **Split File** window, click the button next to "organize output by groups". This will generate separate outputs for men and women. Select *sex* from the variable list and click the arrow to move the variable name into the **Groups Based On** window. Click **OK**, and all procedures requested will be done separately for men and women. To restore the full sample, call up the **Split File** window again and click the **Reset** button. For now, click **Statistics, Correlate,** and **Bivariate** and rerun the correlation matrix for *prestg80, papres80,* and *educ.* The output will look like this:

FOR MALES: Correlations(a)

		HIGHEST YEAR OF SCHOOL COMPLETED	HIGHEST YEAR SCHOOL COMPLETED, FATHER	RS OCCUPATIONAL PRESTIGE SCORE (1980)
HIGHEST YEAR OF SCHOOL COMPLETED	Pearson Correlation	1.000	.382(**)	.571(**)
	Sig. (2-tailed)		.000	.000
	N	656	506	626
HIGHEST YEAR SCHOOL COMPLETED, FATHER	Pearson Correlation	.382(**)	1.000	.217(**)
	Sig. (2-tailed)	.000		.000
	N	506	506	481
RS OCCUPATIONAL PRESTIGE SCORE (1980)	Pearson Correlation	.571(**)	.217(**)	1.000
	Sig. (2-tailed)	.000	.000	
	N	626	481	626

*Correlation is significant at the 0.05 level (2-tailed).
**Correlation is significant at the 0.01 level (2-tailed).
ªRESPONDENTS' SEX = MALE

FOR FEMALES: Correlations(a)

		HIGHEST YEAR OF SCHOOL COMPLETED	HIGHEST YEAR SCHOOL COMPLETED, FATHER	RS OCCUPATIONAL PRESTIGE SCORE (1980)
HIGHEST YEAR OF SCHOOL COMPLETED	Pearson Correlation	1.000	.454(**)	.551(**)
	Sig. (2-tailed)		.000	.000
	N	759	553	702
HIGHEST YEAR SCHOOL COMPLETED, FATHER	Pearson Correlation	.454(**)	1.000	.266(**)
	Sig. (2-tailed)	.000		.000
	N	553	553	514
RS OCCUPATIONAL PRESTIGE SCORE (1980)	Pearson Correlation	.551(**)	.266(**)	1.000
	Sig. (2-tailed)	.000	.000	
	N	702	514	702

*Correlation is significant at the 0.05 level (2-tailed).
**Correlation is significant at the 0.01 level (2-tailed).
ªRESPONDENTS' SEX = FEMALE

The correlations for both groups are statistically significant and positive. The present social class of females is a little more affected by father's prestige (social class of family of origin) but the correlation between prestige and education is almost exactly the same value for both genders. At least for this sample, these results suggest that gender makes little difference for these relationships.

Exercises

13.1 Run the analysis in Demonstration 13.1 again with *income98* rather than *prestg80* as the indicator of social class standing. Use the same independent variables (*paeduc* and *educ*) and see if the patterns are similar to those identified in Demonstration 13.1. (Ignore the fact that *income98* is only ordinal in level of measurement.) Write up your conclusions.

13.2 Run the analysis in Demonstration 13.2 again but substitute *income98* for *prestg80*. Divide the sample by sex as before. Is there a gender difference in the correlates of *income98* as there was with *prestg80?*

13.3 Switch topics entirely and see if you can explain the variation in television viewing habits. What are the correlates of *tvhours?* Choose three or four possible independent variables (including perhaps *age* and *educ*). Write up your results.

Part IV

Multivariate Techniques

Chapter 14 introduces multivariate analytical techniques or statistics that allow us to analyze the relationships among more than two variables at a time. These statistics are extremely useful for probing possible causal relationships among variables and are commonly reported in the professional literature. In particular, the chapter introduces regression analysis, which is the basis for many of the most popular and powerful statistical techniques in use today. These techniques are designed to be used with variables measured at the interval-ratio level of measurement. The mathematics underlying these statistics can become very complicated, so the chapter focuses on the simplest possible applications and stresses interpretation.

14

Partial Correlation and Multiple Regression and Correlation

LEARNING OBJECTIVES By the end of this chapter, you will be able to:

1. Compute and interpret partial correlation coefficients.
2. Find and interpret the least-squares multiple regression equation with partial slopes.
3. Calculate and interpret the multiple correlation coefficient (R^2).
4. Explain the limitations of partial and multiple regression analysis.

14.1 INTRODUCTION

Very few (if any) worthwhile research questions can be answered through a statistical analysis of only two variables. Social science research is, by nature, multivariate and often involves the simultaneous analysis of scores of variables. Some of the most powerful and widely used statistical tools for multivariate analysis are introduced in this chapter. We will cover techniques that are used to analyze causal relationships and to make predictions, both crucial endeavors in any science.

These techniques are based on Pearson's r (see Chapter 13). We will first consider partial correlation analysis, a technique that allows us to examine a bivariate relationship while controlling for a third variable. The second technique involves multiple regression and correlation and allows the researcher to assess the effects, separately and in combination, of more than one independent variable on the dependent variable.

Throughout this chapter, we will focus on research situations involving three variables. This is the least complex application of these techniques, but extensions to situations involving four or more variables are relatively straightforward. To deal efficiently with the computations required by the more complex applications, I refer you to any of the computerized statistical packages probably available on your campus.

14.2 PARTIAL CORRELATION

In Chapter 13, we used Pearson's r to measure the strength and direction of bivariate relationships. To provide an example, we looked at the relationship between husband's contribution to housework (the dependent or Y variable) and number of children (the independent or X variable) for a sample of 12 families. We found a positive relationship of moderate strength ($r = +0.50$) and concluded that husbands tend to make a larger contribution to housework as the number of children increases.

You might wonder, as researchers commonly do, if this relationship always holds true for *all* types of families? For example, might husbands in strongly religious families respond differently from those in less religious families? Would husbands from families that were politically conservative behave differently from husbands in more liberal families? How about more educated husbands? Would they respond differently from less educated husbands? Or perhaps husbands in some ethnic or racial groups would have different responses than husbands in other groups. We can address these kinds of issues by a technique called **partial correlation** in which we observe how the bivariate relationship changes when a third variable like religiosity, education, or ethnicity is introduced. Third variables are often referred to as Z variables or **control variables.**

Partial correlation proceeds by first computing Pearson's r for the bivariate (or zero-order) relationship and then computing the partial (or first-order) correlation coefficient. If the partial correlation coefficient differs from the zero-order correlation coefficient, we conclude that the third variable does affect the bivariate relationship. If, for example, well-educated husbands respond differently to an additional child from less well-educated husbands, the partial correlation coefficient will differ in strength (and, perhaps, in direction) from the bivariate correlation coefficient.

Before considering matters of computation, we'll consider the relationships between the partial and bivariate correlation coefficients and what they might mean. We will examine each of three possible patterns in turn.

Types of Relationships

Direct Relationship.   One possible outcome is that the partial correlation coefficient is essentially the same value as the bivariate coefficient. Imagine, for example, that after we controlled for husband's education, we found a partial correlation coefficient of $+0.49$ compared to the zero-order Pearson's r of $+0.50$. This would mean that the third variable (husband's education) has no effect on the relationship between number of children and husband's hours of housework. In other words, regardless of their education, husbands respond in a similar way to additional children. This outcome is consistent with the conclusion that there is a **direct** or casual relationship (see Figure 14.1) between X and Y and that the third variable (Z) is irrelevant to the investigation. In this case, the researcher would discard that particular Z variable from further consideration but might well run additional tests with other likely control variables (e.g., the religion or ethnicity of the family).

FIGURE 14.1 A DIRECT RELATIONSHIP BETWEEN TWO VARIABLES

$X \longrightarrow Y$

Spurious and Intervening Relationships.   A second possible outcome occurs when the partial correlation coefficient is much weaker than the bivariate correlation, perhaps even dropping to zero. This outcome is consistent with two different relationships between the variables. The first is called a **spurious** relationship: the control variable (Z) is a cause of both the independent (X) and dependent (Y) variable (see Figure 14.2). This outcome would mean that there is no actual relationship between X and Y and that they appear to be related only because both depend on a common cause (Z). Once Z is taken into account, the apparent relationship between X and Y disappears.

What would a spurious relationship look like? Imagine that we controlled for the political ideology of the parents in our 12-family sample and found that the partial correlation coefficient was much weaker than the bivariate Pearson's r. This would indicate that the number of children does not actually change the husband's contribution to housework (that is, the relationship between X and Y is not direct). Rather, political ideology is the mutual cause of both of the other

FIGURE 14.2 A SPURIOUS RELATIONSHIP BETWEEN X AND Y

FIGURE 14.3 AN INTERVENING RELATIONSHIP BETWEEN X AND Y

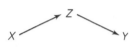

variables: More conservative families would be more likely to follow traditional gender role patterns (in which husbands contribute less to housework) and have more children.

This pattern (partial correlation much weaker than the bivariate correlation) is also consistent with an **intervening** relationship between the variables (see Figure 14.3). In this situation, X and Y are not linked directly but are causally connected through the Z variable. Again, once Z is controlled, the apparent relationship between X and Y disappears.

How can we tell the difference between spurious and intervening relationships? This distinction cannot be made on statistical grounds: Spurious and intervening relationships look exactly the same in terms of statistics. The researcher may be able to distinguish between these two relationships in terms of the time order of the variables (that is, which came first) or theoretical grounds, but not on statistical grounds.

Interaction. I'll mention a final possible relationship between variables even though it *cannot* be detected by partial correlation analysis. This relationship, called **interaction,** occurs when the relationship between X and Y changes markedly under the various values of Z. For example, if we controlled for social class and found that husbands in middle-class families increased their contribution to housework as the number of children increased while husbands in working-class families did just the reverse, we would conclude that there was interaction among these three variables. In other words, there would be a positive relationship between X and Y for one category of Z (middle class) and a negative relationship for the other category (working class), as Figure 14.4 illustrates.

FIGURE 14.4 AN INTERACTIVE RELATIONSHIP BETWEEN X, Y AND Z

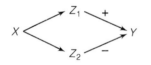

Computing and Interpreting the Partial Correlation Coefficient

Terminology and Formula. The formula for partial correlation requires some new terminology. We will be dealing with more than one bivariate relationship and need to differentiate among them with subscripts. Thus, the symbol r_{yx} will refer to the correlation coefficient between variable Y and variable X, r_{yz} will refer to the correlation coefficient between Y and Z, and r_{xz} to the correlation coefficient between X and Z. Recall that correlation coefficients calculated for bivariate relationships are often referred to as **zero-order correlations.**

Partial correlation coefficients or first-order partials are symbolized as $r_{yx.z}$. The variable to the right of the dot is the control variable. Thus, $r_{yx.z}$ refers to the partial correlation coefficient that measures the relationship between variables X and Y while controlling for variable Z. Here is the formula for the first-order partial:

FORMULA 14.1

$$r_{yx.z} = \frac{r_{yx} - (r_{yz})(r_{xz})}{\sqrt{1 - r_{yz}^2}\,\sqrt{1 - r_{xz}^2}}$$

TABLE 14.1   SCORES ON THREE VARIABLES FOR 12 DUAL-WAGE-EARNER FAMILIES
AND ZERO-ORDER CORRELATIONS

Family	Husband's Housework (Y)	Number of Children (X)	Husband's Years of Education (Z)
A	1	1	12
B	2	1	14
C	3	1	16
D	5	1	16
E	3	2	18
F	1	2	16
G	5	3	12
H	0	3	12
I	6	4	10
J	3	4	12
K	7	5	10
L	4	5	16

TABLE 14.2   ZERO-ORDER CORRELATIONS

	Husband's Housework (Y)	Number of Children (X)	Husband's Years of Education (Z)
Husband's Housework (Y)	1.00	0.50	−0.30
Number of Children (X)		1.00	−0.47
Husband's Years of Education (Z)			1.00

Note that you must first calculate the zero-order coefficients between all possible pairs of variables (variables X and Y, X and Z, and Y and X) before solving this formula.

Computation.   To illustrate the computation of a first-order partial, we will return to the relationship between number of children (X) and husband's contribution to housework (Y) for 12 dual-career families. The zero-order r between these two variables ($r_{yx} = 0.50$) indicated a moderate, positive relationship (as number of children increased, husbands tended to contribute more to housework). Suppose the researcher wanted to investigate the possible effects of husband's education on the bivariate relationship. Table 14.1 presents the original data (from Table 13.1) and the scores of the 12 families on the new variable.

The zero-order correlations, as presented in Table 14.2, indicate that the husband's contribution to housework is positively related to number of children ($r_{yx} = 0.50$), that better-educated husbands tend to do less housework ($r_{yz} = -0.30$), and that families with better-educated husbands have fewer children ($r_{xz} = -0.47$).

Is the relationship between husband's housework and number of children affected by husband's years of education? Substituting the zero-order correlations into Formula 14.1, we would have:

$$r_{yx.z} = \frac{r_{yx} - (r_{yz})(r_{xz})}{\sqrt{1 - r_{yz}^2}\sqrt{1 - r_{xz}^2}}$$

$$r_{yx.z} = \frac{(0.50) - (-0.30)(-0.47)}{\sqrt{1 - (-0.30)^2}\sqrt{1 - (-0.47)^2}}$$

STEP BY STEP	Computing and Interpreting Partial Correlations

Computation

Step 1: Compute Pearson's r for all pairs of variables. Be clear about which variable is independent (X), which is dependent (Y), and which is the control (Z).

Step 2: To compute the partial correlation coefficient, solve Formula 14.1:

$$r_{yx.z} = \frac{r_{yx} - (r_{yz})(r_{xz})}{\sqrt{1 - r_{yz}^2}\,\sqrt{1 - r_{xz}^2}}$$

a. Multiply r_{yz} by r_{xz}.
b. Subtract the value you found in step a from r_{yx}.
c. Square the value of r_{yz}.
d. Subtract the quantity you found in step c from 1.
e. Take the square root of the quantity you found in step d.
f. Square the value of r_{xz}.
g. Subtract the quantity you found in step f from 1.
h. Take the square root of the quantity you found in step g.
i. Multiply the quantity you found in step h by the quantity you found in step e.
j. Divide the quantity you found in step b by the quantity you found in step i.

Interpretation

Step 3: Compare the value of r_{yx} with $r_{yx.z}$. Choose the scenario below that comes closest to describing the relationship between the two values:

a. The partial correlation coefficient is roughly the same value (no less than, say, 0.10 lower) as the bivariate correlation. This suggests that the control variable (Z) has no effect and that the relationship between X and Y is direct.
b. The partial correlation coefficient is much less (say, at least 0.10 lower)—than the bivariate correlation. This suggests that the control variable (Z) changes the relationship between X and Y. The relationship between X and Y is either spurious (Z causes both X and Y) or intervening (X and Y are linked by Z).
c. Be aware that X, Y, and Z may have an interactive relationship in which the relationship between X and Y changes for each category of Z. Partial correlation analysis cannot detect interactive relationships.

$$r_{yx.z} = \frac{(0.50) - (0.14)}{\sqrt{1 - 0.09}\sqrt{1 - 0.22}}$$

$$r_{yx.z} = \frac{0.36}{\sqrt{0.91}\sqrt{0.78}}$$

$$r_{yx.z} = \frac{0.36}{(0.95)(0.88)}$$

$$r_{yx.z} = \frac{0.36}{0.84}$$

$$r_{yx.z} = .43$$

Interpretation. The first-order partial ($r_{yx.z} = 0.43$), which measures the strength of the relationship between husband's housework (Y) and number of children (X) while controlling for husband's education (Z), is lower in value than the zero-order coefficient ($r_{yx} = 0.50$), but the difference in the two values is not great. This result suggests a direct relationship between variables X and Y. That is, when controlling for husband's education, the statistical relationship between husband's housework and number of children is essentially unchanged. Regardless of education, husband's hours of housework increase with the number of children.

Our next step in statistical analysis would probably be to select another control variable. The more the bivariate relationship retains its strength across a series of controls for third variables (Zs), the stronger the evidence for a direct relationship between X and Y. *(For practice in computing and interpreting partial correlation coefficients, see problems 14.1 to 14.3.)*

14.3 MULTIPLE REGRESSION: PREDICTING THE DEPENDENT VARIABLE

In Chapter 13, the least-squares regression line was introduced as a way of describing the overall linear relationship between two interval-ratio variables and of predicting scores on Y from scores on X. This line was the best-fitting line to summarize the bivariate relationship and was defined by the formula:

FORMULA 14.2

$$Y = a + bX$$

where a = the Y intercept
b = the slope

The least-squares regression line can be modified to include (theoretically) any number of independent variables. This technique is called **multiple regression.** For ease of explication, we will confine our attention to the case involving two independent variables. The least-squares multiple regression equation for two independent variables is:

FORMULA 14.3

$$Y = a + b_1X_1 + b_2X_2$$

where b_1 = the partial slope of the linear relationship between the first independent variable and Y.

b_2 = the partial slope of the linear relationship between the second independent variable and Y.

Some new notation and some new concepts are introduced in this formula. First, while the dependent variable is still symbolized as Y, the independent variables are differentiated by subscripts. Thus, X_1 identifies the first independent variable and X_2 the second. The symbol for the slope (b) is also subscripted to identify the independent variable with which it is associated.

Computing the Regression Coefficients

Partial Slopes. A major difference between the multiple and bivariate regression equations concerns the slopes (bs). In the case of multiple regression, the bs are called **partial slopes,** and they show the amount of change in Y for a unit change in one independent variable while controlling for the effects of the other independent variable(s) in the equation. The partial slopes are thus analogous to partial correlation coefficients and represent the direct effect of the associated independent variable on Y.

Computing Partial Slopes. We use Formula 14.4 and Formula 14.5 to determine the partial slopes for the independent variables.[1] The subscripts attached

---

[1]Partial slopes can be computed from zero-order slopes but Formulas 14.4 and 14.5 are somewhat easier to use.

to the symbols in the formula (b, s, r) identify the variables: "Y" is the dependent variable, "1" refers to the first independent variable, and "2" represents the second independent variable.

FORMULA 14.4

$$b_1 = \left(\frac{s_y}{s_1}\right)\left(\frac{r_{y1} - r_{y2}r_{12}}{1 - r_{12}^2}\right)$$

FORMULA 14.5

$$b_2 = \left(\frac{s_y}{s_2}\right)\left(\frac{r_{y2} - r_{y1}r_{12}}{1 - r_{12}^2}\right)$$

where b_1 = the partial slope of X_1 on Y.

   b_2 = the partial slope of X_2 on Y.

   s_y = the standard deviation of Y

   s_1 = the standard deviation of the first independent variable (X_1)

   s_2 = the standard deviation of the second independent variable (X_2)

   r_{y1} = the bivariate correlation between Y and X_1

   r_{y2} = the bivariate correlation between Y and X_2

   r_{12} = the bivariate correlation between X_1 and X_2

To illustrate the computation of the partial slopes, we will assess the combined effects of number of children (X_1) and husband's education (X_2) on husband's contribution to housework. All the relevant information, calculated from Table 14.1, appears here:

Husband's Housework	Number of Children	Husband's Education
$\overline{Y} = 3.3$	$\overline{X}_1 = 2.7$	$\overline{X}_2 = 13.7$
$s_y = 2.1$	$s_1 = 1.5$	$s_2 = 2.6$

Zero-Order Correlations
$r_{y1} = 0.50$
$r_{y2} = -0.30$
$r_{12} = -0.47$

The partial slope for the first independent variable, number of children or X_1, is:

$$b_1 = \left(\frac{s_y}{s_1}\right)\left(\frac{r_{y1} - r_{y2}r_{12}}{1 - r_{12}^2}\right)$$

$$b_1 = \left(\frac{2.1}{1.5}\right)\left(\frac{0.50 - (-0.30)(-0.47)}{1 - (-0.47)^2}\right)$$

$$b_1 = (1.4)\left(\frac{0.50 - 0.14}{1 - 0.22}\right)$$

$$b_1 = (1.4)\left(\frac{0.36}{0.78}\right)$$

$$b_1 = (1.4)(0.46)$$

$$b_1 = 0.65$$

For the second independent variable, husband's education or X_2, the partial slope is:

$$b_2 = \left(\frac{s_y}{s_2}\right)\left(\frac{r_{y2} - r_{y1}r_{12}}{1 - r_{12}^2}\right)$$

$$b_2 = \left(\frac{2.1}{2.6}\right)\left(\frac{-0.30 - (0.50)(-0.47)}{1 - (-0.47)^2}\right)$$

$$b_2 = (0.81)\left(\frac{-0.30 - (-0.24)}{1 - 0.22}\right)$$

$$b_2 = (0.81)\left(\frac{-0.30 + 0.24}{0.78}\right)$$

$$b_2 = (0.81)\left(\frac{-0.06}{0.78}\right)$$

$$b_2 = (0.81)(-0.08)$$

$$b_2 = -0.07$$

Finding the Y Intercept.   Once we've determined the partial slopes for both independent variables, we can find the Y intercept (a). Note that a is calculated from the mean of the dependent variable (symbolized as $\bar{Y}$) and the means of the two independent variables ($\bar{X}_1$ and $\bar{X}_2$):

FORMULA 14.6

$$a = \bar{Y} - b_1\bar{X}_1 - b_2\bar{X}_2$$

Substituting the proper values for the example problem at hand, we have:

$$a = \bar{Y} - b_1\bar{X}_1 - b_2\bar{X}_2$$

$$a = 3.3 - (0.65)(2.7) - (-0.07)(13.7)$$

$$a = 3.3 - (1.8) - (-1.0)$$

$$a = 3.3 - 1.8 + 1.0$$

$$a = 2.5$$

Regression and Prediction

The Least Squares Multiple Regression Line and Predicting Y'.   For our example problem, the full least-squares multiple regression equation is:

$$Y = a + b_1X_1 + b_2X_2$$

$$Y = 2.5 + (0.65)X_1 + (-0.07)X_2$$

As was the case with the bivariate regression line, we can use this formula to predict scores on the dependent variable from scores on the independent variables. For example, what would be our best prediction of husband's housework (Y') for a family of four children ($X_1 = 4$) where the husband had completed 11 years of schooling ($X_2 = 11$)? Substituting these values into the least-squares formula, we would have:

$$Y' = 2.5 + (0.65)(4) + (-0.07)(11)$$

$$Y' = 2.5 + 2.6 - 0.8$$

$$Y' = 4.3$$

Our prediction would be that this husband would contribute 4.3 hours per week to housework. This prediction is, of course, a kind of "educated guess," which is unlikely to be perfectly accurate. However, we will make fewer errors

---

STEP BY STEP	Computing the Regression Coefficients (*b* and *a*)

Computing Partial Slope (*b*)

Step 1: To compute the partial slope of the first independent variable, solve Formula 14.4:

$$b_1 = \left(\frac{s_y}{s_1}\right)\left(\frac{r_{y1} - r_{y2}r_{12}}{1 - r_{12}^2}\right)$$

a. Divide the standard deviation of Y (s_y) by the standard deviation of the first independent variable (s_1).
b. Multiply r_{y2} by r_{12}.
c. Subtract the value you found in step b from r_{y1}.
d. Square r_{12}.
e. Subtract the quantity you found in step d from 1.
f. Divide the quantity you found in step c by the value you found in step e.
g. Multiply the quantity you found in step f by the quantity you found in step a.

Step 2: To compute the partial slope of the second independent variable, solve Formula 14.5:

$$b_2 = \left(\frac{s_y}{s_2}\right)\left(\frac{r_{y2} - r_{y1}r_{12}}{1 - r_{12}^2}\right)$$

a. Divide the standard deviation of Y (s_y) by the standard deviation of the second independent variable (s_2).
b. Multiply r_{y1} by r_{12}.
c. Subtract the value you found in step b from r_{y2}.
d. Square r_{12}.
e. Subtract the quantity you found in step d from 1.
f. Divide the quantity you found in step c by the value you found in step e.
g. Multiply the quantity you found in step f by the quantity you found in step a.

Computing the *Y* intercept (*a*)

Step 3: To compute *a*, solve Formula 14.6:
$$a = \bar{Y} - b_1\bar{X}_1 - b_2\bar{X}_2$$

a. Multiply the slope (b_1) by the mean ($\bar{X}_1$) of the first independent variable.
b. Multiply the slope (b_2) by the mean ($\bar{X}_2$) of the second independent variable.
c. Subtract the quantity you found in step b from the quantity you found in step a.
d. Subtract the quantity you found in step c from the mean of Y ($\bar{Y}$).

---

of prediction using the least-squares line (and, thus, incorporating information from the independent variables) than we would using any other method of prediction (assuming, of course, that there is a linear association between the independent and the dependent variables). *(For practice in predicting Y scores and in computing slopes and the Y intercept, see problems 14.1 to 14.6.)*

14.4 MULTIPLE REGRESSION: ASSESSING THE EFFECTS OF THE INDEPENDENT VARIABLES

The least-squares multiple regression equation (Formula 14.3) is used to isolate the separate effects of the independent variables and to predict scores on the dependent variable. However, in many situations, using this formula to determine the relative importance of the various independent variables will be awkward—especially when the independent variables differ in terms of units of measurement (e.g., number of children versus years of education). When the independent variables are measured in different units, a comparison of the partial slopes will not necessarily tell us which independent variable has the strongest effect and is thus the most important. Comparing the partial slopes of variables that differ in units of measurement is like comparing apples and oranges.

We can make it easier to compare the effects of the independent variables by converting all variables in the equation to a common scale or unit of measurement. This will eliminate variations in the values of the partial slopes that are solely a function of differences in units of measurement. We can, for example, standardize the independent variables by changing their scores to Z scores, as we did in Chapter 5. Each distribution of scores would then have a mean of 0 and a standard deviation of 1, making comparisons between the independent variables much more meaningful.

Computing the Standardized Regression Coefficients

Beta-Weights.   To standardize the variables to the normal curve, we could actually convert all scores into the equivalent Z scores and then recompute the slopes and the Y intercept. This would require a good deal of work. Fortunately, a shortcut is available for computing the slopes of the standardized scores directly. These **standardized partial slopes** are called **beta-weights** and are symbolized b^*. The beta-weights show the amount of change in the standardized scores of Y for a one-unit change in the standardized scores of each independent variable while controlling for the effects of all other independent variables.

Formulas and Computation for Beta-Weights. When we have two independent variables, we find the beta-weight for each by using Formula 14.7 and Formula 14.8:

FORMULA 14.7

$$b_1^* = b_1\left(\frac{s_1}{s_y}\right)$$

FORMULA 14.8

$$b_2^* = b_2\left(\frac{s_2}{s_y}\right)$$

We can now compute the beta-weights for our sample problem to see which of the two independent variables (number of children and husband's education) has the stronger effect on the dependent variable (husband's hours of housework). For the first independent variable, number of children (X_1):

$$b_1^* = b_1\left(\frac{s_1}{s_y}\right)$$

$$b_1^* = (0.65)\left(\frac{1.5}{2.1}\right)$$

$$b_1^* = (0.65)(0.71)$$

$$b_1^* = 0.46$$

For the second independent variable, husband's years of education (X_2):

$$b_2^* = b_2\left(\frac{s_2}{s_y}\right)$$

$$b_2^* = (-0.07)\left(\frac{2.6}{2.1}\right)$$

$$b_2^* = (-0.07)(1.24)$$

$$b_2^* = -0.09$$

STEP BY STEP	Computing Beta-Weight (b^*)

Step 1: To compute the standardized partial slope of the first independent variable (b_1^*), solve Formula 14.7:

$$b_1^* = b_1\left(\frac{s_1}{s_y}\right)$$

a. Divide the standard deviation of the first independent variable (s_1) by the standard deviation of the dependent variable (s_y).

b. Multiply the quantity you found in step a by b_1.

Step 2: To compute the standardized partial slope of the second independent variable (b_2^*), solve Formula 14.8:

$$b_2^* = b_2\left(\frac{s_2}{s_y}\right)$$

a. Divide the standard deviation of the second independent variable (s_2) by the standard deviation of the dependent variable (s_y).

b. Multiply the quantity you found in step a by b_2.

Interpreting Beta-Weights. Comparing the value of the beta-weights for our example problem, we see that number of children ($b_1^* = 0.46$) has a stronger effect than husband's education ($b_2^* = -0.09$) on husband's housework. Furthermore, the net effect (after controlling for the effect of education) of the first independent variable is positive while the net effect of the second independent variable (after controlling for the effect of number of children) is negative. We can conclude that number of children is the more important of the two variables and that husband's contribution to housework increases as number of children increases regardless of husband's years of education.

The Standardized Least-Squares Regression Line. Using standardized scores, we can write the least-squares regression equation:

FORMULA 14.9
$$Z_y = a_z + b_1^* Z_1 + b_2^* Z_2$$

where Z indicates that all scores have been standardized to the normal curve

The standardized regression equation can be further simplified by dropping the term for the Y intercept because this term will always be zero when scores have been standardized. Remember that a is the point where the regression line crosses the Y axis and is equal to the mean of Y when all independent variables equal 0. By substituting 0 for all independent variables in Formula 14.6, we can see this relationship:

$$a = \overline{Y} - b_1\overline{X_1} - b_2\overline{X_2}$$
$$a = \overline{Y} - b_1(0) - b_2(0)$$
$$a = \overline{Y}$$

Because the mean of any standardized distribution of scores is zero, the mean of the standardized Y scores will be zero and the Y intercept will also be zero ($a = \overline{Y} = 0$). Thus, Formula 14.9 simplifies to:

FORMULA 14.10
$$Z_y = b_1^* Z_1 + b_2^* Z_2$$

The standardized regression equation, with beta-weights noted, would be:

$$Z_y = (0.46)Z_1 + (-0.09)Z_2$$

and it is immediately obvious that the first independent variable has a much stronger direct effect on Y than the second independent variable does.

Summary. Multiple regression analysis permits the researcher to summarize the linear relationship among two or more independents and a dependent variable. The unstandardized regression equation (Formula 14.3) permits values of Y to be predicted from the independent variables in the original units of the variables. The standardized regression equation (Formula 14.10) allows the researcher to easily assess the relative importance of the various independent variables by comparing the beta-weights. *(For practice in computing and interpreting beta-weights, see any of the problems at the end of this chapter. It is probably a good idea to start with problem 14.1, which has the smallest data set and the least complex computations.)*

14.5 MULTIPLE CORRELATION

We use the multiple regression equations to disentangle the separate direct effects of each independent variable on the dependent. Using **multiple correlation** techniques, we can also ascertain the *combined* effects of all independents on the dependent variable. We do so by computing the **multiple correlation coefficient (R)** and the **coefficient of multiple determination (R^2).** The value of the latter statistic represents the proportion of the variance in Y that is explained by all the independent variables combined.

In terms of zero-order correlation, we have seen that number of children (X_1) explains a proportion of .25 of the variance in Y ($r_{y1}^2 = (.50)^2 = .25$) by itself and that husband's education explains a proportion of .09 of the variance in Y ($r_{y2}^2 = (-.30)^2 = .09$). The zero-order correlations cannot be simply added together to ascertain their combined effect on Y, because the two independents are also correlated with each other. Therefore, they will "overlap" in their effects on Y and explain some of the same variance. Formula 14.11 eliminates this overlap:

FORMULA 14.11

$$R^2 = r_{y1}^2 + r_{y2.1}^2(1 - r_{y1}^2)$$

where R^2 = the coefficient of multiple determination

r_{y1}^2 = the zero order correlation between Y and X_1, the quantity squared

$r_{y2.1}^2$ = the partial correlation of Y and X_2, while controlling for X_1, the quantity squared

The first term in this formula (r_{y1}^2) is the coefficient of determination for the bivariate relationship between Y and X_1. It represents the amount of variation in Y explained by X_1 by itself. To this quantity we add the amount of the variation remaining in Y (given by $1 - r_{y1}^2$) that can be explained by X_2 after the effect of X_1 is controlled ($r_{y2.1}^2$). Basically, Formula 14.11 allows X_1 to explain as much of Y as it can and then adds in the effect of X_2 after X_1 is controlled (thus eliminating the "overlap" in the variance of Y that X_1 and X_2 have in common).

Computing and Interpreting R and R^2. To observe the combined effects of number of children (X_1) and husband's years of education (X_2) on husband's housework (Y), we need two quantities. The correlation between X_1 and Y ($r_{y1} = .50$) has already been found. Before we can solve Formula 14.11, we

STEP BY STEP	Computing R^2

Step 1: Before finding R^2, we must first find the partial correlation coefficient between Y and X_2 while controlling for X_1. See the earlier step-by-step box (p. 334) on computing partial correlations.

Step 2: To compute R^2, solve Formula 14.11:

$$R^2 = r_{y1}^2 + r_{y2.1}^2(1 - r_{y1}^2)$$

a. Square the value of r_{y1}.
b. Subtract the quantity you found in step a from 1.
c. Square the value of $r_{y2.1}$.
d. Multiply the quantity you found in step c by the quantity you found in step b.
e. Add the quantity you found in step d to the quantity you found in step a.

must first calculate the partial correlation of Y and X_2 while controlling for X_1 ($r_{y2.1}$):

$$r_{y2.1} = \frac{r_{y2} - (r_{y1})(r_{12})}{\sqrt{1 - r_{y1}^2} \sqrt{1 - r_{12}^2}}$$

$$r_{y2.1} = \frac{(-0.30) - (0.50)(-0.47)}{\sqrt{1 - (0.50)^2} \sqrt{1 - (-0.47)^2}}$$

$$r_{y2.1} = \frac{(-0.30) - (-0.24)}{\sqrt{0.75} \sqrt{0.78}}$$

$$r_{y2.1} = \frac{-0.06}{0.77}$$

$$r_{y2.1} = -0.08$$

Formula 14.11 can now be solved for our sample problem:

$$R^2 = r_{y1}^2 + r_{y2.1}^2(1 - r_{y1}^2)$$

$$R^2 = (0.50)^2 + (-0.08)^2(1 - 0.50^2)$$

$$R^2 = 0.25 + (0.006)(1 - 0.25)$$

$$R^2 = 0.25 + 0.005$$

$$R^2 = 0.255$$

The first independent variable (X_1), number of children, explains 25% of the variance in Y by itself. To this total, the second independent (X_2), SES, adds only a half a percent, for a total explained variance of 25.5% in the dependent variable. *(For practice in computing and interpreting R and R^2, see any of the problems at the end of this chapter. I suggest starting with problem 14.1, which has the smallest data set and the least complex computations.)*

14.6 THE LIMITATIONS OF MULTIPLE REGRESSION AND CORRELATION

Partial correlation and multiple regression and correlation are very powerful tools for analyzing the interrelationships among three or more variables. The techniques presented in this chapter permit the researcher to predict scores on one variable from two or more other variables, to distinguish between

Application 14.1

The following table presents information on three variables for a small sample of eight nations. We will take abortion rate as the dependent variable and examine its relationship with two variables: One measures women's status and power and the other measures religiosity. In this analysis, we will focus on R^2 and the beta-weights only.

Our expectation is that the rate of abortion will have a positive relationship with women's status ("abortion rate will increase as women's power and freedom of choice increase") and a negative relationship with religiosity ("the greater the strength of traditional value systems, including religion, the lower the rate of abortion").

Nation	Abortion Rate[1] (Y)	Women's Status[2] (X_1)	Religiosity[3] (X_2)
Canada	165	.76	74
Chile	10	.45	93
Denmark	400	.71	48
Germany	208	.75	67
Italy	389	.54	82
Japan	379	.52	87
United Kingdom	207	.67	67
United States	428	.74	90
Mean	273.25	.64	76.00
Std. deviation	138.86	.11	14.11

[1] Number of abortions per 1,000 live births.

[2] This variable is a measure of the status of women relative to men across a variety of areas including politics and education. The higher the score, the higher the status of women.

[3] The percentage of respondents who say they pray "at least sometimes."

The zero-order correlations for these variables appear in the correlation matrix below.

	Abortion Rate	Women's Status	Religiosity
Abortion Rate	1.00	0.22	−0.17
Women's Status		1.00	−0.56
Religiosity			1.00

Consistent with our expectations, there is a positive but fairly weak relationship between abortion rate and women's status. The relationship between abortion rate and religiosity is negative, as expected, and also weak. The relationship between the two independent variables (women's status and religiosity) is moderate to strong and negative, indicating that women in more religious nations have lower status.

We find the combined effect of women's status and religiosity on abortion rate by computing R^2:

$$R^2 = r_{y1}^2 + r_{y2.1}^2(1 - r_{y1}^2)$$
$$R^2 = (0.22)^2 + (-0.06)^2(1 - .22^2)$$
$$R^2 = 0.05 + (0.0036)(0.95)$$
$$R^2 = 0.05 + (0.003)$$
$$R^2 = 0.053$$

By itself, women's status explains 5% of the variance in abortion rates. To this, religiosity adds another .3%—a very minimal amount—for a total of 5.3%. This leaves about 95% of the variance unexplained, a sizeable proportion but not unusually large in social science research.

To assess the separate effects of the two independent variables, we must calculate the beta-weights. We need values for the unstandardized partial slopes to compute beta-weights, and we will simply report the values as 271.97 for X_1 and −0.67 for X_2.

For the first independent variable (women's status):

$$b_1^* = b_1\left(\frac{s_1}{s_y}\right)$$

$$b_1^* = (271.97)\left(\frac{0.11}{138.86}\right)$$

$$b_1^* = (271.97)(0.00079)$$

$$b_1^* = 0.22$$

(continued next page)

independent variables in terms of the importance of their direct effects on a dependent, and to ascertain the total effect of a set of independent variables on a dependent variable. They are some of the most powerful and flexible statistical tools available to social science researchers.

Application 14.1 *(continued)*

For the second independent variable (religiosity):

$$b_2^* = b_2 \left(\frac{s_2}{s_y} \right)$$

$$b_2^* = (-0.67) \left(\frac{14.11}{138.86} \right)$$

$$b_2^* = (-0.67)(0.102)$$

$$b_2^* = -0.07$$

Recall that the beta-weights show the effect of each independent variable on the dependent variable while controlling for the other independent variables in the equation. In this case, women's status has the stronger effect and the relationship is positive. The effect of religiosity is negative.

In summary, for these eight nations, abortion rate has weak relationships with women's status and with religiosity. Taken together, the independent variables explain 5.3% of the variation in abortion rates. As expected, abortion rates increase as women gain status and power and decrease as religiosity increases.

Powerful tools are not cheap. They demand high-quality data, and measurement at the interval-ratio level is often difficult to accomplish. Furthermore, these techniques assume that the interrelationships among the variables follow a particular form. First, they assume that each independent variable has a linear relationship with the dependent variable. Scattergrams enable us to quickly check how well a given set of variables meets this assumption.

Second, the techniques presented in this chapter assume that the effects of the independent variables are *additive*. This means that we must assume that the best prediction of the dependent variable (Y) can be obtained by simply adding up the scores of the independent variables, as reflected by the plus signs in Formula 14.3. Not all combinations of variables will conform to this assumption. For example, some variables have *interactive* relationships with each other. These are relationships in which some of the scores of the variables combine in unusual, nonadditive ways. For example, consider a survey of neighborhoods that found positive and linear bivariate relationships between poverty (X_1), racial residential segregation (X_2), and crime rates (Y). When combined in regression analysis, however, certain combinations of scores on the independent variables (for example, high levels of poverty combined with high levels of segregation) produced especially high crimes rates. In other words, poverty and segregation interact with each other, and neighborhoods that were high in poverty *and* segregation had much higher crime rates than areas that were impoverished but not racially segregated or areas that were segregated but not impoverished.

If there is interaction among the variables, we cannot accurately estimate or predict the dependent variable by simply adding the effects of the independent variables. The techniques for identifying and handling interaction among variables are beyond the scope of this text.

Third, the techniques of multiple regression and correlation assume that the independent variables are uncorrelated with each other. Strictly speaking, this condition means that the zero-order correlation among all pairs of independents should be zero. But in practice, we act as if this assumption has been met if the intercorrelations among the independents are low.

STATISTICS IN EVERYDAY LIFE: Beta-Weights and Baseball

Baseball is paradise for statistics junkies. Virtually every aspect of what happens on the field of play is recorded, and there are decades of data—hits, runs, stolen bases, strikeouts, assists, passed balls, and so on—for fans to analyze and argue about. Who has the most hits against left-handed pitchers in August for teams no longer in the pennant race? What team has the highest winning percentage in May? What pitcher older than 30 has the highest ratio of strikes to balls when pitching on the East Coast in night games? No matter how specific or unusual the question, there is almost certainly data to provide an answer.

However, by themselves, baseball facts are not particularly useful or interesting (unless you really are a statistics junkie). How can these data be put to work? Can baseball facts be used to improve the performance of a team or are they mere curiosities suitable only for arcane discussions among baseball aficionados?

According to *Moneyball,* a recent best seller by Michael Lewis, the answer to these questions is a resounding yes. Lewis chronicles the efforts of Billy Beane, General Manager of the Oakland A's, to apply principles of scientific management to the sport. Beane's challenge was that the A's had one of the smallest budgets and lowest payrolls in the sport and simply could not afford star players. How could he field a competitive team when there was no money to hire the best players?

Beane succeeded by discarding conventional baseball wisdom and, along with some statistical analysts he hired, finding new ways to evaluate players. He was able to identify talents that were undervalued in the marketplace and focused on players who had those "cheaper" talents. How did he do this? According to Lewis, Beane and his team of analysts ran a series of regression analyses with won-lost record as the dependent variable and every measure of offensive, defensive, and pitching prowess as independent variables. They found that some hallowed statistics (e.g., batting average) had weaker relationships (lower beta-weights) with winning than expected, after other variables had been entered in the equation. They also discovered some measures of performance that had surprisingly strong relationships (higher beta-weights) with winning, even though they were not widely regarded as important by baseball people. Among these was the percentage of times a batter got on base regardless of method, including not only hits but also walks, fielder's choices, hit by pitch, and errors.

Beane and his management team used the findings of their regression analyses to identify a number of high school, college, and minor league players who ranked high on their new measures of performance (such as "on-base percentage") but did not have impressive "conventional" statistics (like batting average) to which all the other teams paid attention. Furthermore, some of these players did not have the physique (broad shoulders and bulging biceps) commonly associated with great athletic promise. In other words, these players were undervalued in the baseball marketplace, and Beane was able to sign them for comparatively low salaries, staying within his limited budget while still giving his team a chance to win.

Did it work? The A's continued a long-term losing streak during Beane's first two years (1997–98) but, in the next seven years, they won 58% of their games, finished first in their division three times, and were second in three other years. They have not been to the World Series under Beane's leadership, but they have made the postseason playoffs four times in the seven seasons from 1999 to 2005. This is a remarkable record for a small-budget team and a convincing demonstration of the power of statistical analysis for sports and for everyday life.

To the extent that these assumptions are violated, the regression coefficients (especially partial and standardized slopes) and the coefficient of multiple determination (R^2) become less and less trustworthy and the techniques less and less useful. A careful inspection of the bivariate scattergrams will help to assess the reasonableness of the assumptions of partial and multiple correlation and regression.

Finally, we should note that we have covered only the simplest applications of partial correlation and multiple regression and correlation. In terms of logic and interpretation, the extensions to situations involving more independent variables are relatively straightforward. However, the computations for these situations are extremely complex. If you are faced with a situation involving more than three variables, turn to one of the computerized statistical packages that are commonly available on college campuses (e.g., SPSS and MicroCase). These programs require minimal computer literacy and can handle complex calculations in, literally, the blink of an eye. Efficient use of these packages will enable you to avoid drudgery and will free you to do what social scientists everywhere enjoy doing most: pondering the meaning of your results and, by extension, the nature of social life.

SUMMARY

1. Partial correlation involves controlling for the effect of a third variable (Z) on a bivariate relationship. Partial correlations permit the detection of direct and spurious or intervening relationships among X, Y, and Z.
2. Multiple regression includes statistical techniques by which we can make predictions of the dependent variable from more than one independent variable (by using partial slopes and the multiple regression equation) and by which we can disentangle the relative importance of the independent variables (by using standardized partial slopes).
3. The coefficient of multiple determination (R^2) summarizes the combined effects of all independent

variables on the dependent variable in terms of the proportion of the total variation in Y that is explained by all of the independent variables.
4. Partial correlation and multiple regression and correlation are some of the most powerful tools available to the researcher and demand high-quality measurement and relationships among the variables that are linear and noninteractive. Further, correlations among the independents must be low (preferably zero). Although the price is high, these techniques pay considerable dividends in the volume of precise and detailed information they generate about the interrelationships among the variables.

SUMMARY OF FORMULAS

Partial correlation coefficient	14.1	$r_{yx.z} = \dfrac{r_{yx} - (r_{yz})(r_{xz})}{\sqrt{1 - r_{yz}^2}\,\sqrt{1 - r_{xz}^2}}$
Least-squares regression line (bivariate)	14.2	$Y = a + bX$
Least-squares multiple regression line	14.3	$Y = a + b_1 X_1 + b_2 X_2$
Partial slope for X_1	14.4	$b_1 = \left(\dfrac{s_y}{s_1}\right)\left(\dfrac{r_{y1} - r_{y2} r_{12}}{1 - r_{12}^2}\right)$
Partial slope for X_2	14.5	$b_2 = \left(\dfrac{s_y}{s_2}\right)\left(\dfrac{r_{y2} - r_{y1} r_{12}}{1 - r_{12}^2}\right)$
Y intercept	14.6	$a = \overline{Y} - b_1 \overline{X}_1 - b_2 \overline{X}_2$
Standardized partial slope (bet.a-weight) for X_1	14.7	$b_1^* = b_1\left(\dfrac{s_1}{s_y}\right)$

Standardized partial slope (beta-weight) for X_2	14.8	$b_2^* = b_2 \left(\dfrac{s_2}{s_y} \right)$
Standardized least-squares regression line	14.9	$Z_y = a_z + b_1^* Z_1 + b_2^* Z_2$
Standardized least-squares regression line (simplified)	14.10	$Z_y = b_1^* Z_1 + b_2^* Z_2$
Coefficient of multiple determination	14.11	$R^2 = r_{y1}^2 + r_{y2.1}^2 (1 - r_{y1}^2)$

GLOSSARY

Beta-weights (b^*). See standardized partial slope.

Coefficient of multiple determination (R^2). A statistic that equals the total variation explained in the dependent variable by all independent variables combined.

Control variable. A "third variable" (Z) that might affect a bivariate relationship.

Direct relationship. A multivariate relationship in which the control variable has no effect on the bivariate relationship.

Interaction. A multivariate relationship in which a bivariate relationship changes across the categories of the control variable.

Intervening relationship. A multivariate relationship in which a bivariate relationship becomes substantially weaker after controlling for a third variable. The independent (X) and dependent (Y) variables are linked primarily through the control variable (Z).

Multiple correlation. A multivariate technique for examining the combined effects of more than one independent variable on a dependent variable.

Multiple correlation coefficient (R). A statistic that indicates the strength of the correlation between a dependent variable and two or more independent variables.

Multiple regression. A multivariate technique that breaks down the separate effects of the independent variables on the dependent variable; used to make predictions of the dependent variable.

Partial correlation. A multivariate technique for examining a bivariate relationship while controlling for other variables.

Partial correlation coefficient. A statistic that shows the relationship between two variables while controlling for other variables; $r_{yx.z}$ is the symbol for the partial correlation coefficient when controlling for one variable.

Partial slope. In a multiple regression equation, the slope of the relationship between a particular independent variable and the dependent variable while controlling for all other independents in the equation.

Spurious relationship. A multivariate relationship in which a bivariate relationship becomes substantially weaker after controlling for a third variable. The independent (X) and dependent (Y) variables are not causally linked. Rather, both are caused by the control variable (Z).

Standardized partial slope (beta-weight). The slope of the relationship between a particular independent variable and the dependent variable when all scores have been normalized.

Zero-order correlation. Correlation coefficient for bivariate relationships.

MULTIMEDIA RESOURCES

The Wadsworth Sociology Resource Center: Virtual Society at
http://www.thomsonedu.com/sociology

Visit the companion website for the seventh edition of *Statistics: A Tool for Social Research* to access a wide range of student resources. Begin by clicking on the Student Resources section of the book's website to access the following study tools:

- Basic math review
- Flash cards

- Additional chapter problems
- Internet links
- Table of random numbers
- MicroCase and SPSS examples and exercises
- Hypothesis testing for variables measured at the ordinal level

PROBLEMS

14.1 PS Problem 13.1 gave data regarding voter turnout in five cities. For the sake of convenience, the data for three of the variables are presented again here along with descriptive statistics and zero-order correlations.

City	Turnout	Unemployment Rate	% Negative Ads
A	55	5	60
B	60	8	63
C	65	9	55
D	68	9	53
E	70	10	48
Mean =	63.6	8.2	55.8
s =	5.5	1.7	5.3

	Unemployment Rate	% Negative Ads
Turnout	0.95	−0.87
Unemployment Rate		−0.70

a. Compute the partial correlation coefficient for the relationship between turnout (Y) and unemployment (X) while controlling for the effect of negative advertising (Z). What effect does this control variable have on the bivariate relationship? Is the relationship between turnout and unemployment direct? (*HINT: Use Formula 14.1 and see Section 14.2.*)

b. Compute the partial correlation coefficient for the relationship between turnout (Y) and negative advertising (X) while controlling for the effect of unemployment (Z). What effect does this have on the bivariate relationship? Is the relationship between turnout and negative advertising direct? (*HINT: Use Formula 14.1 and see Section 14.2. You will need this partial correlation to compute the multiple correlation coefficient.*)

c. Find the unstandardized multiple regression equation with unemployment (X_1) and negative ads (X_2) as the independent variables. What turnout would be expected in a city in which the unemployment rate was 10% and 75% of the campaign ads were negative? (*HINT: Use Formulas 14.4 and 14.5 to compute the partial slopes and then use Formula 14.6 to*

find *a*, the Y intercept. The regression line is stated in Formula 14.3. Substitute 10 for X_1 and 75 for X_2 to compute predicted Y.)

d. Compute beta-weights for each independent variable. Which has the stronger impact on turnout? (*HINT: Use Formulas 14.7 and 14.8 to calculate the beta-weights.*)

e. Compute the multiple correlation coefficient (R) and the coefficient of multiple determination (R^2). How much of the variance in voter turnout is explained by the two independent variables combined? (*HINT: Use Formula 14.11. You calculated the partial correlation coefficient in part b of this problem.*)

f. Write a paragraph summarizing your conclusions about the relationships among these three variables.

14.2 SOC A scale measuring support for increases in the national defense budget has been administered to a sample. The respondents have also been asked to indicate how many years of school they have completed and how many years, if any, they served in the military. Take "support" as the dependent variable.

Case	Support	Years of School	Years of Service
A	20	12	2
B	15	12	4
C	20	16	20
D	10	10	10
E	10	16	20
F	5	8	0
G	8	14	2
H	20	12	20
I	10	10	4
J	20	16	0

a. Compute r for all three relationships (Support and Years of School, Support and Years of Service, and Years of School and Years of Service). Compute the partial correlation coefficient for the relationship between support (Y) and years of school (X) while controlling for the effect of years of service (Z). What effect does the third variable (years of service or Z) have on the bivariate relationship? Is the relationship between support and years of school direct?

b. Compute the partial correlation coefficient for the relationship between support (Y) and years of service (X) while controlling for the effect of years of school (Z). What effect does Z have on the bivariate relationship? Is the relationship between support and years of service direct? (*HINT: You will need this partial correlation to compute the multiple correlation coefficient.*)

c. Find the unstandardized multiple regression equation with school (X_1) and service (X_2) as the independent variables. What level of support would be expected in a person with 13 years of school and 15 years of service?

d. Compute beta-weights for each independent variable. Which has the stronger impact on turnout?

e. Compute the multiple correlation coefficient (R) and the coefficient of multiple determination (R^2). How much of the variance in support do the two independent variables explain? (*HINT: You calculated the partial correlation coefficient in part b of this problem.*)

f. Write a paragraph summarizing your conclusions about the relationships among these three variables.

14.3 SOC   Data on civil strife (number of incidents), unemployment, and urbanization have been gathered for 10 nations. Take "civil strife" as the dependent variable. Compute the zero-order correlations among all three variables.

Number of Incidents of Civil Strife	Unemployment Rate	Percentage of Population Living in Urban Areas
0	5.3	60
1	1.0	65
5	2.7	55
7	2.8	68
10	3.0	69
23	2.5	70
25	6.0	45
26	5.2	40
30	7.8	75
53	9.2	80

a. Compute the partial correlation coefficient for the relationship between strife (Y) and unemployment (X) while controlling for the effect of urbanization (Z). What effect does Z have on the bivariate relationship? Is the relationship between strife and unemployment direct?

b. Compute the partial correlation coefficient for the relationship between strife (Y) and urbanization (X) while controlling for the effect of unemployment (Z). What effect does Z have on the bivariate relationship? Is the relationship between strife and urbanization direct? (*HINT: You will need this partial correlation to compute the multiple correlation coefficient.*)

c. Find the unstandardized multiple regression equation with unemployment (X_1) and urbanization (X_2) as the independent variables. What level of strife would be expected in a nation in which the unemployment rate was 10% and 90% of the population lived in urban areas?

d. Compute beta-weights for each independent variable. Which has the stronger impact on strife?

e. Compute the multiple correlation coefficient (R) and the coefficient of multiple determination (R^2). How much of the variance in strife is explained by the two independent variables?

f. Write a paragraph summarizing your conclusions about the relationships among these three variables.

14.4 SOC/CJ   In problem 13.5, crime and population data were presented for each of 10 states. The data are reproduced below. Take the three crime variables as the dependent variables (one at a time) and:

a. Find the multiple regression equations (unstandardized) with growth and urbanization as independent variables.

State	Crime Rates			Population		
	Homicide	Robbery	Car Theft	Growth	Density	Urban
Maine	1	19	104	3.8	41.7	52.6
New York	5	214	286	5.5	402.7	92.1
Ohio	4	138	344	4.7	277.8	81.2
Iowa	2	37	184	5.4	52.3	45.3
Virginia	6	89	252	14.4	181.5	78.1
Kentucky	5	81	230	9.6	102.3	48.8
Texas	6	145	447	22.8	81.5	84.8
Arizona	7	146	842	40.0	46.7	88.2
Washington	3	99	594	21.1	90.0	83.1
California	6	178	537	13.6	221.2	96.7

b. Make a prediction for each crime variable for a state with a 5% growth rate and a population that is 90% urban.

c. Compute beta-weights for each independent variable in each equation and compare their relative effect on each dependent.

d. Compute R and R^2 for each crime variable, using the population variables as independent variables.

e. Write a paragraph summarizing your findings.

14.5 PS Problem 13.4 presented data on 10 precincts. The information is reproduced here. Take "voter turnout" as the dependent variable.

Precinct	Percent Democrat	Percent Minority	Voter Turnout
A	50	10	56
B	45	12	55
C	56	8	52
D	78	15	60
E	13	5	89
F	85	20	25
G	62	18	64
H	33	9	88
I	25	0	42
J	49	9	36

a. Find the multiple regression equations (unstandardized).

b. Predict the turnout for a precinct in which 0% of the voters were Democrats and 5% were minorities.

c. Compute beta-weights for each independent variable and compare their relative effect on turnout. Which was the more important factor?

d. Compute R and R^2.

e. Write a paragraph summarizing your findings.

14.6 SW Twelve families have been referred to a counselor, and she has rated each of them on a cohesiveness scale. Also, she has information on family income and number of children currently living at home. Take "family cohesion" as the dependent variable.

Family	Cohesion Score	Income (in dollars)	Number of Children
A	10	30,000	5
B	10	70,000	4
C	9	35,000	4
D	5	25,000	0
E	1	55,000	3
F	7	40,000	0
G	2	60,000	2
H	5	30,000	3
I	8	50,000	5
J	8	25,000	4
K	2	45,000	3
L	4	50,000	0

a. Find the multiple regression equations (unstandardized).

b. Predict the level of cohesion in a family with an income of $20,000 and 6 children.

c. Compute beta-weights for each independent variable and compare their relative effect on cohesion. Which was the more important factor?

d. Compute R and R^2.

e. Write a paragraph summarizing your findings.

14.7 Problem 13.8 presented per capita expenditures on education for 15 states, along with

State	Per Capita Expenditures on Education (in dollars)	Percent High School Graduates*	Rank in Per Capita Income, 2001
Arkansas	1,102	82	48
Colorado	1,339	90	7
Connecticut	1,907	88	1
Florida	1,171	84	25
Illinois	1,621	86	9
Kansas	1,276	88	24
Louisiana	1,159	81	45
Maryland	1,412	86	5
Michigan	1,487	86	18
Mississippi	1,041	80	50
Nebraska	1,194	90	22
New Hampshire	1,262	88	6
North Carolina	1,163	79	32
Pennsylvania	1,223	86	15
Wyoming	1,549	90	20

rank on income per capita and the percentage of the population that has graduated from high school. The data are reproduced here. Take "educational expenditures" as the dependent variable.

a. Compute beta-weights for each independent variable and compare their relative effect on expenditures. Which was the more important factor?

b. Compute R and R^2.

c. Write a paragraph summarizing your findings.

14.8 SOC The scores on four variables for 20 individuals are reported here: hours of TV (average number of hours of TV viewing each day), occupational prestige (higher scores indicate greater prestige), number of children, and age. Take "TV viewing" as the dependent variable and select two of the remaining variables as independents.

a. Compute beta-weights for each of the independent variables you selected and compare their relative effect on the hours of television watching. Which was the more important factor?

b. Compute R and R^2.

c. Write a paragraph summarizing your findings.

Hours of TV	Occupational Prestige	Number of Children	Age
4	50	2	43
3	36	3	58
3	36	1	34
4	50	2	42
2	45	2	27
3	50	5	60
4	50	0	28
7	40	3	55
1	57	2	46
3	33	2	65
1	46	3	56
3	31	1	29
1	19	2	41
0	52	0	50
2	48	1	62
4	36	1	24
3	48	0	25
1	62	1	87
5	50	0	45
1	27	3	62

SPSS for Windows

Using *SPSS for Windows* for Regression Analysis

SPSS DEMONSTRATION 14.1 Another Look at the Correlates of Occupational Prestige

In Demonstration 13.1, we used the **Correlate** procedure to calculate zero-order correlation coefficients among *prestg80, paeduc,* and *educ.* In this section, we will use a significantly more complex and flexible procedure called **Regression** to analyze the effects of *paeduc* and *educ* on *prestg80.* The **Regression** procedure permits the user to control many aspects of the regression formula, and it can produce a much greater volume of output than the **Correlate** procedure can. Among other things, **Regression** displays the slope (*b*) and the *Y* intercept (*a*), so we can use this procedure to find least-squares regression lines. This demonstration represents a very sparing use of the power of this command and an extremely economical use of all the options available. I urge you to explore some of the variations and capabilities of this powerful data-analysis procedure.

With the 2004 GSS loaded, click **Analyze, Regression,** and **Linear,** and the **Linear Regression** window will appear. Move *prestg80* into the **Dependent** box and *educ* and *paeduc* into the **Independent(s)** box. If you wish, you can click the **Statistics** button and then click **Descriptives** to get zero-order correlations, means, and standard deviations for the variables. Click **Continue** and **OK,** and the following output will appear (descriptive information about the variables and the zero-order correlations are omitted here to conserve space).

Model Summary(a)

Model	R	R Square	Adjusted R Square	Std. Error of the Estimate
1	.575(a)	.331	.329	11.684

aPredictors: (Constant), HIGHEST YEAR OF SCHOOL COMPLETED, HIGHEST YEAR SCHOOL COMPLETED, FATHER

ANOVA(b)

Model		Sum of Squares	df	Mean Square	F	Sig.
1	Regression	66,949.621	2	33474.810	245.229	.000
	Residual	135,412.235	992	136.504		
	Total	202,361.855	994			

bDependent variable: RS OCCUPATIONAL PRESTIGE SCORE (1980)

Coefficients(c)

Model		Unstandardized Coefficients		Standardized Coefficients		
		B	Std. Error	Beta	t	Sig.
1	(Constant)	3.532	1.978		1.786	.074
	HIGHEST YEAR SCHOOL COMPLETED, FATHER	−.011	.102	−.003	−.111	.912
	HIGHEST YEAR OF SCHOOL COMPLETED	3.025	.150	.577	20.144	.000

cDependent variable: RS OCCUPATIONAL PRESTIGE SCORE (1980)

The Model Summary block reports the multiple R (.575) and R square (.331). The ANOVA output block shows the significance of the relationship (Sig. = .000). So far, we know that the independent variables explain about 33% of the variance in *prestg80* and that this result is statistically significant. In the last output block, we see the slopes (reported under B in the "Unstandardized Coefficients" column) of the independent variables on *prestg80*, the standardized partial slopes (reported under Beta in the "Standardized Coefficients" column), and the Y intercept (reported as a constant of 3.532 under the "Unstandardized Coefficients" column). From this information, we can build a regression equation to predict scores on *prestg80*. The beta for *educ* (.577) is much greater than the beta for *paeduc* (-.003). The beta for *paeduc* suggests that father's education has virtually no effect on prestige once we control for the educational level of the respondent. At any rate, education is the more important independent variable.

What does all this mean? At least for this sample, a person's occupational prestige is much more affected by education than by the social class of his or her family of origin.

SPSS DEMONSTRATION 14.2 What Are the Correlates of Sexual Activity?

The 2004 GSS includes an item (*sexfreq*) that asks respondents about the frequency of their sexual activity over the past year. What causes variations in sexual activity? Some obvious correlates include gender and age: We commonly suppose that men are more sexually active than women and younger people more active than older. What are the effects of social class? Does the level of sexual activity—as with so many other aspects of the social world—vary by a person's economic status?

Gender is a nominal-level variable, and its inclusion in a regression procedure may be surprising. Actually, nominal variables with only two categories—sometimes called "dummy variables"—are commonly used in regression. Click **Analyze, Regression,** and **Linear** and name *sexfreq* as the dependent variable and *age, income98,* and *sex* as the independent variables. Request descriptive statistics, if you wish, by clicking the **Statistics** button and making the appropriate selection. Confining our attention only to R^2 and the regression coefficients, the output will look like this:

Model Summary(a)

Model	R	R Square	Adjusted R Square	Std. Error of the Estimate
1	.450(a)	.202	.200	1.723

[a]Predictors: (Constant), RESPONDENTS' SEX, AGE OF RESPONDENT, TOTAL FAMILY INCOME

Coefficients(b)

Model		Unstandardized Coefficients		Standardized Coefficients		
		B	Std. Error	Beta	t	Sig.
1	(Constant)	4.551	.299		15.202	.000
	AGE OF RESPONDENT	−.051	.004	−.418	−14.582	.000
	TOTAL FAMILY INCOME	.057	.010	.167	5.752	.000
	RESPONDENT'S SEX	−.202	.112	−.052	−1.805	.071

[b]Dependent variable: FREQUENCY OF SEX DURING LAST YEAR

The R^2 statistics indicate that the independent variables account for about 20% of the variation in frequency of sexual activity. The slopes (B) indicate that the dependent variable increases with income (.057) and decreases with age (−.051). The slope for gender is also negative (−.202). Because a male is coded as 1 and a female as 2, this indicates that the frequency of sex is greater for men. In other words, the frequency of sex increases as gender "decreases"—its rate is higher for people with the "lower" score on gender. The beta-weights (Beta) suggest that age has the greatest influence on frequency of sexual activity, followed by income and gender. The rate of sexual activity increases as affluence increases, decreases as age increases, and is more associated with males than females.

Exercises

14.1 Conduct the analysis in Demonstration 14.1 again with *income98* as the dependent variable. Compare your conclusions with those that you made in Demonstration 13.1. Write a paragraph summarizing the relationships.

14.2 Conduct the analysis in Demonstration 14.2 again with *partnrs5* (number of different sexual partners over the past five years) as the dependent variable. Compare and contrast with the analysis of *sexfreq*.

14.3 Conduct a regression analysis of *tvhours, childs,* or *attend.* Choose two or three potential independent variables and follow the instructions in Demonstrations 14.1 and 14.2. Ideally, your independent variables should be interval-ratio in level of measurement, but ordinal variables with a broad range of scores and nominal variables with only two scores will work as well.

Area Under the Normal Curve

Column (a) lists Z scores from 0.00 to 4.00. Only positive scores are displayed, but, since the normal curve is symmetrical, the areas for negative scores will be exactly the same as areas for positive scores. Column (b) lists the proportion of the total area between the Z score and the mean. Figure A.1 displays areas of this type. Column (c) lists the proportion of the area beyond the Z score, and Figure A.2 displays this type of area.

FIGURE A.1   AREA BETWEEN MEAN AND Z **FIGURE A.2**   AREA BEYOND Z

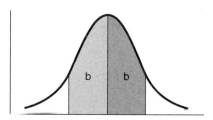 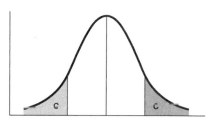

(a)	(b) Area Between	(c) Area Beyond	(a)	(b) Area Between	(c) Area Beyond
Z	Mean and Z	Z	Z	Mean and Z	Z
0.00	0.0000	0.5000	0.26	0.1026	0.3974
0.01	0.0040	0.4960	0.27	0.1064	0.3936
0.02	0.0080	0.4920	0.28	0.1103	0.3897
0.03	0.0120	0.4880	0.29	0.1141	0.3859
0.04	0.0160	0.4840	0.30	0.1179	0.3821
0.05	0.0199	0.4801			
0.06	0.0239	0.4761	0.31	0.1217	0.3783
0.07	0.0279	0.4721	0.32	0.1255	0.3745
0.08	0.0319	0.4681	0.33	0.1293	0.3707
0.09	0.0359	0.4641	0.34	0.1331	0.3669
0.10	0.0398	0.4602	0.35	0.1368	0.3632
			0.36	0.1406	0.3594
0.11	0.0438	0.4562	0.37	0.1443	0.3557
0.12	0.0478	0.4522	0.38	0.1480	0.3520
0.13	0.0517	0.4483	0.39	0.1517	0.3483
0.14	0.0557	0.4443	0.40	0.1554	0.3446
0.15	0.0596	0.4404			
0.16	0.0636	0.4364	0.41	0.1591	0.3409
0.17	0.0675	0.4325	0.42	0.1628	0.3372
0.18	0.0714	0.4286	0.43	0.1664	0.3336
0.19	0.0753	0.4247	0.44	0.1700	0.3300
0.20	0.0793	0.4207	0.45	0.1736	0.3264
			0.46	0.1772	0.3228
0.21	0.0832	0.4168	0.47	0.1808	0.3192
0.22	0.0871	0.4129	0.48	0.1844	0.3156
0.23	0.0910	0.4090	0.49	0.1879	0.3121
0.24	0.0948	0.4052	0.50	0.1915	0.3085
0.25	0.0987	0.4013			

(a)	(b) Area Between Mean and Z	(c) Area Beyond Z	(a)	(b) Area Between Mean and Z	(c) Area Beyond Z
Z			Z		
0.51	0.1950	0.3050	1.03	0.3485	0.1515
0.52	0.1985	0.3015	1.04	0.3508	0.1492
0.53	0.2019	0.2981	1.05	0.3531	0.1469
0.54	0.2054	0.2946	1.06	0.3554	0.1446
0.55	0.2088	0.2912	1.07	0.3577	0.1423
0.56	0.2123	0.2877	1.08	0.3599	0.1401
0.57	0.2157	0.2843	1.09	0.3621	0.1379
0.58	0.2190	0.2810	1.10	0.3643	0.1357
0.59	0.2224	0.2776	1.11	0.3665	0.1335
0.60	0.2257	0.2743	1.12	0.3686	0.1314
0.61	0.2291	0.2709	1.13	0.3708	0.1292
0.62	0.2324	0.2676	1.14	0.3729	0.1271
0.63	0.2357	0.2643	1.15	0.3749	0.1251
0.64	0.2389	0.2611	1.16	0.3770	0.1230
0.65	0.2422	0.2578	1.17	0.3790	0.1210
0.66	0.2454	0.2546	1.18	0.3810	0.1190
0.67	0.2486	0.2514	1.19	0.3830	0.1170
0.68	0.2517	0.2483	1.20	0.3849	0.1151
0.69	0.2549	0.2451	1.21	0.3869	0.1131
0.70	0.2580	0.2420	1.22	0.3888	0.1112
0.71	0.2611	0.2389	1.23	0.3907	0.1093
0.72	0.2642	0.2358	1.24	0.3925	0.1075
0.73	0.2673	0.2327	1.25	0.3944	0.1056
0.74	0.2703	0.2297	1.26	0.3962	0.1038
0.75	0.2734	0.2266	1.27	0.3980	0.1020
0.76	0.2764	0.2236	1.28	0.3997	0.1003
0.77	0.2794	0.2206	1.29	0.4015	0.0985
0.78	0.2823	0.2177	1.30	0.4032	0.0968
0.79	0.2852	0.2148	1.31	0.4049	0.0951
0.80	0.2881	0.2119	1.32	0.4066	0.0934
0.81	0.2910	0.2090	1.33	0.4082	0.0918
0.82	0.2939	0.2061	1.34	0.4099	0.0901
0.83	0.2967	0.2033	1.35	0.4115	0.0885
0.84	0.2995	0.2005	1.36	0.4131	0.0869
0.85	0.3023	0.1977	1.37	0.4147	0.0853
0.86	0.3051	0.1949	1.38	0.4162	0.0838
0.87	0.3078	0.1922	1.39	0.4177	0.0823
0.88	0.3106	0.1894	1.40	0.4192	0.0808
0.89	0.3133	0.1867	1.41	0.4207	0.0793
0.90	0.3159	0.1841	1.42	0.4222	0.0778
0.91	0.3186	0.1814	1.43	0.4236	0.0764
0.92	0.3212	0.1788	1.44	0.4251	0.0749
0.93	0.3238	0.1762	1.45	0.4265	0.0735
0.94	0.3264	0.1736	1.46	0.4279	0.0721
0.95	0.3289	0.1711	1.47	0.4292	0.0708
0.96	0.3315	0.1685	1.48	0.4306	0.0694
0.97	0.3340	0.1660	1.49	0.4319	0.0681
0.98	0.3365	0.1635	1.50	0.4332	0.0668
0.99	0.3389	0.1611	1.51	0.4345	0.0655
1.00	0.3413	0.1587	1.52	0.4357	0.0643
1.01	0.3438	0.1562	1.53	0.4370	0.0630
1.02	0.3461	0.1539	1.54	0.4382	0.0618

(a)	(b) Area Between	(c) Area Beyond	(a)	(b) Area Between	(c) Area Beyond
Z	Mean and Z	Z	Z	Mean and Z	Z
1.55	0.4394	0.0606	2.07	0.4808	0.0192
1.56	0.4406	0.0594	2.08	0.4812	0.0188
1.57	0.4418	0.0582	2.09	0.4817	0.0183
1.58	0.4429	0.0571	2.10	0.4821	0.0179
1.59	0.4441	0.0559	2.11	0.4826	0.0174
1.60	0.4452	0.0548	2.12	0.4830	0.0170
1.61	0.4463	0.0537	2.13	0.4834	0.0166
1.62	0.4474	0.0526	2.14	0.4838	0.0162
1.63	0.4484	0.0516	2.15	0.4842	0.0158
1.64	0.4495	0.0505	2.16	0.4846	0.0154
1.65	0.4505	0.0495	2.17	0.4850	0.0150
1.66	0.4515	0.0485	2.18	0.4854	0.0146
1.67	0.4525	0.0475	2.19	0.4857	0.0143
1.68	0.4535	0.0465	2.20	0.4861	0.0139
1.69	0.4545	0.0455	2.21	0.4864	0.0136
1.70	0.4554	0.0446	2.22	0.4868	0.0132
1.71	0.4564	0.0436	2.23	0.4871	0.0129
1.72	0.4573	0.0427	2.24	0.4875	0.0125
1.73	0.4582	0.0418	2.25	0.4878	0.0122
1.74	0.4591	0.0409	2.26	0.4881	0.0119
1.75	0.4599	0.0401	2.27	0.4884	0.0116
1.76	0.4608	0.0392	2.28	0.4887	0.0113
1.77	0.4616	0.0384	2.29	0.4890	0.0110
1.78	0.4625	0.0375	2.30	0.4893	0.0107
1.79	0.4633	0.0367	2.31	0.4896	0.0104
1.80	0.4641	0.0359	2.32	0.4898	0.0102
1.81	0.4649	0.0351	2.33	0.4901	0.0099
1.82	0.4656	0.0344	2.34	0.4904	0.0096
1.83	0.4664	0.0336	2.35	0.4906	0.0094
1.84	0.4671	0.0329	2.36	0.4909	0.0091
1.85	0.4678	0.0322	2.37	0.4911	0.0089
1.86	0.4686	0.0314	2.38	0.4913	0.0087
1.87	0.4693	0.0307	2.39	0.4916	0.0084
1.88	0.4699	0.0301	2.40	0.4918	0.0082
1.89	0.4706	0.0294	2.41	0.4920	0.0080
1.90	0.4713	0.0287	2.42	0.4922	0.0078
1.91	0.4719	0.0281	2.43	0.4925	0.0075
1.92	0.4726	0.0274	2.44	0.4927	0.0073
1.93	0.4732	0.0268	2.45	0.4929	0.0071
1.94	0.4738	0.0262	2.46	0.4931	0.0069
1.95	0.4744	0.0256	2.47	0.4932	0.0068
1.96	0.4750	0.0250	2.48	0.4934	0.0066
1.97	0.4756	0.0244	2.49	0.4936	0.0064
1.98	0.4761	0.0239	2.50	0.4938	0.0062
1.99	0.4767	0.0233	2.51	0.4940	0.0060
2.00	0.4772	0.0228	2.52	0.4941	0.0059
2.01	0.4778	0.0222	2.53	0.4943	0.0057
2.02	0.4783	0.0217	2.54	0.4945	0.0055
2.03	0.4788	0.0212	2.55	0.4946	0.0054
2.04	0.4793	0.0207	2.56	0.4948	0.0052
2.05	0.4798	0.0202	2.57	0.4949	0.0051
2.06	0.4803	0.0197	2.58	0.4951	0.0049

(a) Z	(b) Area Between Mean and Z	(c) Area Beyond Z	(a) Z	(b) Area Between Mean and Z	(c) Area Beyond Z
2.59	0.4952	0.0048	3.09	0.4990	0.0010
2.60	0.4953	0.0047	3.10	0.4990	0.0010
2.61	0.4955	0.0045	3.11	0.4991	0.0009
2.62	0.4956	0.0044	3.12	0.4991	0.0009
2.63	0.4957	0.0043	3.13	0.4991	0.0009
2.64	0.4959	0.0041	3.14	0.4992	0.0008
2.65	0.4960	0.0040	3.15	0.4992	0.0008
2.66	0.4961	0.0039	3.16	0.4992	0.0008
2.67	0.4962	0.0038	3.17	0.4992	0.0008
2.68	0.4963	0.0037	3.18	0.4993	0.0007
2.69	0.4964	0.0036	3.19	0.4993	0.0007
2.70	0.4965	0.0035	3.20	0.4993	0.0007
2.71	0.4966	0.0034	3.21	0.4993	0.0007
2.72	0.4967	0.0033	3.22	0.4994	0.0006
2.73	0.4968	0.0032	3.23	0.4994	0.0006
2.74	0.4969	0.0031	3.24	0.4994	0.0006
2.75	0.4970	0.0030	3.25	0.4994	0.0006
2.76	0.4971	0.0029	3.26	0.4994	0.0006
2.77	0.4972	0.0028	3.27	0.4995	0.0005
2.78	0.4973	0.0027	3.28	0.4995	0.0005
2.79	0.4974	0.0026	3.29	0.4995	0.0005
2.80	0.4974	0.0026	3.30	0.4995	0.0005
2.81	0.4975	0.0025	3.31	0.4995	0.0005
2.82	0.4976	0.0024	3.32	0.4995	0.0005
2.83	0.4977	0.0023	3.33	0.4996	0.0004
2.84	0.4977	0.0023	3.34	0.4996	0.0004
2.85	0.4978	0.0022	3.35	0.4996	0.0004
2.86	0.4979	0.0021	3.36	0.4996	0.0004
2.87	0.4979	0.0021	3.37	0.4996	0.0004
2.88	0.4980	0.0020	3.38	0.4996	0.0004
2.89	0.4981	0.0019	3.39	0.4997	0.0003
2.90	0.4981	0.0019	3.40	0.4997	0.0003
2.91	0.4982	0.0018	3.41	0.4997	0.0003
2.92	0.4982	0.0018	3.42	0.4997	0.0003
2.93	0.4983	0.0017	3.43	0.4997	0.0003
2.94	0.4984	0.0016	3.44	0.4997	0.0003
2.95	0.4984	0.0016	3.45	0.4997	0.0003
2.96	0.4985	0.0015	3.46	0.4997	0.0003
2.97	0.4985	0.0015	3.47	0.4997	0.0003
2.98	0.4986	0.0014	3.48	0.4997	0.0003
2.99	0.4986	0.0014	3.49	0.4998	0.0002
3.00	0.4986	0.0014	3.50	0.4998	0.0002
3.01	0.4987	0.0013	3.60	0.4998	0.0002
3.02	0.4987	0.0013	3.70	0.4999	0.0001
3.03	0.4988	0.0012			
3.04	0.4988	0.0012	3.80	0.4999	0.0001
3.05	0.4989	0.0011	3.90	0.4999	<0.0001
3.06	0.4989	0.0011			
3.07	0.4989	0.0011	4.00	0.4999	<0.0001
3.08	0.4990	0.0010			

Distribution of *t*

Degrees of Freedom (df)	Level of Significance for One-tailed Test					
	.10	.05	.025	.01	.005	.0005
	Level of Significance for Two-tailed Test					
	.20	.10	.05	.02	.01	.001
1	3.078	6.314	12.706	31.821	63.657	636.619
2	1.886	2.920	4.303	6.965	9.925	31.598
3	1.638	2.353	3.182	4.541	5.841	12.941
4	1.533	2.132	2.776	3.747	4.604	8.610
5	1.476	2.015	2.571	3.365	4.032	6.859
6	1.440	1.943	2.447	3.143	3.707	5.959
7	1.415	1.895	2.365	2.998	3.499	5.405
8	1.397	1.860	2.306	2.896	3.355	5.041
9	1.383	1.833	2.262	2.821	3.250	4.781
10	1.372	1.812	2.228	2.764	3.169	4.587
11	1.363	1.796	2.201	2.718	3.106	4.437
12	1.356	1.782	2.179	2.681	3.055	4.318
13	1.350	1.771	2.160	2.650	3.012	4.221
14	1.345	1.761	2.145	2.624	2.977	4.140
15	1.341	1.753	2.131	2.602	2.947	4.073
16	1.337	1.746	2.120	2.583	2.921	4.015
17	1.333	1.740	2.110	2.567	2.898	3.965
18	1.330	1.734	2.101	2.552	2.878	3.922
19	1.328	1.729	2.093	2.539	2.861	3.883
20	1.325	1.725	2.086	2.528	2.845	3.850
21	1.323	1.721	2.080	2.518	2.831	3.819
22	1.321	1.717	2.074	2.508	2.819	3.792
23	1.319	1.714	2.069	2.500	2.807	3.767
24	1.318	1.711	2.064	2.492	2.797	3.745
25	1.316	1.708	2.060	2.485	2.787	3.725
26	1.315	1.706	2.056	2.479	2.779	3.707
27	1.314	1.703	2.052	2.473	2.771	3.690
28	1.313	1.701	2.048	2.467	2.763	3.674
29	1.311	1.699	2.045	2.462	2.756	3.659
30	1.310	1.697	2.042	2.457	2.750	3.646
40	1.303	1.684	2.021	2.423	2.704	3.551
60	1.296	1.671	2.000	2.390	2.660	3.460
120	1.289	1.658	1.980	2.358	2.617	3.373
∞	1.282	1.645	1.960	2.326	2.576	3.291

Source: Table III of Fisher & Yates: *Statistical Tables for Biological, Agricultural and Medical Research*, published by Longman Group Ltd., London (1974), 6th edition (previously published by Oliver & Boyd Ltd., Edinburgh).

Distribution of Chi Square

df	.99	.98	.95	.90	.80	.70	.50	.30	.20	.10	.05	.02	.01	.001
1	$.0^3157$	$.0^3628$	.00393	.0158	.0642	.148	.455	1.074	1.642	2.706	3.841	5.412	6.635	10.827
2	.0201	.0404	.103	.211	.446	.713	1.386	2.408	3.219	4.605	5.991	7.824	9.210	13.815
3	.115	.185	.352	.584	1.005	1.424	2.366	3.665	4.642	6.251	7.815	9.837	11.341	16.268
4	.297	.429	.711	1.064	1.649	2.195	3.357	4.878	5.989	7.779	9.488	11.668	13.277	18.465
5	.554	.752	1.145	1.610	2.343	3.000	4.351	6.064	7.289	9.236	11.070	13.388	15.086	20.517
6	.872	1.134	1.635	2.204	3.070	3.828	5.348	7.231	8.558	10.645	12.592	15.033	16.812	22.457
7	1.239	1.564	2.167	2.833	3.822	4.671	6.346	8.383	9.803	12.017	14.067	16.622	18.475	24.322
8	1.646	2.032	2.733	3.490	4.594	5.527	7.344	9.524	11.030	13.362	15.507	18.168	20.090	26.125
9	2.088	2.532	3.325	4.168	5.380	6.393	8.343	10.656	12.242	14.684	16.919	19.679	21.666	27.877
10	2.558	3.059	3.940	4.865	6.179	7.267	9.342	11.781	13.442	15.987	18.307	21.161	23.209	29.588
11	3.053	3.609	4.575	5.578	6.989	8.148	10.341	12.899	14.631	17.275	19.675	22.618	24.725	31.264
12	3.571	4.178	5.226	6.304	7.807	9.034	11.340	14.011	15.812	18.549	21.026	24.054	26.217	32.909
13	4.107	4.765	5.892	7.042	8.634	9.926	12.340	15.119	16.985	19.812	22.362	25.472	27.688	34.528
14	4.660	5.368	6.571	7.790	9.467	10.821	13.339	16.222	18.151	21.064	23.685	26.873	29.141	36.123
15	5.229	5.985	7.261	8.547	10.307	11.721	14.339	17.322	19.311	22.307	24.996	28.259	30.578	37.697
16	5.812	6.614	7.962	9.312	11.152	12.624	15.338	18.418	20.465	23.542	26.296	29.633	32.000	39.252
17	6.408	7.255	8.672	10.085	12.002	13.531	16.338	19.511	21.615	24.769	27.587	30.995	33.409	40.790
18	7.015	7.906	9.390	10.865	12.857	14.440	17.338	20.601	22.760	25.989	28.869	32.346	34.805	42.312
19	7.633	8.567	10.117	11.651	13.716	15.352	18.338	21.689	23.900	27.204	30.144	33.687	36.191	43.820
20	8.260	9.237	10.851	12.443	14.578	16.266	19.337	22.775	25.038	28.412	31.410	35.020	37.566	45.315
21	8.897	9.915	11.591	13.240	15.445	17.182	20.337	23.858	26.171	29.615	32.671	36.343	38.932	46.797
22	9.542	10.600	12.338	14.041	16.314	18.101	21.337	24.939	27.301	30.813	33.924	37.659	40.289	48.268
23	10.196	11.293	13.091	14.848	17.187	19.021	22.337	26.018	28.429	32.007	35.172	38.968	41.638	49.728
24	10.856	11.992	13.848	15.659	18.062	19.943	23.337	27.096	29.553	33.196	36.415	40.270	42.980	51.179
25	11.524	12.697	14.611	16.473	18.940	20.867	24.337	28.172	30.675	34.382	37.652	41.566	44.314	52.620
26	12.198	13.409	15.379	17.292	19.820	21.792	25.336	29.246	31.795	35.563	38.885	42.856	45.642	54.052
27	12.879	14.125	16.151	18.114	20.703	22.719	26.336	30.319	32.912	36.741	40.113	44.140	46.963	55.476
28	13.565	14.847	16.928	18.939	21.588	23.647	27.336	31.391	34.027	37.916	41.337	45.419	48.278	56.893
29	14.256	15.574	17.708	19.768	22.475	24.577	28.336	32.461	35.139	39.087	42.557	46.693	49.588	58.302
30	14.953	16.306	18.493	20.599	23.364	25.508	29.336	33.530	36.250	40.256	43.773	47.962	50.892	59.703

Source: Table IV of Fisher & Yates: *Statistical Tables for Biological, Agricultural and Medical Research*, published by Longman Group Ltd., London (1974), 6th edition (previously published by Oliver & Boyd Ltd., Edinburgh). Reprinted by permission of Addison Wesley Longman Ltd.

Distribution of *F*

p = .05

n_1 / n_2	1	2	3	4	5	6	8	12	24	∞
1	161.4	199.5	215.7	224.6	230.2	234.0	238.9	243.9	249.0	254.3
2	18.51	19.00	19.16	19.25	19.30	19.33	19.37	19.41	19.45	19.50
3	10.13	9.55	9.28	9.12	9.01	8.94	8.84	8.74	8.64	8.53
4	7.71	6.94	6.59	6.39	6.26	6.16	6.04	5.91	5.77	5.63
5	6.61	5.79	5.41	5.19	5.05	4.95	4.82	4.68	4.53	4.36
6	5.99	5.14	4.76	4.53	4.39	4.28	4.15	4.00	3.84	3.67
7	5.59	4.74	4.35	4.12	3.97	3.87	3.73	3.57	3.41	3.23
8	5.32	4.46	4.07	3.84	3.69	3.58	3.44	3.28	3.12	2.93
9	5.12	4.26	3.86	3.63	3.48	3.37	3.23	3.07	2.90	2.71
10	4.96	4.10	3.71	3.48	3.33	3.22	3.07	2.91	2.74	2.54
11	4.84	3.98	3.59	3.36	3.20	3.09	2.95	2.79	2.61	2.40
12	4.75	3.88	3.49	3.26	3.11	3.00	2.85	2.69	2.50	2.30
13	4.67	3.80	3.41	3.18	3.02	2.92	2.77	2.60	2.42	2.21
14	4.60	3.74	3.34	3.11	2.96	2.85	2.70	2.53	2.35	2.13
15	4.54	3.68	3.29	3.06	2.90	2.79	2.64	2.48	2.29	2.07
16	4.49	3.63	3.24	3.01	2.85	2.74	2.59	2.42	2.24	2.01
17	4.45	3.59	3.20	2.96	2.81	2.70	2.55	2.38	2.19	1.96
18	4.41	3.55	3.16	2.93	2.77	2.66	2.51	2.34	2.15	1.92
19	4.38	3.52	3.13	2.90	2.74	2.63	2.48	2.31	2.11	1.88
20	4.35	3.49	3.10	2.87	2.71	2.60	2.45	2.28	2.08	1.84
21	4.32	3.47	3.07	2.84	2.68	2.57	2.42	2.25	2.05	1.81
22	4.30	3.44	3.05	2.82	2.66	2.55	2.40	2.23	2.03	1.78
23	4.28	3.42	3.03	2.80	2.64	2.53	2.38	2.20	2.00	1.76
24	4.26	3.40	3.01	2.78	2.62	2.51	2.36	2.18	1.98	1.73
25	4.24	3.38	2.99	2.76	2.60	2.49	2.34	2.16	1.96	1.71
26	4.22	3.37	2.98	2.74	2.59	2.47	2.32	2.15	1.95	1.69
27	4.21	3.35	2.96	2.73	2.57	2.46	2.30	2.13	1.93	1.67
28	4.20	3.34	2.95	2.71	2.56	2.44	2.29	2.12	1.91	1.65
29	4.18	3.33	2.93	2.70	2.54	2.43	2.28	2.10	1.90	1.64
30	4.17	3.32	2.92	2.69	2.53	2.42	2.27	2.09	1.89	1.62
40	4.08	3.23	2.84	2.61	2.45	2.34	2.18	2.00	1.79	1.51
60	4.00	3.15	2.76	2.52	2.37	2.25	2.10	1.92	1.70	1.39
120	3.92	3.07	2.68	2.45	2.29	2.17	2.02	1.83	1.61	1.25
∞	3.84	2.99	2.60	2.37	2.21	2.09	1.94	1.75	1.52	1.00

Values of n_1 and n_2 represent the degrees of freedom associated with the between and within estimates of variance, respectively.

Source: Table V of Fisher and Yates: *Statistical Tables for Biological, Agricultural and Medical Research*, published by Longman Group Ltd., London (1974), 6th edition (previously published by Oliver and Boyd Ltd., Edinburgh). Reprinted by permission of Addison Wesley Longman Ltd.

$p = .01$

n_1 n_2	1	2	3	4	5	6	8	12	24	∞
1	4052	4999	5403	5625	5764	5859	5981	6106	6234	6366
2	98.49	99.01	99.17	99.25	99.30	99.33	99.36	99.42	99.46	99.50
3	34.12	30.81	29.46	28.71	28.24	27.91	27.49	27.05	26.60	26.12
4	21.20	18.00	16.69	15.98	15.52	15.21	14.80	14.37	13.93	13.46
5	16.26	13.27	12.06	11.39	10.97	10.67	10.27	9.89	9.47	9.02
6	13.74	10.92	9.78	9.15	8.75	8.47	8.10	7.72	7.31	6.88
7	12.25	9.55	8.45	7.85	7.46	7.19	6.84	6.47	6.07	5.65
8	11.26	8.65	7.59	7.01	6.63	6.37	6.03	5.67	5.28	4.86
9	10.56	8.02	6.99	6.42	6.06	5.80	5.47	5.11	4.73	4.31
10	10.04	7.56	6.55	5.99	5.64	5.39	5.06	4.71	4.33	3.91
11	9.65	7.20	6.22	5.67	5.32	5.07	4.74	4.40	4.02	3.60
12	9.33	6.93	5.95	5.41	5.06	4.82	4.50	4.16	3.78	3.36
13	9.07	6.70	5.74	5.20	4.86	4.62	4.30	3.96	3.59	3.16
14	8.86	6.51	5.56	5.03	4.69	4.46	4.14	3.80	3.43	3.00
15	8.68	6.36	5.42	4.89	4.56	4.32	4.00	3.67	3.29	2.87
16	8.53	6.23	5.29	4.77	4.44	4.20	3.89	3.55	3.18	2.75
17	8.40	6.11	5.18	4.67	4.34	4.10	3.79	3.45	3.08	2.65
18	8.28	6.01	5.09	4.58	4.25	4.01	3.71	3.37	3.00	2.57
19	8.18	5.93	5.01	4.50	4.17	3.94	3.63	3.30	2.92	2.49
20	8.10	5.85	4.94	4.43	4.10	3.87	3.56	3.23	2.86	2.42
21	8.02	5.78	4.87	4.37	4.04	3.81	3.51	3.17	2.80	2.36
22	7.94	5.72	4.82	4.31	3.99	3.76	3.45	3.12	2.75	2.31
23	7.88	5.66	4.76	4.26	3.94	3.71	3.41	3.07	2.70	2.26
24	7.82	5.61	4.72	4.22	3.90	3.67	3.36	3.03	2.66	2.21
25	7.77	5.57	4.68	4.18	3.86	3.63	3.32	2.99	2.62	2.17
26	7.72	5.53	4.64	4.14	3.82	3.59	3.29	2.96	2.58	2.13
27	7.68	5.49	4.60	4.11	3.78	3.56	3.26	2.93	2.55	2.10
28	7.64	5.45	4.57	4.07	3.75	3.53	3.23	2.90	2.52	2.06
29	7.60	5.42	4.54	4.04	3.73	3.50	3.20	2.87	2.49	2.03
30	7.56	5.39	4.51	4.02	3.70	3.47	3.17	2.84	2.47	2.01
40	7.31	5.18	4.31	3.83	3.51	3.29	2.99	2.66	2.29	1.80
60	7.08	4.98	4.13	3.65	3.34	3.12	2.82	2.50	2.12	1.60
120	6.85	4.79	3.95	3.48	3.17	2.96	2.66	2.34	1.95	1.38
∞	6.64	4.60	3.78	3.32	3.02	2.80	2.51	2.18	1.79	1.00

Values of n_1 and n_2 represent the degrees of freedom associated with the between and within estimates of variance, respectively.

Using Statistics: Ideas for Research Projects

This appendix presents outlines for four research projects, each of which requires the use of *SPSS for Windows* to analyze the 2004 General Social Survey, the data set used throughout this text. The research projects should be completed at various intervals during the course, and each project gives students many choices. The first project stresses description and should be done after completing Chapters 2–4. The second involves estimation and should be completed in conjunction with Chapter 6. The third project uses inferential statistics and should be done after completing Part II, and the fourth combines inferential statistics with measures of association (with an option for multivariate analysis) and should be done after Part III (or IV).

PROJECT 1—DESCRIPTIVE STATISTICS

1. Open the 2004 General Social Survey (GSS). Select five variables from the database *(NOTE: your instructor may specify a different number of variables)* and use the **Frequencies** command to get frequency distributions and summary statistics; click the **Statistics** button and request the mean, median, mode, standard deviation, and range. See the SPSS demonstrations at the end of Chapters 3 and 4 for guidelines and examples. Make a note of all relevant information when it appears on screen or make a hard copy. See Appendix G for a list of variables available in the 2004 GSS.

2. Inspect the frequency distributions and choose appropriate measures of central tendency and, for ordinal- and interval-ratio-level variables, dispersion. Also, for interval-ratio and ordinal variables with many scores, check for skew both by using the line chart and by comparing the mean and median. Write a sentence or two describing each variable, being careful to include a description of the overall shape of the distribution (see Chapter 2), the central tendency (Chapter 3), and the dispersion (Chapter 4). For nominal- and ordinal-level variables, be sure to explain any arbitrary numerical codes. For example, on the variable *class* in the 2004 GSS (see Appendix G), a 1 is coded as "lower class," a 2 indicates "working class," and so forth. This is an ordinal-level variable, so you might choose to report the median as a measure of central tendency. If the median score on *class* were 2.45, for example, you might place that value in context by reporting that "the median is 2.45, about halfway between 'working class' and 'middle class'."

3. Below are examples of *minimal* summary sentences, using fictitious data: *For a nominal-level variable* (e.g., marital status), report the mode and some detail about the overall distribution. For example: "Most respondents were married (57.5%), but divorced (17.4%) and single (21.3%) individuals were also common." *For an ordinal-level variable* (e.g., occupational prestige), use the median (and, perhaps, the mode) and the range. For example: "The median prestige score was 44.3, and the range extended from 34 to 87. The most common score was 42."

For an interval-ratio level variable (e.g., population density for 124 nations), use the mean (and, perhaps, the median or mode) and the standard deviation (and, perhaps, the range). For example: "For these nations, population density (population per square mile) ranged from a low of 4.00 to a high of 12,000. The standard deviation was 1,000. The nations averaged 351 people per square mile, and the median density was 193. The distribution is positively skewed, and some nations have very high population density."

PROJECT 2—ESTIMATION

In this exercise, you will use the 2004 GSS sample to estimate characteristics of the U.S. population. You will generate the sample statistics by using SPSS and then use either Formula 6.2 or 6.3 to find the confidence interval and state each interval in words.

A. Estimating Means

1. The 2004 GSS has relatively few interval-ratio variables. For this part of the project, you may use ordinal variables that have *at least* four categories or scores. Choose a total of three variables that fit this description. *(NOTE: your instructor may specify a different number of variables.)*

2. Use the **Descriptives** command to get means, standard deviations, and sample size (N), and use this information to construct 95% confidence intervals for the first of your variables. Make a note of the mean, standard deviation, and sample size or keep a hard copy. Use Formula 6.2 to find the confidence interval. Repeat this procedure for the remaining variables.

3. For each variable, write a summary sentence reporting the variable, the interval itself, the confidence level, and the sample size. Write in plain English, as if you were reporting results in a newspaper. Most importantly, make it clear that you are estimating characteristics of the population of the entire United States. For example, a summary sentence might look like this: "Based on a random sample of 1,231, I estimate at the 95% level that U.S. drivers average between 64.46 and 68.22 miles per hour when driving on interstate highways."

B. Estimating Proportions

4. Choose three variables that are nominal or ordinal and have *two or three* categories or scores. *(NOTE: your instructor may specify a different number of variables.)*

5. Use the **Frequencies** command to get the percentage of the sample in the various categories of each variable. Change the percentages (remember to use the "valid percents" column) to find proportions and construct confidence intervals for one category of each variable (e.g., the % female for *sex*) using Formula 6.3.

6. For each variable, write a summary sentence reporting the variable, the interval, the confidence level, and the sample size. Write in plain English, as if you were reporting results in a newspaper. Remember to make it clear that you are estimating a characteristic of the U.S. population.

7. For any one of the intervals you constructed, identify each of the following concepts and terms and briefly explain their role in estimation: sample, population, statistic, parameter, EPSEM, representative, confidence level.

**PROJECT 3—
SIGNIFICANCE TESTING**

Use the 2004 General Social Survey data set for this project.

A. Two-Sample *t* Test (Chapter 8)

1. Choose two different dependent variables from the interval-ratio or ordinal variables that have three or more scores. Choose independent variables that might logically be a cause of your dependent variables. Remember that, for a *t* test, independent variables can have *only* two categories. Independent variables can be any level of measurement, and you may use the same independent variable for both tests.

2. Click **Analyze, Compare Means,** and then **Independent Samples T Test.** Name your dependent variable(s) in the **Test Variable** window and your independent variable in the **Grouping Variable** window. You will also need to specify the scores used to define the groups on the independent variable. Take notes of the test results (group means, obtained *t* score, significance, sample size) or keep a hard copy. Repeat the procedure for the second dependent variable.

3. Write up the results of the test. At a minimum, your report should clearly identify the independent and dependent variables, the sample statistics, the value of the test statistic (step 4), the results of the test (step 5), and the alpha level you used.

B. Analysis of Variance (Chapter 9)

1. Choose two different dependent variables from the interval-ratio or ordinal variables that have three or more scores. Choose independent variables that might logically be a cause of your dependent variables and that have between three and five categories. You may use the same independent variables for both tests.

2. Click **Analyze, Compare Means,** and then **One-way ANOVA.** The **One-way ANOVA** window will appear. Find your dependent variable in the variable list on the left, and click the arrow to move the variable name into the **Dependent List** box. Note that you can request more than one dependent variable at a time. Next, find the name of your independent variable and move it to the **Factor** box. Click **Options** and then click the box next to **Descriptive** in the **Statistics** box to request means and standard deviations. Click **Continue** and **OK.** Make a note of the test results or keep a hard copy. Repeat for your second dependent variable.

3. Write up the results of the test. At a minimum, your report should clearly identify the independent and dependent variables, the sample statistics (category means), the value of the test statistic (step 4), the results of the test (step 5), the degrees of freedom, and the alpha level you used.

C. Chi Square (Chapter 10)

1. Choose two different dependent variables of any level of measurement with five or fewer (preferably two or three) scores. For each dependent variable, choose an independent variable that might logically be a cause. Independent variables can be any level of measurement as long as they have two to five categories. Output will be easier to analyze if you use variables with few categories. You may use the same independent variable for both tests.

2. Click **Analyze, Descriptive Statistics,** and then **Crosstabs.** The **Crosstabs** dialog box will appear. Highlight your first dependent variable and move it into the **Rows** box. Next, highlight your independent variable and move it into the **Columns** box. Click the **Statistics** button at the bottom of the window and click the box next to chi square. Click **Continue** and **OK.** Record the results or get a hard copy. Repeat for your second dependent variable.

3. Write up the test results. At a minimum, your report should clearly identify the independent and dependent variables, the value of the test statistic (step 4), the results of the test (step 5), the degrees of freedom, and the alpha level you used. It is almost always desirable to also report the column percentages.

PROJECT 4—ANALYZING THE STRENGTH AND SIGNIFICANCE OF RELATIONSHIPS

A. Using Bivariate Tables

1. From the 2004 GSS data set, select either:
   a. One dependent variable and three independent variables (possible causes), or
   b. One independent variable and three possible dependent variables (possible effects)

   Variables can be from any level of measurement but must have only a few (two to five) categories or "scores." Develop research questions or hypotheses about the relationships between variables. Make sure that the causal links you suggest are sensible and logical.

2. Use the **Crosstabs** procedure to generate bivariate tables. Click **Analyze, Descriptive Statistics,** and **Crosstabs** and place your dependent variable(s) in the rows and independent variable(s) in the columns. On the **Crosstabs** dialog box, click the **Statistics** button and choose chi square, phi or *V*, lambda, and gamma for every table you request. On the **Crosstabs** dialog box, click the **Cells** button and get column percentages for every table you request. Note results as they appear on the screen or obtain hard copies.

3. Write a report that presents and analyzes these relationships. Be clear about which variables are dependent and which are independent. For each combination of variables, report the test of significance and a measure of association. In addition, for each relationship, report and discuss column percentages, pattern or direction of the relationship, and strength of the relationship.

B. Using Interval-Ratio Variables

1. From the 2004 GSS, select either:
   a. One dependent variable and three independent variables (possible causes), or
   b. One independent variable and three possible dependent variables (possible effects)

   Variables should be interval-ratio in level of measurement, but you may use ordinal-level variables if they have more than (preferably many more than) five scores. Develop research questions or hypotheses about the relationships between variables. Make sure that the causal links you suggest are sensible and logical.

2. Use the **Regression** and **Scatterplot** (click **Graphs** and then **Scatter**) procedures to analyze the bivariate relationships. Make a note of results (including r, r^2, slope, beta-weights, and a) as they appear on the screen or get hard copies.

3. Write a report that presents and analyzes these relationships. Be clear about which variables are dependent and which are independent. For each combination of variables, report the significance of the relationship (if relevant) and the strength and direction of the relationship. Include r, r^2, and the beta-weights in your report.

4. *OPTIONAL MULTIVARIATE ANALYSIS:* Pick one of the bivariate relationships you produced in step 2 and find another logical independent variable. Run **Regression** again with both independent variables and analyze the results. How much improvement is there in the explained variance after the second independent variable is included? Which independent variable has the stronger effect (higher beta weight)? Note the direction of relationships and compare to the bivariate relationships. Write up the results of this analysis and include them in your summary paper for this project.

An Introduction to *SPSS* for Windows

Computers have affected virtually every aspect of human society and, as you would expect, their impact on the conduct of social research has been profound. Researchers routinely use computers to organize data and compute statistics—activities that humans often find dull, tedious, and difficult but which computers accomplish with accuracy and ease. This division of labor allows social scientists to spend more time on analysis and interpretation—activities that humans typically enjoy but which are beyond the power of computers (so far, at least).

These days, the skills needed to use computers successfully are readily accessible, even for people with little or no experience. This appendix will prepare you to use a statistics program called *SPSS for Windows* (SPSS stands for Statistical Package for the Social Sciences). If you have used a mouse to "point and click" and run a computer program, you are ready to learn how to use SPSS. Even if you are completely unfamiliar with computers, you will find this program accessible. After you finish this appendix, you will be ready to do the exercises at the end of most chapters of this text.

A word of caution before we begin: This appendix is intended only as an *introduction* to SPSS. It will give you an overview of the program and enough information so that you can complete the assignments in the text. It is unlikely, however, that this appendix will answer all your questions or provide solutions to all the problems you might encounter. So, keep in mind that SPSS has an extensive and easy-to-use "help" facility that will provide assistance as you request it. You should familiarize yourself with this feature and use it as needed. To get help, simply click on the **Help** command on the toolbar across the top of the screen.

SPSS is a statistical package (or statpak) or a set of computer programs that work with data and compute statistics as requested by the user (you). Once you have entered the data for a particular group of observations, you can easily and quickly produce an abundance of statistical information without doing any computations yourself. Also, you can access the power of the computer without having to write computer programs yourself.

Why bother to learn this technology? The truth is that the laborsaving capacity of computers is sometimes exaggerated, and there are research situations in which they are unnecessary. If you are working with a small number of observations or need only a few, uncomplicated statistics, then statistical packages are probably not going to be helpful. However, as the number of cases increases and as your requirements for statistics become more sophisticated, computers and statpaks will become more and more useful.

An example should clarify this point. Suppose you have gathered a sample of 150 respondents and the *only* thing you want to know about these people is their average age. To compute an average, as you know, you add the scores and divide by the number of cases. How long do you think it would take you to add 150 two-digit numbers (ages) with a hand calculator? (Don't even think about doing it with paper and pencil.) If you entered the scores at the (pretty fast) rate of one per second, or 60 scores a minute, it would take about three or four minutes to enter the ages and get the average. Even if you worked slowly and carefully and did the addition a second and third time to check your math, you could probably complete all calculations in less than 20 minutes. If this were all the information you needed, computers and statpaks would not save you any time.

But such a simple research project is not very realistic. Typically, researchers deal with not one but scores or even hundreds of variables, and samples have hundreds or thousands of cases. Although you could add 150 numbers in perhaps several minutes, how long would it take to add the scores for 1,500 cases? What are the chances of processing 1,500 numbers without making significant errors of arithmetic? The more complex the research situation, the more valuable and useful statpaks become. SPSS can produce statistical information in a few keystrokes or clicks of the mouse that might take you minutes, hours, or even days to produce with a hand calculator.

Clearly, this is technology worth mastering by any social researcher. With SPSS, you can avoid the drudgery of mere computation, spend more time on analysis and interpretation, and conduct research projects with very large data sets. Mastery of this technology might be very handy indeed in your senior-level courses, in a wide variety of jobs, or in graduate school.

F.1 GETTING STARTED—DATABASES AND COMPUTER FILES

Before statistics can be calculated, SPSS must first have data to process. A database is an organized collection of related information such as the responses to a survey. For purposes of computer analysis, a database is organized into a file: a collection of information that is stored under the same name in the memory of the computer, on a compact disc, or in some other medium. Words as well as numbers can be saved in files. If you've ever used a word processing program to type a term paper, you probably saved your work in a file so that you could update it or make corrections at a later time. Data can be stored in files indefinitely. Because conducting a thorough data analysis can take months, the ability to save a database is another advantage of using computers.

For the SPSS exercises in this text, we will use a database that contains some of the results of the General Social Survey (GSS) for 2004. This database contains the responses of a sample of adult Americans to questions about a wide variety of social issues. The GSS has been conducted regularly since 1972 and has been the basis for hundreds of research projects by professional social researchers. It is a rich source of information about public opinion in the United States and includes data on everything from attitudes about abortion to opinions on assisted suicide.

The GSS is especially valuable because the respondents are chosen so that the sample as a whole is representative of the entire U.S. population. A representative sample reproduces, in miniature form, the characteristics of the population from which it was taken (see Chapters 1 and 6). So, when you analyze the 2004 General Social Survey database, you are in effect analyzing U.S. society as of 2004. The data are real, and the relationships you will analyze reflect some of the most important and sensitive issues in American life.

The complete General Social Survey for 2004 includes hundreds of items of information (age, sex, opinion about issues such as capital punishment, and so forth) for almost 3,000 respondents. Some of you will be using a student version of *SPSS for Windows,* which is limited in the number of cases and variables it can process. To accommodate these limits, I have reduced the database to about 50 items of information and fewer than 1,500 respondents.

Appendix G summarizes the GSS data file. Please turn to this appendix and familiarize yourself with it. Note that the variables are listed alphabetically by their variable names. In SPSS, the names of variables must be no more than eight characters long. In many cases, the resultant need for brevity is not a problem, and variable names (e.g., *age*) are easy to figure out. In other cases, the eight-character limit necessitates extreme abbreviation, and some variable names (like *abany* and *abhlth*) are not so obvious. Appendix G also shows the wording of the item that generated the variable. For example, the *abany* variable consists of responses to a question about legal abortion: Should a woman be able to have an abortion for "any reason"? Note that the variable name is formed from the question: Should an *ab*ortion be possible for *any* reason? The *abhlth* variable consists of responses to a different scenario: Should legal abortion be possible if the woman's health is seriously endangered? Appendix G serves as a codebook for the database: It lists all the codes (or scores) for the survey items along with their meanings.

Notice that some of the possible responses to *abany* and *abhlth* and other variables in Appendix G are labeled "Not applicable," NA, or DK. The first of these codes (NAP) means that the item in question was not given to a respondent. The full GSS is very long and, to keep the time frame for completing the survey reasonable, not all respondents are asked every question. NA stands for "No Answer" and means that the respondent was asked the question but refused to answer. DK stands for "Don't Know," which means that the respondent did not have the requested information. All three of these scores are *missing values* and, as "noninformation" should be eliminated from statistical analysis. Missing values are common on surveys and, as long as they are not too numerous, they are not a particular problem.

It's important that you understand the difference between a statpak (such as SPSS) and a database (the GSS) and what we are ultimately after here. A database consists of information. A statpak organizes the information in the database and produces statistics. Our goal is to send the database to the statpak and produce output (for example, statistics and graphs) that we can analyze and use to answer questions. Figure F.1 suggests the process.

Statpaks like SPSS are general research tools that you can use to analyze databases of all sorts; they are not limited to the 2004 GSS. In the same way, the 2004 GSS could be analyzed with statpaks other than SPSS. Other widely used statpaks include MicroCase, SAS, and Stata—each of which may be available on your campus.

FIGURE F.1   THE DATA ANALYSIS PROCESS

Database ⟶ **Statpak** ⟶ **Output** ⟶ **Analysis**

(raw information) (computer programs) (statistics and graphs) (interpretation)

F.2 STARTING *SPSS FOR WINDOWS* AND LOADING THE 2004 GSS

If you are using the complete, professional version of *SPSS for Windows,* you will probably be working in a computer lab, and you can begin running the program immediately. If you are using the student version of the program on your personal computer, the first thing you need to do is install the software. Follow the instructions that came with the program and return to this appendix when installation is complete.

To start *SPSS for Windows,* find the icon (or picture) on the screen of your monitor that has an "SPSS" label. Use the computer mouse to move the arrow on the monitor screen over this icon and then double-click the left button on the mouse. This will start up the SPSS program.

After a few seconds, the *SPSS for Windows* screen will appear and ask, at the top of the screen, "What would you like to do?" Of the choices listed, the button next to "Open an existing file" will be checked (or preselected), and there will probably be a number of data sets listed in the window at the bottom of the screen. Find the 2004 General Social Survey data set, probably labeled *GSS04.sav* or something similar. If you have the data set on a compact disc, you will need to specify the correct drive. Check with your instructor to make sure that you know where to find the version of the 2004 GSS being used for your course.

Once you've located the data set, click on the name of the file with the left-hand button on the mouse, and SPSS will load the data. The next screen you will see is the SPSS Data Editor screen. Figure F.2 (p. 372) shows the screen with the GSS 2004 loaded.

Note that there is a list of commands across the very top of the screen. These commands begin with **File** at the far left and end with **Help** at the far right. This is the main menu bar for SPSS. When you click any of these words, a menu of commands and choices will drop down. Basically, you tell SPSS what to do by clicking on your desired choices from these menus. Sometimes, submenus will appear, and you will need to specify your choices further.

SPSS provides a variety of options for displaying information about the data file and output on the screen. I recommend that you tell the program to display lists of variables by name (e.g., *age, abany*) rather than labels (e.g, AGE OF RESPONDENT, ABORTION IF WOMAN WANTS FOR ANY REASON). Lists displayed this way will be easier to read and compare to Appendix G. To do this, click **Edit** on the main menu bar and then click **Options** from the drop-down submenu. A dialog box labeled "Options" will appear with a series of "tabs" along the top. The "General" options should be displayed but, if not, click on this tab. On the "General" screen, find the box labeled "Variable Lists" and, if they are not already selected, click "Display names" and "alphabetical" and then click **OK.** Depending on which version of SPSS you are using, a message may appear on the screen that tells you that changes will take effect the next time a data file is opened.

In this section, you learned how to start up *SPSS for Windows,* load a data file, and set some of the display options for this program. Table F.1 (p. 372) summarizes these procedures.

F.3 WORKING WITH DATABASES

Note that in the SPSS Data Editor window, the data are organized into a two-dimensional grid with columns running up and down (vertically) and rows running across (horizontally). Each column is a variable or item of information from the survey. The names of the variables are listed at the tops of the columns.

FIGURE F.2   THE SPSS DATA EDITOR SCREEN WITH THE 2004 GSS LOADED

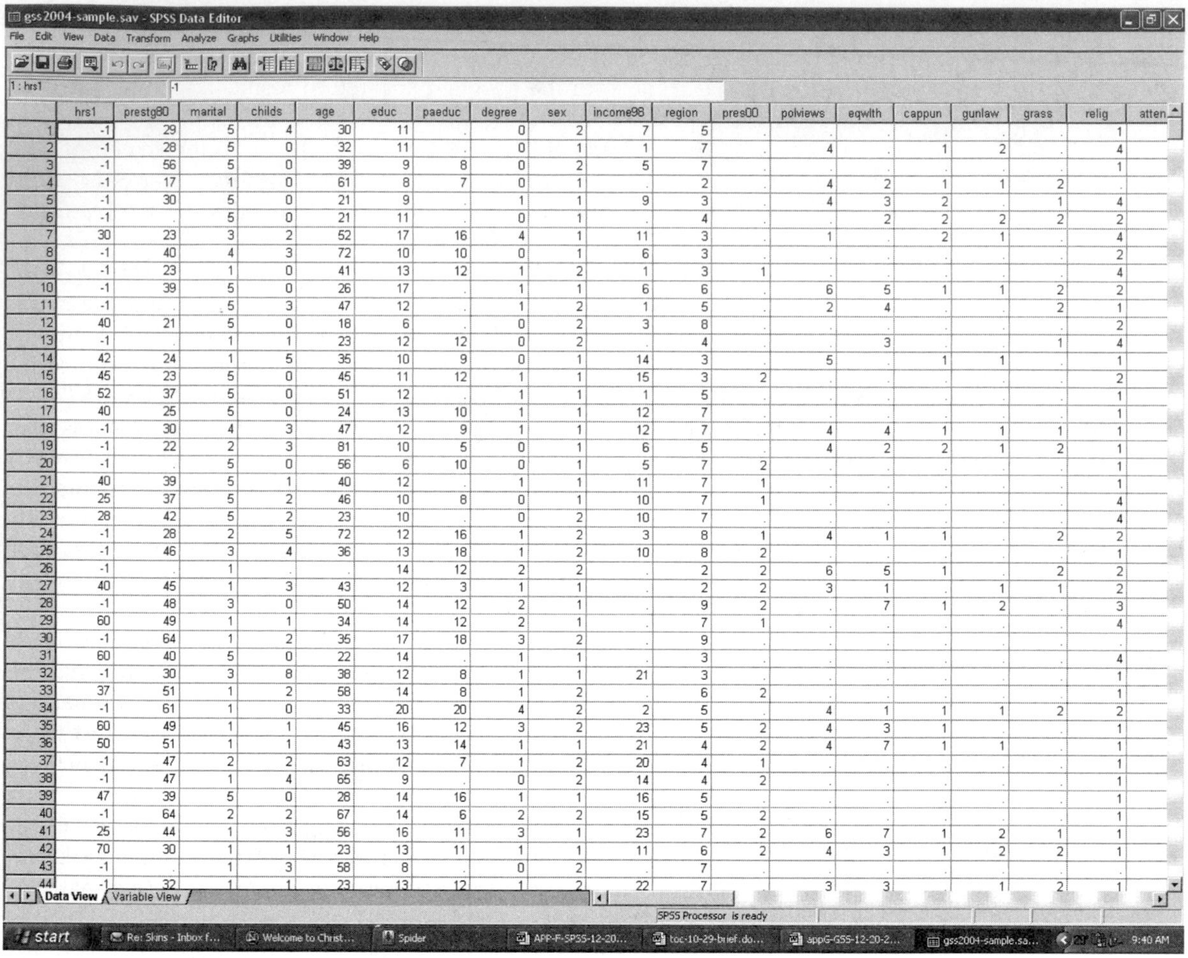

	hrs1	prestg80	marital	childs	age	educ	paeduc	degree	sex	income98	region	pres00	polviews	eqwlth	cappun	gunlaw	grass	relig	atten
1	-1	29	5	4	30	11		0	2	7	5								1
2	-1	28	5	0	32	11		0	1	1	7		4		1	2		4	
3	-1	56	5	0	39	9	8	0	2	5	7							1	
4	-1	17	1	0	61	8	7	0	1		2		4	2	1	1	2		
5	-1	30	5	0	21	9		1	1	9	3		4	3	2		1	4	
6	-1		5	0	21	11		0	1		4			2	2	2	2	2	
7	30	23	3	2	52	17	16	4	1	11	3		1		2	1		4	
8	-1	40	4	3	72	10	10	0	1	6	3							2	
9	-1	23	1	0	41	13	12	1	2	1	3	1						4	
10	-1	39	5	0	26	17		1	1	6	6		6	5	1	1	2	2	
11	-1		5	3	47	12		1	2	1	5		2	4			2	1	
12	40	21	5	0	18	6		0	2	3	8							2	
13	-1		1	1	23	12	12	0	2		4		3				1	4	
14	42	24	1	5	35	10	9	0	1	14	3		5		1	1		1	
15	45	23	5	0	45	11	12	1	1	15	3	2						2	
16	52	37	5	0	51	12		1	1	1	5							1	
17	40	25	5	0	24	13	10	1	1	12	7							1	
18	-1	30	4	3	47	12	9	1	1	12	7		4	4	1	1	1	1	
19	-1	22	2	3	81	10	5	0	1	6	5		4	2	2	1	2	1	
20	-1		5	0	56	6	10	0	1	5	7	2						1	
21	40	39	5	1	40	12		1	1	11	7	1						1	
22	25	37	5	2	46	10	8	0	1	10	7	1						4	
23	28	42	5	2	23	10		0	2	10	7							4	
24	-1	28	2	5	72	12	16	1	2	3	8	1	4	1	1		2	2	
25	-1	46	3	4	36	13	18	1	2	10	8	2						1	
26	-1		1		14	12		2	2		2	2	6	5	1		2	2	
27	40	45	1	3	43	12	3	1	1		2	2	3	1		1	1	2	
28	-1	48	3	0	50	14	12	2	1		9	2		7	1	2		3	
29	60	49	1	1	34	14	12	2	1		7	1						4	
30	-1	64	1	2	35	17	18	3	2		9								
31	60	40	5	0	22	14		1	1		3							4	
32	-1	30	3	8	38	12	8	1	1	21	3							1	
33	37	51	1	2	58	14	8	1	2		6	2						1	
34	-1	61	1	0	33	20	20	4	2	2	5		4	1	1	1	2	2	
35	60	49	1	1	45	16	12	3	2	23	5	2	4	3	1	1		1	
36	50	51	1	1	43	13	14	1	1	21	4	2	4	7	1	1		1	
37	-1	47	2	2	63	12	7	1	2	20	4	1						1	
38	-1	47	1	4	65	9		0	2	14	4	2						1	
39	47	39	5	0	28	14	16	1	1	16	5							1	
40	-1	64	2	2	67	14	6	2	2	15	5	2						1	
41	25	44	1	3	56	16	11	3	1	23	7	2	6	7	1	2	1	1	
42	70	30	1	1	23	13	11	1	1	11	6	2	4	3	1	2	2	1	
43	-1		1	3	58	8		0	2		7								
44	-1	32	1	1	23	13	12	1	2	22	7		3	3		1	2	1	

TABLE F.1   SUMMARY OF COMMANDS TO START SPSS AND LOAD THE 2004 GSS

To start *SPSS for Windows*	Click the SPSS icon on the screen of the computer monitor.
To open a data file	Double-click on the data file name.
To set display options for lists of variables	Click **Edit** from the main menu bar, then click **Options**. On the "General" tab, make sure that "Display names" and "alphabetical" are selected and then click **OK**.

Remember that you can find the meaning of these variable names in the GSS 2004 codebook in Appendix G.

Another way to decipher the meaning of variable names is to click **Utilities** on the menu bar and then click **Variables**. The Variables window opens. This window has two parts. On the left is a list of all variables in the database arranged in alphabetical order with the first variable highlighted. On the right is the Variable

TABLE F.2 SUMMARY OF COMMANDS TO VIEW THE DATA EDITOR WINDOW AND TO GET INFORMATION ABOUT VARIABLES

To move around in the Data Editor window	1. Click the cell you want to highlight or 2. Use the arrow keys on your keyboard or 3. Move the slider buttons or 4. Click the arrows on the right-hand and bottom margins.
To get information about a variable	1. From the menu bar, click **Utilities** and then click **Variables.** Scroll through the list of variable names until you highlight the name of the variable in which you are interested. Variable information will appear in the window on the right. 2. See Appendix G.

Information window with information about the highlighted variable. The first variable is listed as *abany*. The Variable Information window displays a fragment of the question that was actually asked during the survey ("ABORTION IF WOMAN WANTS FOR ANY REASON") and shows the possible scores on this variable (a score of 1 = yes and a score of 2 = no) along with some other information.

The same information can be displayed for any variable in the data set. For example, find the variable *marital* in the list. You can do this by using the arrow keys on your keyboard or the slider bar on the right of the variable list window. You can also move through the list by typing the first letter of the variable name you are interested in. For example, type "m" and you will be moved to the first variable name in the list that begins with that letter. Now you can see that the variable measures marital status and that a score of "1" indicates that the respondent was married, and so forth. What do *prestg80* and *class* measure? Close this window by clicking the **Close** button at the bottom of the window.

Examine the window displaying the 2004 GSS a little more. Each row of the window (reading across or from left to right) contains the scores of a particular respondent on all the variables in the database. Note that the upper-left-hand cell is highlighted (outlined in a darker border than the other cells). This cell contains the score of respondent #1 on the first variable. The second row contains the scores of respondent #2, and so forth. You can move around in this window with the arrow keys on your keyboard. The highlight moves in the direction of the arrow, one cell at a time.

In this section, you learned to read information in the data display window and to decipher the meaning of variable names and scores. These commands are summarized in Table F.2. We are now prepared to actually perform some statistical operations with the 2004 GSS database.

F.4 PUTTING SPSS TO WORK: PRODUCING STATISTICS

At this point, the database on the screen is just a mass of numbers with little meaning for you. That's okay because you will not have to actually read any information from this screen. Virtually all of the statistical operations you will conduct will begin by clicking the **Analyze** (or **Statistics** for version 8.0 or earlier) command from the menu bar, selecting a procedure and statistics, and then naming the variable or variables you would like to process.

To illustrate, let's have *SPSS for Windows* produce a frequency distribution for the variable *sex*. Frequency distributions are tables that display the number of times each score of a variable occurred in the sample (see Chapter 2). So, when we complete this procedure, we will know the number of males and females in the 2004 GSS sample.

With the 2004 GSS loaded, begin by clicking the **Analyze** command on the menu bar. From the menu that drops down, click **Descriptive Statistics** and then **Frequencies.** The Frequencies window appears with the variables listed in alphabetical order in the box on the left. The first variable (*abany*) will be highlighted. Use the slider button or the arrow keys on the right-hand margin of this box to scroll through the variable list until you highlight the variable *sex,* or type "s" to move to the approximate location. Once the variable you want to process has been highlighted, click the arrow button in the middle of the screen to move the variable name to the box on the right-hand side of the screen. SPSS will produce frequency distributions for all variables listed in this box, but, for now, we will confine our attention to sex. Figure F.3 shows what the screen should look like at this point.

FIGURE F.3 REQUESTING A FREQUENCIES DISTRIBUTION FOR *sex*

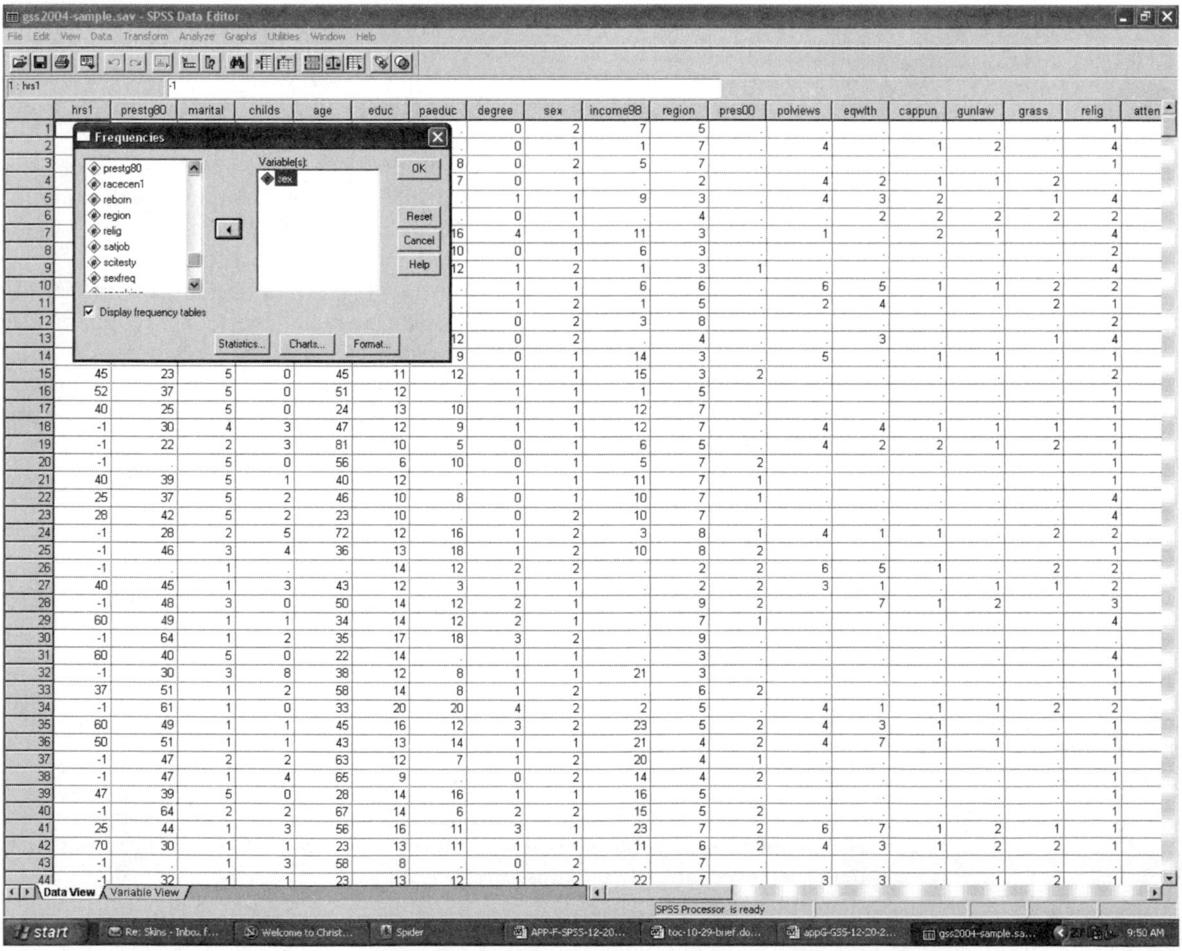

TABLE F.3   AN EXAMPLE OF SPSS OUTPUT: RESPONDENTS' *sex*

		Frequency	Percent	Valid Percent	Cumulative Percent
Valid	MALE	656	46.4	46.4	46.4
	FEMALE	759	53.6	53.6	100.0
	Total	1,460	100.0	100.0	

Click the **OK** button in the upper-right-hand corner of the Frequencies window and, in seconds, a frequency distribution will be produced.

SPSS sends all tables and statistics to the Output window or SPSS viewer. This window is now "closest" to you, and the Data Editor window is "behind" the Output window. If you wanted to return to the Data Editor, click on any visible part of it, and it will move to the "front" and the Output window will be "behind" it. To display the Data Editor window if it is not visible, minimize the Output window by clicking the "–" box in the upper-right-hand corner.

Frequencies. The output from SPSS, slightly modified, is reproduced as Table F.3. What can we tell from this table? The score labels (male and female) are printed at the left with the number of cases (frequency) in each category of the variable one column to the right. As you can see, there are 656 males and 759 females in the sample. The next two columns give information about percentages, and the last column to the right displays cumulative percentages. We will defer a discussion of this last column until a later exercise.

One of the percentage columns is labeled Percent and the other is labeled Valid Percent. The difference between these two columns lies in the handling of missing values. The Percent column is based on all cases, including people who did not respond to the item (NA) and people who said they did not have the requested information (DK). The Valid Percent column excludes all missing scores. Because we will almost always want to ignore missing scores, we will pay attention only to the Valid Percent column. Note that for *sex,* there are no missing scores (the interviewer determined gender), and the two columns are identical.

F.5 PRINTING AND SAVING OUTPUT

Once you've gone to the trouble of producing statistics, a table, or a graph, you will probably want to keep a permanent record. There are two ways to do this. First, you can print a copy of the contents of the Output window to take with you. To do this, click on **File** and then click **Print** from the **File** menu. Alternatively, find the icon of a printer (third from the left) in the row of icons just below the menu bar and click on it.

The other way to create a permanent record of SPSS output is to save the Output window to the computer's memory or to a diskette or other storage device. To do this, click **Save** from the **File** menu. The Save dialog box opens. Give the output a name (some abbreviation such as "freqsex" might do) and, if necessary, specify the name of the drive in which your storage device is located. Click **OK,** and the table will be permanently saved.

F.6 ENDING YOUR *SPSS FOR WINDOWS* SESSION

Once you have saved or printed your work, you may end your SPSS session. Click on **File** from the menu bar and then click **Exit.** If you haven't already done so, you will be asked if you want to save the contents of the Output window. You may save the frequency distribution at this point if you wish. Otherwise, click **NO.** The program will close, and you will be returned to the screen from which you began

Codebook for the General Social Survey, 2004

The General Social Survey (GSS) is a public opinion poll that the National Opinion Research Council has conducted regularly. A version of the 2004 GSS is available at the website for this text and is used for all end-of-chapter SPSS exercises. Our version of the 2004 GSS includes about 50 variables for a randomly selected subsample of about half of the original respondents. This codebook lists each item in the data set. The variable names are those used in the data files. The questions have been reproduced exactly as they were asked (with a few exceptions to conserve space), and the numbers beside each response are the scores recorded in the data file.

The data set includes variables that measure demographic or background characteristics of the respondents, including sex, age, race, religion, and several indicators of socioeconomic status. Also included are items that measure opinion on such current and controversial topics as abortion, capital punishment, and homosexuality.

Most variables in the data set have codes for "missing data." These codes are italicized in the listings below for easy identification. The codes refer to various situations in which the respondent does not or cannot answer the question and are excluded from all statistical operations. The codes are NAP or "Not Applicable" (the respondent was not asked the question), DK or "Don't Know" (the respondent didn't have the requested information), and NA or "No Answer" (the respondent refused to answer).

Please tell me if you think it should be possible for a woman to get a legal abortion if . . .

abany

She wants it for any reason.
1. Yes
2. No
0. NAP, 8. DK, 9. NA.

abhlth

The woman's health is seriously endangered.
(Same scoring as abany)

affrmact

Some people say that because of past discrimination, blacks should be given preference in hiring and promotion. Others say that such preference . . . is wrong because it discriminates against whites. What about your opinion—are you for or against preferential hiring and promotion of blacks?
1. Strongly supports preferences
2. Supports preferences
3. Opposes preferences

4. Strongly opposes preferences
8. DK, 9. NA

age   Age of respondent
18–89. Actual age in years
99. NA

attend   How often do you attend religious services?
0. Never	1. Less than once per year
2. Once or twice a year	3. Several times per year
4. About once a month	5. 2–3 times a month
6. Nearly every week	7. Every week
8. Several times a week	
9. DK or NA	

cappun   Do you favor or oppose the death penalty for persons convicted of murder?
1. Favor
2. Oppose
0. NAP, 8. DK, 9. NA

childs   How many children have you ever had? Please count all that were born alive at any time (including any from a previous marriage).
0–7. Actual number
8. Eight or more
9. NA

class   Subjective class identification
1. Lower class	2. Working class
3. Middle class	4. Upper class
0. NAP, 8. DK, 9. NA	

closeblk In general, on a scale of 1 to 9, how close do you feel to blacks?
1. Not at all close
5. Neither one feeling nor the other
9. Very close
98. DK, 99. NA

degree   Respondent's highest degree
0. Less than HS
1. High school
2. Assoc./Junior college
3. Bachelor's
4. Graduate
7. NAP, 8. DK, 9. NA

demtoday On the whole, on a scale of 1 to 10, how well does democracy work in America today?
0. Very poorly
10. Very well
−1 NAP, 98. Can't choose, 99. NA

educ   Highest year of school completed
0–20. Actual number of years
97. NAP, 98. DK, 99. NA

eqwlth   Some people think that the government in Washington ought to reduce the income differences between the rich and poor. . . . Others think that the government should not concern itself with [this]. On a scale of 1 to 7, what score comes closest to the way you feel?
1. Government should reduce differences
7. Government should not reduce differences
8. DK, 9. NA

fefam   It is much better for everyone involved if the man is the achiever outside the home and the woman takes care of the home and family.
1. Strongly agree
2. Agree
3. Disagree
4. Strongly disagree
0. NAP, 8. DK, 9. NA

fepresch   A preschool child is likely to suffer if his or her mother works.
1. Strongly agree
2. Agree
3. Disagree
4. Strongly disagree
0. NAP, 8. DK, 9. NA

goodlife   The way things are in America, people like me and my family have a good chance of improving our standard of living.
1. Strongly agree
2. Agree
3. Neither agree nor disagree
4. Disagree
5. Strongly disagree
8. Can't choose, 9. NA

grass   Do you think the use of marijuana should be made legal or not?
1. Should
2. Should not
0. NAP, 8. DK, 9. NA

gunlaw   Would you favor or oppose a law which would require a person to obtain a police permit before he or she could buy a gun?
1. Favor
2. Oppose
8. DK, 9. NA

hapmar   Taking things all together, how would you describe your marriage? Would you say that your marriage is very happy, pretty happy, or not too happy?
1. Very happy
2. Pretty happy

3. Not too happy
0. NAP, 8. DK, 9. NA

happy

Taken all together, how would you say things are these days—would you say that you are very happy, pretty happy, or not too happy?
1. Very happy
2. Pretty happy
3. Not too happy
0. NAP, 8. DK, 9. NA

helpful

Would you say that most of the time people try to be helpful, or are they mostly just looking out for themselves?
1. Try to be helpful
2. Just look out for themselves
3. Depends
0. NAP, 8. DK, 9. NA

hrs1

How many hours did you work last week?
1–89. Actual hours
−1. NAP, 98. DK, 99. NA

immameco

There are different opinions about immigrants . . . living in America. How much do you agree or disagree that immigrants are generally good for America's economy?
1. Agree strongly
2. Agree
3. Neither agree nor disagree
4. Disagree
5. Disagree strongly
8. Can't choose, 9. NA

income98

Respondent's total family income (in dollars) from all sources

1. Less than 1,000	2. 1,000 to 2,999
3. 3,000 to 3,999	4. 4,000 to 4,999
5. 5,000 to 5,999	6. 6,000 to 6,999
7. 7,000 to 7,999	8. 8,000 to 9,999
9. 10,000 to 12,499	10. 12,500 to 14,999
11. 15,000 to 17,499	12. 17,500 to 19,999
13. 20,000 to 22,499	14. 22,500 to 24,999
15. 25,000 to 29,999	16. 30,000 to 34,999
17. 35,000 to 39,999	18. 40,000 to 49,999
19. 50,000 to 59,999	20. 60,000 to 74,999
21. 75,000 to 89,999	22. 90,000 to 109,999
23. 110,000 or more	

98. DK, 99. NA

letdie1

When a person has a disease that cannot be cured, do you think doctors should be allowed by law to end the patient's life by some painless means if the patient and his family request it?
1. Yes

2. No
0. NAP, 8. DK, 9. NA

marblk
[Would you favor or oppose] . . . having a close relative or family member marrying a black person?
1. Strongly favor
2. Favor
3. Neither favor nor oppose
4. Oppose
5. Strongly oppose
0. NAP, 8. DK, 9. NA

marhomo
Do you agree or disagree: Homosexual couples should have the right to marry one another?
1. Strongly agree
2. Agree
3. Neither agree nor disagree
4. Disagree
5. Strongly disagree
8. Can't choose, 9. NA

marital
Are you currently married, widowed, divorced, separated, or have you never been married?
1. Married
2. Widowed
3. Divorced
4. Separated
5. Never married
9. NA

news
How often do you read the newspaper?
1. Every day
2. A few times a week
3. Once a week
4. Less than once a week
5. Never
0. NAP, 8. DK, 9. NA

paeduc
Father's highest year of school completed
0–20. Actual number of years
97. NAP, 98. DK, 99, NA

partnrs5
How many sex partners have you had over the past five years?
0. No partners
1. 1 partner
2. 2 partners
3. 3 partners
4. 4 partners
5. 5–10 partners
6. 11–20 partners

7. 21–100 partners
8. More than 100 partners
9. *1 or more, don't know the number, 95. Several, 98. DK, 99. NA, −1. NAP*

pillok
Do you strongly agree, agree, disagree, or strongly disagree that it is acceptable to make birth control devices available to teenagers, age 14–16?
1. Strongly agree
2. Agree
3. Disagree
4. Strongly disagree
0. NAP, 8. DK, 9. NA

polviews
I'm going to show you a seven-point scale on which the political views that people might hold are arranged from extremely liberal to extremely conservative. Where would you place yourself on this scale?
1. Extremely liberal
2. Liberal
3. Slightly liberal
4. Moderate
5. Slightly conservative
6. Conservative
7. Extremely conservative
0. NAP, 8. DK, 9. NA

premarsx
There's been a lot of discussion about the way morals and attitudes about sex are changing in this country. If a man and a woman have sex relations before marriage, do you think it is always wrong, almost always wrong, wrong only sometimes, or not wrong at all?
1. Always wrong
2. Almost always wrong
3. Wrong only sometimes
4. Not wrong at all
0. NAP, 8. DK, 9. NA

pres00
In 2000, you remember that Gore ran for president on the Democratic ticket against Bush for the Republicans and Nader as an Independent. Did you vote for Gore, Bush, or Nader? (Includes only those who said they voted in this election)
1. Gore
2. Bush
3. Nader
4. Other, 6. No presidential vote, 0. NAP, 8. DK, 9. NA

prestg80
Prestige of respondent's occupation
17–86. Actual score
0. NAP, DK, NA

racecen1
Race of respondent
1. White
2. Black

      3. American Indian or Alaska Native
      4. Asian American and Pacific Islanders
      5. Hispanic
      0. NAP, 98. DK

reborn   Would you say you have been "born again" or have had a "born again" experience—that is, a turning point in your life when you committed yourself to Christ?
      1. Yes
      2. No
      8. DK, 9. NA

region   Region of interview
      1. New England
      2. Mid-Atlantic
      3. East N. Cent.
      4. West N. Cent.
      5. So. Atlantic
      6. East So. Cent.
      7. West So. Cent.
      8. Mountain
      9. Pacific

relig   What is your religious preference? Is it Protestant, Catholic, Jewish, some other religion, or no religion?
      1. Protestant
      2. Catholic
      3. Jewish
      4. None
      5. Other
      8. DK, 9. NA

satjob   All in all, how satisfied would you say you are with your job?
      1. Very satisfied
      2. Moderately satisfied
      3. A little dissatisfied
      4. Very dissatisfied
      0. NAP, 8. DK, 9. NA

scitesty   "Human beings developed from earlier species of animals." This statement is
      1. Definitely true
      2. Probably true
      3. Probably not true
      4. Definitely not true
      8. Can't choose, 9. NA

sex   Respondent's gender
      1. Male
      2. Female

sexfreq
About how many times did you have sex during the last 12 months?
0. Not at all
1. Once or twice
2. About once a month
3. 2 or 3 times a month
4. About once a week
5. 2 or 3 times a week
6. More than 3 times a week
−1. NAP, 8. DK, 9. NA

spanking
Do you strongly agree, agree, disagree, or strongly disagree that it is sometimes necessary to discipline a child with a good, hard spanking?
1. Strongly agree
2. Agree
3. Disagree
4. Strongly disagree
0. NAP, 8. DK, 9. NA

tvhours
On the average day, about how many hours do you personally watch television?
0–22. Actual hours
−1. NAP, 8. DK, 9. NA

wwwhr
Not counting e-mail, about how many hours per week do you use the web?
0–130. Actual hours
998. DK, 999. NA

xmarsex
What is your opinion about a married person having sexual relations with someone other that the marriage partner?
1. Always wrong
2. Almost always wrong
3. Wrong only sometimes
4. Not wrong at all
0. NAP, 8. DK, 9. NA

Glossary of Symbols

The number in parentheses indicates the chapter in which the symbol is introduced.

a	Point at which the regression line crosses the Y axis (13)
ANOVA	The analysis of variance (9)
b	Slope of the regression line (13)
b_i	Partial slope of the linear relationship between the ith independent variable and the dependent variable (14)
b_i^*	Standardized partial slope of the linear relationship between the ith independent variable and the dependent variable (14)
df	Degrees of freedom (7)
f	Frequency (2)
F	The F ratio (9)
f_e	Expected frequency (10)
f_o	Observed frequency (10)
G	Gamma for a sample (12)
H_0	Null hypothesis (7)
H_1	Research or alternate hypothesis (7)
Md	Median (3)
Mo	Mode (3)
N	Number of cases (2)
N_d	Number of pairs of cases ranked in different order on two variables (12)
N_s	Number of pairs of cases ranked in the same order on two variables (12)
%	Percentage (2)
P	Proportion (2)

P_s	A sample proportion (6)
P_u	A population proportion (6)
PRE	Proportional reduction in error (11)
Q	Interquartile range (4)
r	Pearson's correlation coefficient for a sample (13)
r^2	Coefficient of determination (13)
R	Multiple correlation coefficient (14)
R	Range (4)
r_s	Spearman's rho for a sample (12)
$r_{xy.z}$	Partial correlation coefficient (14)
R^2	Coefficient of multiple determination (14)
s	Sample standard deviation (4)
SSB	The sum of squares between (9)
SST	The total sum of squares (9)
SSW	The sum of squares within (9)
s^2	Sample variance (4)
t	Student's t score (7)
V	Cramer's V (11)
X	Any independent variable (11)
$\overline{X}$	Mean of a sample (3)
X_i	Any score in a distribution (3)
Y	Any dependent variable (11)
Y'	A predicted score on Y (13)
Z	A control variable (14)
Z scores	Standard scores (5)

GREEK LETTERS

α	Probability of Type I error (7)
β	Probability of Type II error (7)
λ	Lambda (11)
μ	Mean of a population (3)
μ_p	Mean of a sampling distribution of sample proportions (6)
$\mu_{\overline{X}}$	Mean of a sampling distribution of sample means (6)
σ	Population standard deviation (4)
σ_p	Standard deviation of a sampling distribution of sample proportions (6)
σ_{p-p}	Standard deviation of the sampling distribution of difference in sample proportions (8)

$\sigma_{\overline{X}}$	Standard deviation of a sampling distribution of sample means (6)
$\sigma_{\overline{X}-\overline{X}}$	Standard deviation of the sampling distribution of the difference in sample means (8)
σ^2	Population variance (4)
Σ	"Summation of" (3)
ϕ	Phi (11)
χ^2	Chi square statistic (10)
χ_c^2	Chi square corrected by Yates's correction (10)

Answers to Odd-Numbered Computational Problems

In addition to answers, this section suggests some problem-solving strategies and provides examples of how to interpret some statistics. You should try to solve and interpret the problems on your own before consulting this section.

In solving these problems, I let my calculator or computer do most of the work. I worked with whatever level of precision these devices permitted and, generally, didn't round off until the end or until I had to record an intermediate sum. I always rounded off to two places of accuracy (that is, two places beyond the decimal point, or to 100ths). If you follow these same conventions, your answers will almost always match mine. However, be aware that small discrepancies might occur and that these differences almost always will be trivial. If the difference between your answer and mine doesn't seem trivial, you should double-check to make sure you haven't made an error.

Finally, please allow me a brief disclaimer about mathematical errors in this section. Let me assure you, first of all, that I know how important this section is for most students and that I worked hard to be certain that these answers are correct. Human fallibility being what it is, however, I cannot make absolute guarantees. Should you find any errors, please let me know so I can make corrections in the future.

Chapter 2

2.1　**a.** Complex A: $\left(\dfrac{5}{20}\right) \times 100 = 25.00\%$

Complex B: $\left(\dfrac{10}{20}\right) \times 100 = 50.00\%$

b. Complex A: $4{:}5 = 0.80$
Complex B: $6{:}10 = 0.60$

c. Complex A: $\left(\dfrac{0}{20}\right) = 0.00$

Complex B: $\left(\dfrac{1}{20}\right) = 0.05$

d. $\left(\dfrac{6}{4+6}\right) = \left(\dfrac{6}{10}\right) = 60.00\%$

e. Complex A: $8{:}5 = 1.60$
Complex B: $2{:}10 = 0.20$

2.3　Bank robbery rate

$= \left(\dfrac{47}{211{,}732}\right) \times 100{,}000 = 22.20$

Homicide rate $= \left(\dfrac{13}{211{,}732}\right) \times 100{,}000 = 6.14$

Auto theft rate $= \left(\dfrac{23}{211{,}732}\right) \times 100{,}000 = 10.86$

Chapter 3

3.1　"Region of birth" and "Religion" are nominal-level variables, "support for legalization" and "opinion of food" are ordinal, and "expenses" and "number of movies" are interval-ratio. The mode, the most common score, is the only measure of central tendency available for nominal-level variables. For the two ordinal-level variables, *don't forget to array the scores from high to low* before locating the median. There are 10 freshmen (*N* is even), so the median for freshmen will be the score halfway between the scores of the two middle cases. There are 11 seniors (*N* is odd), so the median for seniors will be the score of the middle case. To find the mean for the interval-ratio variables, add the scores and divide by the number of cases.

Variable	Freshmen	Seniors
Region of birth:	Mode = North	Mode = North
Legalization:	Median = 3	Median = 5
Expenses:	Mean = 48.50	Mean = 63.00
Movies:	Mean = 5.80	Mean = 5.18
Food:	Median = 6	Median = 4
Religion:	Mode = Protestant	Mode = Protestant and None (4 cases each)

3.3

Variable	Level of Measurement	Measure of Central Tendency
Sex	Nominal	Mode = male
Social Class	Ordinal	Median = "medium" (the middle case is in this category)
Number of Years in Party	I-R	Mean = 26.15
Education	Ordinal	Median = high school
Marital Status	Nominal	Mode = married
Number of Children	I-R	Mean = 2.39

3.5

Variable	Level of Measurement	Measure of Central Tendency
Marital Status	Nominal	Mode = married
Race	Nominal	Mode = white
Age	I-R	Mean = 27.53
Attitude on Abortion	Ordinal	Median = 7

3.7

	1999	2002
Mean	676.36	726.55
Median	660	706

3.9

	1995	2005
Mean	54.25	60.3

3.11

	Pretest	Posttest
Mean	9.33	12.93
Median	10	12

3.13 a. Median = 14.5, Mean = 17.78

Chapter 4

4.1 The high score is 50 and the low score is 10 so the range is 50 − 10 or 40. The standard deviation is 12.28.

4.3

	Expenses		Movies		Food	
	Fresh.	Srs.	Fresh.	Srs.	Fresh.	Srs.
Mean	48.5	63.0	5.80	5.18	5.50	4.55
Std. Dev.	9.21	15.19	5.06	4.04	3.35	2.50
Range	32	50	14	14	10	8

4.5

Statistic	1999	2002
Mean	676.36	726.55
Median	660	706
Standard Deviation	79.09	79.58
Range	324	306

In this time period, the mean and median increase. The distributions for both years show a positive skew (the mean is greater than the median). This is caused by the Yukon, which has a much higher average income than the other 10 provinces. The standard deviation and range do not change much over the time period. This indicates that the dispersion in the group (the differences in income from province to province) stayed about the same.

4.7

Variable	Statistic	Males	Females
Labor Force Participation	Mean	77.60	58.40
	Standard deviation	2.73	6.73
% High School Graduate	Mean	69.20	70.20
	Standard deviation	5.38	4.98
Mean Income	Mean	33,896.60	29,462.40
	Standard deviation	4,443.16	4,597.93

Males and females are very similar in terms of educational level, but females are less involved in the labor force and, on the average, earn almost $4,500 less than males per year. The females in these 10 states are much more variable in their labor force participation but are similar to males in dispersion on the other two variables.

4.9

	1973	1975
Mean	12.21	19.06
Standard Deviation	12.20	9.63

Chapter 5

5.1

X_i	Z Score	% Area Above	% Area Below
5	−1.67	95.25	4.75
6	−1.33	90.82	9.18
7	−1.00	84.13	15.87
8	−0.67	74.86	25.14
9	−0.33	62.93	37.07
11	0.33	37.07	62.93
12	0.67	25.14	74.86
14	1.33	9.18	90.82
15	1.67	4.75	95.25
16	2.00	2.28	97.72
18	2.67	0.38	99.62

5.3

	Z Scores	Area
a.	0.10 & 1.10	32.45%
b.	0.60 & 1.10	13.86%
c.	0.60	27.43%
d.	0.90	18.41%
e.	0.60 & −0.40	38.11%
f.	0.10 & −0.40	19.52%
g.	0.10	53.98%
h.	0.30	61.79%
i.	0.60	72.57%
j.	1.10	86.43%

5.5

X_i	Z Score	Number of Students Above	Number of Students Below
60	−2.00	195	5
57	−2.50	199	1
55	−2.83	199	1
67	−0.83	159	41
70	−0.33	126	74
72	0.00	100	100
78	1.00	32	168
82	1.67	9	191
90	3.00	1	199
95	3.83	1	199

NOTE: Number of students has been rounded off to the nearest whole number

5.7

	Z Score	Percentage of Sample
a.	−2.20	1.39%
b.	1.80	96.41%
c.	−0.20 & 1.80	54.34%
d.	0.80 & 2.80	20.93%
e.	−1.20	88.49%
f.	0.80	21.19%

5.9

	Z Score	Probability
a.	−1.00 & 1.50	.7745
b.	0.25 & 1.50	.3345
c.	1.50	.0668
d.	0.25 & −2.25	.5865
e.	−1.00 & −2.25	.1465
f.	−1.00	.1587

5.11 Yes. The raw score of 110 translates into a Z score of +2.88; 99.80% of the area lies below this score, so this individual was in the top 1% on this test.

5.13 For the first event, the probability is .0919 and, for the second, the probability is .0655. The first event is more likely.

Chapter 6

6.1 **a.** 5.2 ± 0.11 **b.** 100 ± 0.71
c. 20 ± 0.40 **d.** 1020 ± 5.41
e. 7.3 ± 0.23 **f.** 33 ± 0.80

6.3

Confidence Level	Alpha	Area Beyond Z	Z Score
95%	.05	.0250	±1.96
94%	.06	.0300	±1.89
92%	.08	.0400	±1.76
97%	.03	.0150	±2.17
98%	.02	.0100	±2.33
99.9%	.001	.0005	±3.32

NOTE: When I had a choice of Z scores, I chose the larger score.

6.5 **a.** 2.30 ± 0.04 **b.** 2.10 ± 0.01, $0.78 \pm .07$
c. 6.00 ± 0.37

6.7 **a.** 178.23 ± 1.97. The estimate is that students spent between \$176.26 and \$180.20 on books.
b. 1.5 ± 0.04. The estimate is that students visited the clinic between 1.46 and 1.54 times on the average.
c. $2.8 \pm .13$
d. $3.5 \pm .19$

6.9 $14 \pm .07$ The estimate is that between 7% and 21% of the population consists of unmarried couples living together.

6.11 **a.** $P_s = (823/1,496) = 0.55$, Confidence interval: 0.55 ± 0.02. Between 53% and 57% of the population agrees with the statement.

b. $P_s = (650/1,496) = 0.44$, Confidence interval: 0.44 ± 0.02

c. $P_s = (375/1,496) = 0.25$, Confidence interval: 0.25 ± 0.02

d. $P_s = (1,023/1,496) = 0.68$, Confidence interval: 0.68 ± 0.02

e. $P_s = (800/1,496) = 0.54$, Confidence interval: 0.54 ± 0.02

6.13

Alpha (α)	Confidence Level	Confidence Interval
0.10	90%	100 ± 0.74
0.05	95%	100 ± 0.88
0.01	99%	100 ± 1.16
0.001	99.9%	100 ± 1.47

6.15 The confidence interval is $.51 \pm .05$. The estimate would be that between 46% and 56% of the population prefer candidate A. The population parameter (P_u) is equally likely to be anywhere in the interval (that is, it's just as likely to be 46% as it is to be 56%), so a winner cannot be predicted.

6.17 The confidence interval is 0.23 ± 0.08. At the 95% confidence level, the estimate would be that between 240 (15%) and 496 (31%) of the 1,600 freshmen would be extremely interested. The estimated numbers are found by multiplying N (1,600) by the upper (.31) and lower (.15) limits of the interval.

Chapter 7

7.1 **a.** For each situation, find Z(critical).

Alpha	Form	Z(Critical)
.05	One-tailed	$+1.65$ or -1.65
.10	Two-tailed	± 1.65
.06	Two-tailed	± 1.89
.01	One-tailed	$+2.33$ or -2.33
.02	Two-tailed	± 2.33

b. For each situation, find the critical t score.

Alpha	Form	N	t(Critical)
.10	Two-tailed	31	± 1.697
.02	Two-tailed	24	± 2.500
.01	Two-tailed	121	± 2.617
.01	One-tailed	31	$+2.457$ or -2.457
.05	One-tailed	61	$+1.671$ or -1.671

c. Using $\alpha = 0.05$ throughout
1. Z(obtained) $= -3.77$ **2.** t(obtained) -2.21
3. Z(obtained) $= -5.74$ **4.** Z(obtained) $= 0.66$
5. Z(obtained) $= -0.77$

7.3 **a.** Z(obtained) $= -41.00$
b. Z(obtained) $= 29.09$

7.5 Z(obtained) $= 6.04$

7.7 **a.** Z(obtained) $= -13.66$
b. Z(obtained) $= 25.50$

7.9 t(obtained) $= 4.50$

7.11 Z(obtained) $= 3.06$

7.13 Z(obtained) $= -1.48$

7.15 **a.** Z(obtained) $= -0.74$
b. Z(obtained) $= 2.19$
c. Z(obtained) $= -8.55$
d. Z(obtained) $= -18.07$
e. Z(obtained) $= 2.09$
f. Z(obtained) $= -53.33$

7.17 t(obtained) $= -1.14$

Chapter 8

8.1 **a.** $\sigma = 1.39$, Z(obtained) $= -2.52$
b. $\sigma = 1.60$, Z(obtained) $= 2.49$

8.3 **a.** Z(obtained) $= 1.70$
b. Z(obtained) $= -2.48$

8.5 **a.** $\sigma = 0.08$, Z(obtained) $= -11.25$
b. Phone calls: $\sigma = 0.12$, Z(obtained) $= -3.33$, E-mail messages: $\sigma = 0.15$, Z(obtained) $= 20.00$

8.7 These are small samples (combined Ns of less than 100), so be sure to use Formulas 8.5 and 8.6 in step 4.
a. $\sigma = 0.12$, t(obtained) $= -1.33$
b. $\sigma = 0.13$, t(obtained) $= 14.85$

8.9 **a.** (France) $\sigma = 0.0095$, Z(obtained) $= -31.58$
b. (Nigeria) $\sigma = 0.0075$, Z(obtained) $= -146.67$
c. (China) $\sigma = 0.0065$, Z(obtained) $= 76.92$
d. (Mexico) $\sigma = 0.0107$, Z(obtained) $= -74.77$
e. (Japan) $\sigma = 0.012$, Z(obtained) $= -41.67$

The large values for the Z scores indicate that the differences are significant at very low alpha levels (that is, they are extremely unlikely to have been caused by random chance alone). Note that women are significantly happier than men in every nation except China.

8.11 $P_u = .45$, $\sigma_p = .06$, Z(obtained) $= 0.67$

8.13 a. $P_u = .46$, $\alpha_p = 0.05$, Z(obtained) $= 2.60$
b. $P_u = .80$, $\alpha_p = 0.07$, Z(obtained) $= 1.43$
c. $P_u = .72$, $\alpha_p = 0.08$, Z(obtained) $= 0.75$

8.15 a. Z(obtained) $= 1.67$
b. Z(obtained) $= 3.50$
c. Z(obtained) $= 4.00$
d. Z(obtained) $= 4.88$
e. Z(obtained) $= -7.40$
f. Z(obtained) $= -5.17$

Chapter 9

9.1

Problem	Grand Mean	SST	SSB	SSW	F ratio
a.	12.17	231.67	173.17	58.5	13.32
b.	6.87	455.73	78.53	377.20	1.25
c.	31.65	8,362.55	5,053.35	3,309.2	8.14

9.3

Problem	Grand Mean	SST	SSB	SSW	F ratio
a.	4.39	86.28	45.78	40.50	8.48
b.	16.44	332.44	65.44	267.200	1.84

For problem 9.3a, with alpha $= 0.05$ and df $= 2, 15$, the critical F ratio would be 3.68. We would reject the null hypothesis and conclude that decision making *does* vary significantly by type of relationship. By inspection of the group means, it seems that the "cohabitational" category accounts for most of the differences.

9.5 SST $= 213.61$, SSB $= 2.11$, SSW $= 211.50$, F(obtained) $= .08$

9.7 SST $= 429.48$, SSW $= 305.42$, SSB $= 124.06$, F(obtained) $= 5.96$

9.9

Nation	Grand Mean	SST	SSB	SSW	F ratio
Mexico	3.78	300.98	154.08	146.90	12.59
Canada	6.88	156.38	20.08	136.30	1.78
U.S.	5.13	286.38	135.28	151.10	10.74

At alpha $= 0.05$ and df $= 3, 36$, the critical F ratio is 2.92. There is a significant difference in support for suicide by class in Mexico and the United States but not in Canada. The category means for Mexico suggest that the upper class accounts for most of the differences. For the United States, there is more variation across the category means, and the working class seems to account for most of the differences. Going beyond the ANOVA test and comparing the grand means, support is highest in Canada and lowest in Mexico.

Chapter 10

10.1 a. 1.11 **b.** 0.00 **c.** 1.52 **d.** 1.46

10.3 A computing table is highly recommended as a way of organizing the computations for chi square:

Computational Table for Problem 10.3

(1) f_o	(2) f_e	(3) $f_o - f_e$	(4) $(f_o - f_e)^2$	(5) $(f_o - f_e)^2 / f_e$
6	5	1	1	.20
7	8	−1	1	.13
4	5	−1	1	.20
9	8	1	1	.13
$N = 26$	$N = 26$	0		χ^2(obtained) $= 0.65$

There is 1 degree of freedom in a 2×2 table. With alpha set at .05, the critical value for the chi square would be 3.841. The obtained chi square is 0.65, so we fail to reject the null hypothesis of independence between the variables. There is no statistically significant relationship between race and services received.

10.5

Computational Table for Problem 10.5

(1) f_o	(2) f_e	(3) $f_o - f_e$	(4) $(f_o - f_e)^2$	(5) $(f_o - f_e)^2 / f_e$
21	17.5	3.5	12.25	.70
29	32.5	−3.5	12.25	.38
14	17.5	−3.5	12.25	.70
36	32.5	3.5	12.25	.38
$N = 100$	$N = 100.0$	0		χ^2(obtained) $= 2.15$

With 1 degree of freedom and alpha set at .05, the critical region will begin at 3.841. The obtained chi square of 2.15 does not fall within this area, so the null hypothesis cannot be rejected. There is no statistically significant relationship between unionization and salary.

10.7   The obtained chi square is 5.12, which is significant (df = 1, alpha = .05).

10.9   The obtained chi square is 6.67.

10.11   The obtained chi square is 12.59.

10.13   The obtained chi square is 19.33.

10.15

Problem	Chi Square	Significant at $\alpha = 0.05$?
a.	25.19	Yes
b.	1.80	No
c.	5.23	Yes
d.	28.43	Yes
e.	14.17	Yes

Chapter 11

Tables display column percentages.

11.1

Efficiency	Authoritarianism	
	Low	High
Low	37.04	70.59
High	62.96	29.41
Totals	100.00	100.00

The conditional distributions change, so there is a relationship between the variables. The change from column to column is large and the maximum difference is (70.59 − 37.04) = 33.55. Using Table 11.5 as a guideline, we can say that this relationship is strong. The phi for this table is .33 and lambda is .32, both values reinforcing the impression that the relationship is strong.

From inspection of the percentages, we can see that efficiency decreases as authoritarianism increases—workers with dictatorial bosses are less productive (or, maybe, bosses become more dictatorial when workers are inefficient), so this relationship is negative in direction.

11.3

2000 Election	1996 Election	
	Democrat	Republican
Democrat	87.31	11.44
Republican	12.69	88.56
Totals	100.00	100.00

The maximum difference for this table is (87.31 − 11.44) or 75.87, phi is .75, and lambda is .71. This is a very strong relationship. People are very consistent in their voting habits.

11.5   The maximum difference is 30.84, the phi is .36, and lambda is .13. This appears to be a strong relationship. Lower turnover is associated with a more experienced director.

11.7   Maximum difference = 60
Phi = .59
Lambda = .55
High isolation is associated with living in an integrated neighborhood.

11.9

	Phi	Lambda
a.	.01	.00
b.	.11	.00
c.	.08	.00
d.	.00	.00
e.	.02	.00

Chapter 12

12.1   **a.** $G = 0.71$ **b.** $G = 0.69$ **c.** $G = -0.88$

These relationships are strong. Facility in English and income increase with length of residence (+0.71, +0.69). Use the percentages to help interpret the direction of a relationship. In the first table, 80% of the "newcomers" were "low" in English facility while 60% of the residents of more than five years were "high." In this relationship, low scores on one variable are associated with low scores on the other, and scores increase together (as one increases, the other increases), so this is a positive relationship. In contrast, contact with the old country decreases with length of residence (−0.88). Most newcomers have higher levels of contact, and most longer-term residents have lower levels.

12.3 $G = .27$

12.5 $G = 0.22$

12.7 $G = -0.14$

12.9   **a.** $G = -0.14$
      b. $G = -0.17$
      c. $G = -0.14$
      d. $G = -0.13$
      e. $G = .39$

12.11 $r_s = -0.46$

12.13 $r_s = 0.33$
For these nations, there is a moderate positive relationship between diversity and inequality. The greater the diversity, the greater the inequality.

Chapter 13

13.1   (*HINT: When finding the slope, remember that "turnout" is the dependent or Y variable.*)

	For Turnout (Y) and		
	Unemployment	Education	Neg. Campaigning
Slope (b)	3.00	12.67	−0.90
Y intercept (a)	39.00	−94.73	114.01
Reg. Eq.	$Y = (39) + (3)X$	$Y = (-94.73) + (12.67)X$	$Y = (114.01) + (-0.90)X$
r	0.95	0.98	−0.87
r^2	0.90	0.96	0.76

13.3   (*HINT: When finding the slope, remember that "number of visitors" is the dependent or Y variable.*)

Slope (b)	−0.37
Y intercept (a)	13.42
r	−0.31
r^2	0.10

13.5

Dependent Variables		Independent Variables		
		Density	Growth	Urbanization
Car Theft	a	417.08	135.47	−215.46
	b	−0.23	17.50	7.96
	r	−0.13	0.89	0.67
	r^2	0.02	0.79	0.45
Robbery	a	59.97	94.38	−96.71
	b	0.37	1.44	2.81
	r	0.72	0.27	0.87
	r^2	0.52	0.07	0.76
Homicide	a	3.87	3.01	−0.58
	b	0.00	0.11	0.07
	r	0.26	0.62	0.65
	r^2	0.07	0.38	0.42

13.5   **c.** For a growth rate of −1, the predicted homicide rate would be 2.90.
      For a population density of 250, the predicted robbery rate would be 152.47.
      For a state with 50% urbanization, the predicted rate of auto theft would be 182.54.

13.7 $b = 0.05$, $a = 53.18$, $r = 0.40$, $r^2 = 0.16$

13.9

Relationship	r	r^2
Prestige and age	−0.30	0.09
Attendance and number of children	−0.39	0.15
Number of children and hours of TV	0.18	0.03
Age and hours of TV	0.16	0.03
Age and number of children	0.67	0.45
Hours of TV and prestige	−0.19	0.04

Chapter 14

14.1   **a.** For turnout (Y) and unemployment (X) while controlling for negative advertising (Z), $r_{yx.z} = 0.95$. The relationship between X and Y is not affected by the control variable Z.
      b. For turnout (Y) and negative advertising (X) while controlling for unemployment (Z), $r_{yx.z} = -0.89$. The bivariate relationship is not affected by the control variable.

c. Turnout $(Y) = 70.25 + (2.09)$ unemployment $(X_1) + (-0.43)$ negative advertising (X_2). For unemployment $(X_1) = 10$ and negative advertising $(X_2) = 75$, turnout $(Y) = 58.90$.

d. For unemployment (X_1): $b_1^* = 0.66$. For negative advertising (X_2): $b_2^* = -0.41$.
Unemployment has a stronger effect on turnout than negative advertising. Note that the independent variables' effects on turnout are in opposite directions.

e. $R^2 = 0.98$

14.3 **a.** For strife (Y) and unemployment (X), controlling for urbanization (Z), $r_{yx.z} = 0.79$.
 b. For strife (Y) and urbanization (X), controlling for unemployment (Z), $r_{yx.z} = 0.20$.

c. Strife $(Y) = (-14.60) + (4.94)$ unemployment $(X_1) + (0.16)$ urbanization (X_2). With unemployment $= 10$ and urbanization $= 90$, strife (Y') would be 49.19.

d. For unemployment (X_1): $b_1^* = 0.78$. For urbanization (X_2): $b_2^* = 0.13$.

e. $R^2 = 0.65$

14.5 **a.** Turnout $(Y) = 83.80 + (-1.16)$ Democrat $(X_1) + (2.89)$ minority (X_2)

b. For $X_1 = 0$ and $X_2 = 5$, $Y' = 98.25$

c. $Z_y = (-1.27)Z_1 + (.84)Z_2$

d. $R^2 = 0.51$

14.7 **a.** $Z_y = (-0.001)$ HS grads $(Z_1) + (-0.71)$ rank (Z_2)

b. $R^2 = 0.51$

Glossary

Each entry includes a brief definition and notes the chapter that introduces the term.

Alpha error. See Type I error. Chapter 7.

Alpha level (α). In inferential statistics, the probability of error. (1) In estimation, the probability that a confidence interval does not contain the population value. Chapter 6 (2) In hypothesis testing, the proportion of the area under the sampling distribution that contains unlikely sample outcomes if the null is true. The probability of Type I error. Chapter 7

Analysis of variance (ANOVA). A test of significance for testing the differences between more than two sample means. Chapter 9

Association. Variables are associated if the distribution of one variable changes for the various categories or scores of the other variable. Chapter 11

Bar chart. A graphic display device for nominal or ordinal variables with few categories. Categories are represented by bars of equal width, the height of each corresponding to the number (or percentage) of cases in the category. Chapter 2

Beta error. See Type II error. Chapter 7

Beta-weights (b^*). See standardized partial slopes. Chapter 14

Bias. A criterion used to select sample statistics as estimators. A statistic is unbiased if the mean of its sampling distribution is equal to the population value of interest. Chapter 6

Bivariate table. A table that displays the joint frequency distributions of two variables. Chapter 10

Cell. The cross-classification categories of the variables in a bivariate table. Chapter 10

Central Limit Theorem. A theorem that specifies the mean, standard deviation, and shape of the sampling distribution, given that the sample is large. Chapter 6

χ^2(critical). The score on the sampling distribution that marks the beginning of the critical region. Chapter 10

χ^2(obtained). The test statistic as computed from sample results. Chapter 10

Chi square test. A test of significance for variables in a bivariate table. Chapter 10

Coefficient of determination (r^2). The proportion of all variation in Y that is explained by X. Chapter 13

Coefficient of multiple determination (R^2). A statistic that equals the total variation explained in the dependent variable by all independent variables combined. Chapter 14

Column. The vertical dimension of a bivariate table. Chapter 10

Column percentages. The percentage of all cases in a particular column of a bivariate table that are in each row. Chapter 10

Conditional distribution of Y. The distribution of scores on the dependent variable for a specific score of the independent variable. Chapter 11

Conditional means of Y. The mean of all scores on Y for a value of X. Chapter 13

Confidence interval. An estimate of a population value in which a range of values is specified. Chapter 6

Confidence level. Another way of expressing alpha. The confidence level is the probability, expressed as a percentage, that an infinite number of confidence intervals constructed over the long run would contain the population parameter. Chapter 6

Control variable. A third variable that might affect the relationship between independent and dependent variables. Chapter 14

Correlation matrix. A table showing the strength and direction of the relationships between all possible pairs of variables. Chapter 13

Cramer's V. A chi square–based measure of association for variables that have been organized into a bivariate table of any number of rows and columns. Chapter 11

Critical region (region of rejection). The area under the sampling distribution that, in advance of the test itself, is defined as including unlikely sample outcomes, given that the null hypothesis is true. Chapter 7

Cumulative frequency. An optional column in a frequency distribution that displays the number of cases within an interval and all preceding intervals. Chapter 2

Cumulative percentage. An optional column in a frequency distribution that displays the percentage of cases within an interval and all preceding intervals. Chapter 2

Data. Information expressed in numerical form. Chapter 1

Data reduction. The process of allowing a few numbers to summarize many numbers. Chapter 1

Dependent variable. A variable that is thought to be caused by another variable. The variable that is taken as the effect. Chapters 1 and 11

Descriptive statistics. Statistics designed to describe a single variable or the relationships between two or more variables. Chapter 1

Deviation. The distance between the scores and the mean. Chapter 4

Direct relationship. A relationship between X and Y in which the third variable (Z) has no effect. Chapter 14

Dispersion. The amount of variety or heterogeneity in a distribution of scores. Chapter 4

Efficiency. The extent to which the sample outcomes are clustered around the mean of the sampling distribution. Chapter 6

EPSEM. A method for selecting random samples. Every element or case in the population must have an equal probability of selection for the sample. Chapter 6

Expected frequency (f_e). The cell frequencies that would be expected if the variables were independent. Chapter 10

Explained variation. The proportion of all variation in Y that is attributed to the effect of X. Chapter 13

F ratio. The test statistic computed in step 4 of the ANOVA test. Chapter 9

Five-step model. A step-by-step framework that organizes decisions and computations for all tests of significance. Chapter 7

Frequency distribution. A table that displays the number of cases in each category of a variable. Chapter 2

Frequency polygon. See line chart. Chapter 2

Gamma (G). A measure of association for "collapsed" ordinal-level variables that have been organized into a bivariate table. Chapter 12

Histogram. A graphic display device for interval-ratio variables. Class intervals are represented by contiguous bars of equal width (equal to the class limits), the height of each corresponding to the number (or percentage) of cases in the interval. Chapter 2

Hypothesis. A statement about the relationship between variables that is derived from a theory. Hypotheses are more specific than theories, and all terms and concepts are fully defined. Chapter 1

Hypothesis test. Statistical test that estimates the probability of sample outcomes if assumptions about the population (the null hypothesis) are true. Chapter 7

Independence. Two variables are independent if the classification of cases on one variable has no effect on the classification of cases on the second variable. Chapter 10

Independent random samples. Samples selected in such a way that the selection of cases for one samples is not connected in any way to the selection of cases for the other sample. Chapter 8

Independent variable. A variable that is thought to cause another variable. The variable that is taken as the cause. Chapters 1 and 11

Inferential statistics. Statistical technique that allows researchers to generalize from samples to populations. Chapter 1

Interaction. A relationship in which X and Y are related differently for different values of Z. Chapter 14

Interquartile range (Q). The distance from the third quartile to the first quartile. Chapter 4

Intervening relationship. A relationship in which X and Y are linked by Z. Chapter 14

Lambda (λ). A PRE measure of association for variables that have been organized into a bivariate table of any number of rows and columns. Chapter 11

Level of measurement. The mathematical relationship of the scores of a variable. Chapter 1

Line chart. A graphic display device for interval-ratio variables. Class intervals are represented by dots placed over the midpoints, the height of each corresponding to the number (or percentage) of cases in the interval. All dots are connected by straight lines. Same as a frequency polygon. Chapter 2

Linear relationship. A relationship between two variables that can be approximated with a straight line on a scattergram. Chapter 13

Marginals. The row and column subtotals in a bivariate table. Chapter 10

Maximum difference. The largest difference between column percentages for any row of a bivariate table. Chapter 11

Mean. The arithmetic average of the scores. $\bar{X}$ represents the mean of a sample, and μ, the mean of a population. Chapter 3

Mean square estimate. An estimate of the variance calculated by dividing the sum of squares within (SSW) or the sum of squares between (SSB) by the proper degrees of freedom. Chapter 9

Measure of association. Statistic that quantifies the strength and direction of the association between variables. Chapters 1 and 11

Measure of central tendency. Statistic that summarizes a distribution of scores by reporting the most typical or representative value of the distribution. Chapter 3

Measure of dispersion. Statistic that indicates the amount of variety or heterogeneity in a distribution of scores. Chapter 3

Median (*Md*). The point in a distribution of scores above and below which exactly half of the cases fall. Chapter 3

Midpoint. The point exactly halfway between the upper and lower limits of a class interval. Chapter 2

Mode. The most common value in a distribution or the largest category of a variable. Chapter 3

μ. The mean of a population. Chapter 6

$\mu_{\bar{x}}$. The mean of a sampling distribution of sample means. Chapter 6

μ_p. The mean of a sampling distribution of sample proportions. Chapter 6

Multiple correlation. A multivariate technique for examining the combined effects of more than one independent variable on a dependent variable. Chapter 14

Multiple correlation coefficient (*R*). A statistic that indicates the strength of the correlation between a dependent variable and two or more independent variables. Chapter 14

Multiple regression. A multivariate technique that separates the effects of the independent variables on the dependent variable. Also used to predict the dependent variable using all independent variables. Chapter 14

N_d. The number of pairs of cases ranked in different order on two variables. Chapter 12

Negative association. A relationship in which the variables vary in opposite directions. Chapter 11

Nonparametric test. A test of significance that does not assume a normal sampling distribution. Chapter 10

Normal curve. A theoretical distribution of scores that is symmetrical, unimodal, and bell shaped. Chapter 5

Normal curve table. Appendix A; a detailed description of the area between a Z score and the mean of any standardized normal distribution. Chapter 5

N_s. The number of pairs of cases ranked in the same order on two variables. Chapter 12

Null hypothesis (*H_0*). In single-sample tests of significance, a statement that says that the population from which the sample was drawn has a certain characteristic or value. Chapter 7

Observed frequency (*f_o*). The cell frequencies actually observed. Chapter 10

One-tailed test. A type of hypothesis test that can be used when (1) the direction of the difference can be predicted or (2) concern is focused on only one tail of the sampling distribution. Chapter 7

One-way analysis of variance. An application of ANOVA in which the effect of a single independent variable on a dependent variable is observed. Chapter 9

Parameter. A characteristic of a population. Chapter 6

Partial correlation. A multivariate technique for examining a bivariate relationship while controlling for a third variable. Chapter 14

Partial correlation coefficient. A statistic that shows the relationship between two variables while controlling for other variables. Chapter 14

Partial slopes. The slope of the relationship between an independent variable and the dependent variable while controlling for all other independent variables in the equation. Chapter 14

Pearson's *r*. A measure of association for interval-ratio-level variables. Chapter 13

Percent change. A statistic that expresses the magnitude of change in a variable from time 1 to time 2. Chapter 2

Percentage. The number of cases in a category of a variable divided by the number of cases in all categories of the variable, the entire quantity multiplied by 100. Chapter 2

Phi (*φ*). A chi square–based measure of association for variables that have been organized into a 2×2 table. Chapter 11

Pie chart. A graphic display device especially for nominal or ordinal variables with few categories.

A circle (the pie) is divided into segments proportional in size to the percentage of cases in each category of the variable. Chapter 2

Pooled estimate. An estimate of the standard deviation of the sampling distribution of the difference in sample means based on the standard deviations of both samples. Chapter 8

Population. All cases in which a researcher is interested. Chapter 1

Positive association. A relationship in which the variables vary in the same direction. Chapter 11

Probability. The likelihood that a defined event will occur. Chapter 5

Proportion. The number of cases in one category of a variable divided by the number of cases in all categories of the variable. Chapter 2

Proportional reduction in error (PRE). The logic behind some measures of association. PRE measures compare the number of errors made in predicting the dependent variable while ignoring the independent variable with the number of errors of prediction made while taking the independent variable into account. Chapter 11

P_s **(P-sub-s).** A sample proportion. Chapter 6

P_u **(P-sub-u).** A population proportion. Chapter 6

Quantitative research. Research focused on numerical information or data. Chapter 1

Range (R). The highest score minus the lowest score. Chapter 4

Rate. The number of actual occurrences of some phenomenon or trait divided by the number of possible occurrences per some unit of time. Chapter 2

Ratio. The number of cases in one category divided by the number of cases in some other category. Chapter 2

Regression line. The single, best-fitting straight line that summarizes the relationship between two variables. Chapter 13

Representative. A sample is representative if it reproduces the major characteristics of the population from which it was drawn. Chapter 6

Research. The systematic and careful gathering of information to answer questions and test theories and hypotheses. Chapter 1

Research hypothesis (H_1). A statement that contradicts the null hypothesis. The specific form varies from test to test. Chapter 7

Row. The horizontal dimension of a bivariate table. Chapter 10

Sample. Carefully chosen subsets of populations. Chapter 1

Sampling distribution. The distribution of a statistic for all possible sample outcomes of a certain size. Chapter 6

Scattergram. A graph that shows the relationship between two variables. Chapter 13

Σ. Uppercase Greek letter sigma. A mathematical operator that stands for "the summation of." Chapter 3

σ_{p-p}. Symbol for the standard deviation of the sampling distribution of the differences in sample proportions. Chapter 8

$\sigma_{\bar{x}-\bar{x}}$. Symbol for the standard deviation of the sampling distribution of the differences in sample means. Chapter 8

Simple random sample. A sample chosen from a list of the population by which every case and every combination of cases has an equal chance of being included. Chapter 6

Skew. The extent to which a distribution of scores has a few scores that are extremely high (positive skew) or extremely low (negative skew). Chapter 3

Slope (b). The amount of change in one variable per unit change in the other. Chapter 13

Spearman's rho (r_s). A measure of association appropriate for "continuous" ordinal-level variables. Chapter 12

Spurious relationship. A relationship in which Z causes both X and Y. Chapter 14

Standard deviation. The most important and useful descriptive measure of dispersion; s represents the standard deviation of a sample and σ represents the standard deviation of a population. Chapter 4

Standard error of the mean. The standard deviation of a sampling distribution of sample means. Chapter 6

Standardized partial slopes (beta-weights). The slope of the relationship between an independent and dependent variable when all scores have been normalized. Chapter 14

Statistics. Mathematical techniques used to organize and manipulate data. Chapter 1

Student's t distribution. A distribution used to find the critical region for tests of sample means when σ is unknown and sample size is small. Chapter 7

Sum of squares between (SSB). The sum of the squared deviations of the sample means from the overall mean, weighted by sample size. Chapter 9

Sum of squares within (SSW). The sum of the squared deviations of scores from the category means. Chapter 9

*t***(critical).** The *t* score that marks the beginning of the critical region of a *t* distribution. Chapter 7

*t***(obtained).** The test statistic computed in step 4 of the five-step model for tests of sample means when *N* is small and population standard deviation is unknown. Chapter 7

Test statistic. The value computed in step 4 of the five-step model that places the sample outcome on the sampling distribution. Chapter 7

Theory. A general explanation of the relationship between phenomena. Chapter 1

Total sum of squares (SST). The sum of the squared deviations of the scores from the overall mean. Chapter 9

Total variation. The spread of the *Y* scores around the mean of *Y*. Chapter 13

Two-tailed test. A significance test in which the researcher does not specify the direction of the difference. Chapter 7

Type I error. The probability of rejecting a null hypothesis that is, in fact, true. Chapter 7

Type II error. The probability of failing to reject a null hypothesis that is, in fact, false. Chapter 7

Unexplained variation. The proportion of the total variation in *Y* that is not accounted for by *X*. Chapter 13

Variable. Any trait that can change value from case to case. Chapter 1

Variance. A measure of dispersion used primarily in inferential statistics and also in correlation and regression techniques. Chapter 4

X_i **(*X*-sub-*i*).** Any score in a distribution. Chapter 3

Y **intercept (*a*).** The point where the regression line crosses the *Y* axis. Chapter 13

*Z***(critical).** The *Z* score that marks the beginnings of the critical region. Chapter 7

*Z***(obtained).** The sample outcomes expressed as a *Z* score. Chapter 7

Z **scores.** Standard scores; the way scores are expressed after they have been standardized to the theoretical normal curve. Chapter 5

Zero-order correlation. Correlation coefficient for bivariate relationships. Chapter 14

Index